Applied Statistics:
Analysis of Variance
and Regression

A WILEY PUBLICATION IN APPLIED STATISTICS

Applied Statistics: Analysis of Variance and Regression

OLIVE JEAN DUNN
Professor of Biostatistics
University of California, Los Angeles

and

VIRGINIA A. CLARK
Associate Professor of Biostatistics
University of California, Los Angeles

John Wiley & Sons

New York · London · Sydney · Toronto

Library of Congress Cataloging in Publication Data:

Dunn, Olive Jean.
Applied statistics: analysis of variance and regression.

(Wiley series in probability and mathematical statistics)
"A Wiley-Interscience publication."
Includes bibiliographies.
1. Analysis of variance. 2. Regression analysis. I. Clark, Virginia A., joint author.
II. Title.

QA297.D87 519.5′35 73-13683
ISBN 0-471-22700-5

Printed in the United States of America

10 9 8 7 6 5 4

Preface

This book has been based on notes originally developed for a one-semester course in analysis of variance, regression, and covariance. Students in the course were assumed to have had a one-year course in basic statistics, including a short introduction to analysis of variance and linear regression. Both the one-year prerequisite course and the one-semester course for which these notes were prepared were given at a mathematical level presupposing only elementary algebra.

We had two general objectives in developing the original set of notes into a book. The first was to have a self-contained textbook, and with this objective in mind, we have prepared four introductory chapters. These chapters are somewhat brief, and so the reader with no previous statistics may well study any of several elementary textbooks in statistics, for example, Dunn, *Basic Statistics: A Primer for the Biomedical Sciences*, John Wiley & Sons, 1964, or Dixon and Massey, *An Introduction to Statistical Analysis*, 3rd edition, McGraw-Hill, 1969. We have also added topics which, we feel, make the notes more usable from a practical point of view. Discussions of outliers, transformations, and departures from underlying assumptions, all fall into this category. With these additions, we think that the book can serve as a textbook in a one-year course; with omissions, it can be used for a one-semester course.

Our second objective, sometimes difficult to reconcile with the first, was to produce a book that might be useful as a reference. We realize that more complete reference books are available. On the other hand, many students go forth into the world of statistical applications after one or two years of statistics, and when any problem arises, the textbooks they have studied are their first hope in seeking a solution. For them, therefore, we include briefly some methods and topics that would otherwise be omitted, and we also give reference lists.

The elementary mathematical level to which we have adhered was chosen deliberately. Although this level is often irritating to the reader who has studied more advanced mathematics, we believe that he often gains insight from seeing statistics presented in this way. In each topic we go through the simplest model in detail. More general, more sophisticated treatments have been relegated to special starred sections, which can be omitted without loss of continuity.

The examples used in this book were obtained in the following way. For any research problem as found in the literature, we drew random samples on a large-scale computer from normally distributed populations with appropriate means and variances. These provide the basic data for most of the problems. Where outliers or data from nonnormal distributions were required, these were also generated by computer.

Chapters 1 through 4 provide the introduction necessary for studying analysis of variance and regression. Chapters 5, 6, and 7 deal with the fixed effects model analysis of variance Model I. Chapter 8, a starred chapter, gives a brief introduction to confounding, still with Model I. Chapter 9, also starred, presents variable effects models (Models II and III). Chapters 10, 11, and 12 introduce linear, multiple, and polynomial regression; Chapter 13 is concerned with covariance analysis. In Chapter 14 we discuss various techniques for screening data before analysis. We consider this material so important that it must be gathered together into a single chapter and placed as conspicuously as possible; clearly it cannot be the first chapter, and so it must be the last.

In the later chapters, where the numerical calculations could entail considerable effort, references are made to the output and use of canned programs. Principally, the BMD programs edited by W. J. Dixon, University of California Press, are referenced because of their wide range of programs and broad distribution. There are occasional references to the IBM Scientific Subroutine Programs and the EPID Programs (a set of descriptive programs available from the Epidemiology Division, School of Public Health, UCLA) for additional output.

We are indebted to our colleagues Drs. A. A. Afifi, Potter C. Chang, and F. Massey, Jr., and to the many students who have used the manuscript for helpful comments, to Dr. Maria Zielezny for help in preparing problems, and finally to Mrs. Ann Eiseman and Mrs. Leona Povondra for their tireless work in preparing the manuscript.

OLIVE JEAN DUNN
VIRGINIA A. CLARK

Los Angeles, California
April 1973

Contents

CHAPTER 1

DESCRIPTIVE STATISTICS

WHAT IS STATISTICS?

Statistics has many possible meanings; here we regard it as a body of methods enabling us to draw reasonable conclusions from data. Statistics is often divided into two general types, *descriptive statistics* and *statistical inference*. With descriptive statistics, we summarize data, making calculations, tables, or graphs that can be comprehended easily. Statistical inference, on the other hand, involves drawing conclusions from the data. It is based on the mathematical theory of probability.

THE DATA

We discuss briefly the nature of data. The data to be analyzed can be in many forms. They are most frequently but not always numerical. Data could consist, as in Table 1.1, of a list of girls in a certain school, with the country of birth, age at last birthday, number of siblings, ranking in class, score on algebra examination, IQ, systolic blood pressure, weight, eye color, and date of last smallpox vaccination recorded for each girl. The 10 pieces of information recorded on each individual are called 10 *variables*.

Variables can be either *discrete* or *continuous*. Number of siblings is an example of a discrete variable; it can only be 0, 1, 2,.... Weight, on the other hand, is a continuous variable; it can have any value within a reasonable range. Both these variables are numerical.

Of the variables listed, country of birth is categorical rather than numerical. We can however, for convenience, use a code such as U.S.A. = 1, Mexico = 2, and so forth. In handling such coded data, we must remember that the numbers are arbitrary and do not represent a scale from small to large. We could use the code Mexico = 1 and U.S.A. = 2 without changing the statistical analysis or interpretation. Eye color could be coded

TABLE 1.1. DATA FOR 50 INDIVIDUALS INCLUDING BOTH CONTINUOUS AND DISCRETE VARIABLES

	Country*	Age	Number of Siblings	Ranking	Examination Score	I.Q.	S.B.P.	Weight	Eye** Color	Year of Vaccination
1	2	10	3	16	70	91	114	67	2	1965
2	1	10	1	8	79	121	90	69	1	1965
3	3	10	4	1	76	111	88	85	1	1964
4	1	11	2	3	85	103	96	83	2	1965
5	1	11	3	7	82	104	113	74	3	1964
6	1	11	2	9	81	99	92	81	2	1965
7	1	11	3	25	64	87	103	97	1	1964
8	3	11	2	10	72	121	123	92	1	1964
9	1	12	4	47	25	70	125	114	2	1964
10	1	12	2	17	88	105	83	85	3	1964
11	1	12	1	31	47	96	110	94	2	1968
12	2	12	3	14	85	107	101	95	1	1964
13	4	13	4	3	88	103	127	106	3	1963
14	1	13	5	34	34	79	94	93	1	1963
15	1	13	3	26	63	91	119	108	2	1963
16	1	13	2	20	71	108	105	104	2	1964
17	5	13	3	15	74	132	97	99	2	1962
18	1	14	3	32	54	105	109	105	1	1962
19	1	14	4	37	52	93	102	112	1	1962
20	1	14	3	22	67	101	115	91	2	1961
21	1	14	2	1	92	138	128	105	3	1962
22	1	14	5	19	67	119	106	138	2	1962
23	2	14	4	37	49	73	105	124	1	1961
24	1	15	3	23	70	100	123	114	2	1961
25	1	15	3	11	81	96	115	117	2	1962
26	1	15	2	14	74	81	81	108	3	1961
27	1	15	4	19	79	118	132	110	2	1961
28	1	15	3	21	66	88	105	123	3	1969
29	1	15	3	26	69	98	101	106	2	1962
30	1	15	2	7	80	100	124	98	1	1965

Country* 1 = U.S.A. 2 = Mexico 3 = Canada 4 = English 5 = Oriental

Eye Color** 1 = Black 2 = Brown 3 = Blue

arbitrarily, or it might be assigned a meaningful number according to its position on a color spectrum. Similarly, class rank, score on algebra examination, and IQ represent attempts at meaningful scales to measure something of interest. They are not necessarily measured in units of equal length, and therefore differences may not have the same meaning in different parts of the scale. For example, the difference between the students ranked 1 and 2 in a class of 50 may be larger than the difference between the students ranked 24 and 25.

TABLE 1.1 (CONTINUED)

	Country*	Age	Number of Siblings	Ranking	Examination Score	I.Q.	S.B.P.	Weight	Eye** Color	Year of Vaccination
31	1	16	1	5	82	127	134	144	1	1960
32	1	16	4	14	81	102	129	124	2	1961
33	1	16	2	6	75	114	119	112	2	1959
34	2	16	3	26	73	102	92	111	1	1968
35	1	16	3	5	89	104	108	106	1	1960
36	1	17	2	2	91	137	143	127	3	1959
37	1	17	3	42	43	96	126	122	2	1958
38	1	17	3	32	62	107	115	108	1	1959
39	1	17	1	4	87	104	127	136	1	1967
40	1	17	6	1	99	146	128	117	2	1959
41	2	17	4	23	56	116	95	114	1	1959
42	1	18	5	36	57	99	132	116	3	1959
43	1	18	4	27	68	109	126	125	1	1958
44	1	18	3	15	76	129	139	119	1	1966
45	1	18	2	12	83	111	119	109	3	1957
46	1	18	1	37	52	83	99	118	1	1957
47	1	18	3	3	93	120	118	120	2	1958
48	1	18	4	9	89	101	130	139	2	1969
49	1	19	5	28	46	97	117	131	1	1965
50	1	19	3	4	95	108	135	126	1	1957

Country* 1 = U.S.A.
2 = Mexico
3 = Canada
4 = English
5 = Oriental

Eye Color** 1 = Black
2 = Brown
3 = Blue

Of the remaining variables, age at last birthday, weight, and systolic blood pressure can be termed measurement data. Measurement data are measured in units of equal length. Blood pressure and weight are continuous variables but, in the measurement process, we usually round off the measurements to the nearest whole number. When we report age, we actually report a continuous variable rounded down to the age at last birthday. Date of last smallpox vaccination could be converted to a continuous measurement scale by changing it to "time since last smallpox vaccination," because time is measured in units of equal length.

Throughout this book we assume that our data consist of measurements on a continuous scale, or that the discrete scale used has such fine gradations that for all practical purposes the scale is continuous. However, the methods presented are frequently used with other types of data. For a more complete discussion of scales of measurement, see the references on classification of measurements at the end of this chapter.

DESCRIPTION OF THE DATA

Let us now consider the description of a set of data on just one variable. As an example, we use blood pressure measurements from Table 1.1. The 50 measurements on systolic blood pressure are too numerous for us to comprehend in a quick glance at Table 1.1, and our object is to summarize them in a form that is easier to grasp. We do this by making a frequency table in which we place blood pressure measurements that are similar in a single class interval.

Class intervals should be recorded in a way that makes clear to the reader exactly how the classes were formed. Here blood pressures were recorded in the data to the nearest unit; therefore, a blood pressure recorded as 80 means that the individual's blood pressure lies between 79.5 and 80.5; similarly, a recorded 89 indicates a blood pressure between 88.5 and 89.5. Thus blood pressure values recorded as 80, 81,...,89 include all pressures lying between 79.5 and 89.5. Note that if the pressures are reported to the nearest whole number, each blood pressure can fall in only one class interval. If class intervals had been reported as 80–90, 90–100, and so on, it would not be clear to the reader of the frequency table where a blood pressure of 90 mm Hg was placed.

The first column of Table 1.2 contains the class intervals. Midpoints of the intervals are given in column 2; these are the averages of the upper and lower limits of the class intervals. For example, $(79.5+89.5)/2=84.5$, the midpoint of the first interval. The third column lists the number of individuals whose blood pressures lie within each class interval; these are called *frequencies*. In column 4 of Table 1.2 we find the frequencies as percentages of 50, the total number of girls. Each relative frequency is computed by dividing the frequency in the class interval by 50, the total frequency, and multiplying by 100. Relative frequencies are often of more

TABLE 1.2. FREQUENCY TABLE OF SYSTOLIC BLOOD PRESSURES

Class Interval	Midpoint	Frequency f	Relative Frequency %
79.5 - 89.5	84.5	3	6
89.5 - 99.5	94.5	8	16
99.5 -109.5	104.5	10	20
109.5 -119.5	114.5	11	22
119.5 -129.5	124.5	11	22
129.5 -139.5	134.5	6	12
139.5 -149.5	144.5	1	2
		50	100%

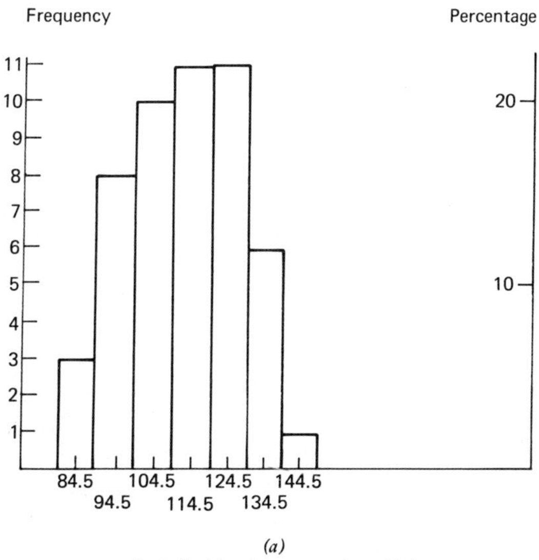

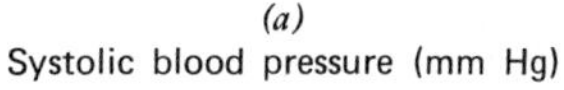

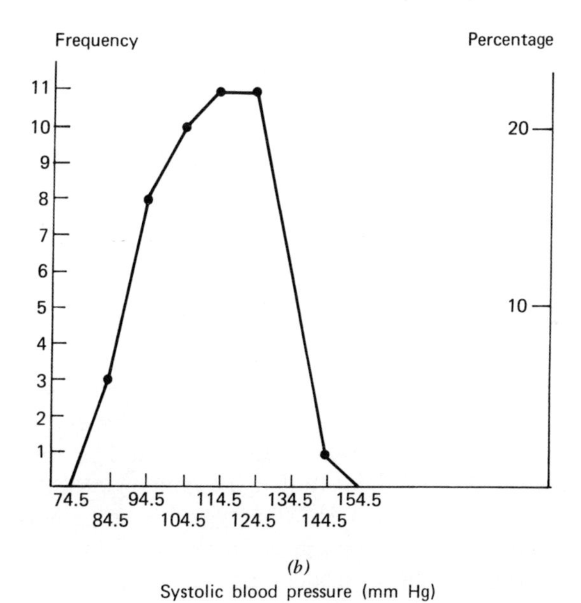

Figure 1.1. (*a*) Histogram of systolic blood pressure; (*b*) frequency polygon of systolic blood pressure.

interest than actual frequencies, particularly when comparing frequency tables based on different numbers of individuals.

Here we have used equal class intervals, which are usually desirable. For more discussion on frequency tables, see Dunn [1.1].

HISTOGRAMS AND FREQUENCY POLYGONS

Histograms or frequency polygons, which are constructed from the frequency tables, show the form of the data even more clearly than do the tables. In both histogram and frequency polygon, class intervals are plotted along the horizontal axis. In plotting a histogram, a rectangle whose height equals the frequency of the class is drawn over each class interval. The histogram drawn from the data in Table 1.2 is shown in Figure 1.1*a*. In Figure 1.1*b* a frequency polygon is plotted from the same data. Here the frequencies are plotted as points above the midpoint of each class interval and the points are joined by line segments. To make the picture more complete, a line segment is also drawn from zero on the abscissa at 74.5 (the midpoint of the class interval preceding the first interval in the table) to 3 at 84.5 (the midpoint of the first interval). Similarly, a line segment joins 1 at 144.5 to zero at 154.5. Often, instead of the frequencies themselves, relative frequencies or percentages are used on the vertical scale. Figure 1.1 includes vertical scales indicating percentages.

Several general remarks can be made about histograms and frequency polygons. First, with a small data set they may look quite different with different choices of class intervals. With a very large set of data and a fairly large number of classes, however, these differences become unimportant.

Second, we can learn to interpret the area of the histogram and the area under the frequency polygon (i.e., the area between the frequency polygon and the horizontal axis). For example, the area of the bars from 79.5 to 109.5 appears to be about one-half the area of all the rectangles in the histogram. Adding the frequencies in the corresponding classes, we find that $3+8+10=21$ of the 50 girls had systolic blood pressures less than 109.5 mm Hg; thus 42% of the individuals have blood pressures between 79.5 and 109.5. We notice at a glance that the number of individuals with blood pressures above 139.5 mm Hg is very small; it actually equals 1. Areas of the rectangles of a histogram are proportional to percentages of the individuals falling into the corresponding classes. The eye has a natural tendency to compare areas.

THE SCATTER DIAGRAM

When two variables such as blood pressure and weight have been measured on the same set of individuals, a simple and effective way of describing them is the *scatter diagram*. Figure 1.2 is a scatter diagram formed from the blood pressure and weight measurements in Table 1. Blood pressure measurements are denoted by Y, weight measurements by X.

The scatter diagram uses two perpendicular axes, one for each variable. Each individual's X and Y measurements are plotted as a point on the diagram, the X value plotted on the horizontal scale, the Y value on the vertical scale. For example, from Table 1.1, the first individual's weight is 67 lb, her blood pressure is 114 mm Hg; that is, $X_1 = 67$ lb, $Y_1 = 114$ mm Hg. The circled point in Figure 1.2 corresponds to this individual's weight and blood pressure.

From the scatter diagram we can often see whether the two variables X and Y are related. In Figure 1.2, for example, we notice a tendency for high blood pressures and high weights to be associated.

NUMERICAL DESCRIPTION OF THE DATA

In addition to summarizing sets of data by means of frequency tables or histograms, we usually calculate certain numbers from the data that are used to characterize the entire set of data. Such numbers are called *statistics*. Many different statistics can be calculated from data, and often are, but here we introduce only those used most frequently—the arithmetic mean, the variance, the standard deviation, the covariance, and the correlation coefficient.

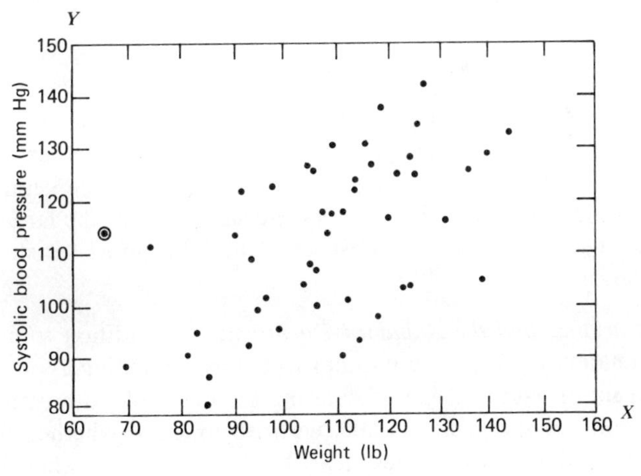

Figure 1.2. Scatter diagram of systolic blood pressure and weight for 50 girls.

The Mean. In our notation, Y denotes a variable for which we have a set of measurements; in our example, Y denotes systolic blood pressure. Also, Y_1 designates the measurement on the first girl, in Table 1.1 $Y_1 = 114$. We use n for the number of measurements; here $n = 50$. A more general notation is Y_i; in this example, i can take on values from 1 to 50 or, generally, from 1 to n.

The arithmetic mean of a set of data is denoted by $\overline{Y}$; it is simply the arithmetic average of all the observations. That is,

$$\overline{Y} = \frac{Y_1 + Y_2 + Y_3 + \cdots + Y_n}{n}.$$

A general notation for summing observations is the symbol $\sum$, which represents the phrase "the sum of." Instead of writing $Y_1 + Y_2 + Y_3 + \cdots + Y_{50}$, we use the expression $\sum_{i=1}^{50} Y_i$. In this example, we are taking the sum of Y_i from $i = 1$ (written below the $\sum$ sign) to $i = 50$ (written above the $\sum$ sign). In general, we sum over n measurements, and we write $\sum_{i=1}^{n} Y_i$ to signify the summation sign; $\overline{Y}$ can be written

$$\overline{Y} = \sum_{i=1}^{n} \frac{Y_i}{n}. \tag{1.1}$$

From our example, we calculate

$$\overline{Y} = \frac{114 + 90 + \cdots + 135}{50}$$

$$= 113.1.$$

The arithmetic mean is familiar to us all; it can be thought of as a measure of central tendency for a set of numbers. Other averages are sometimes appropriate for this purpose—the median, the geometric mean, the harmonic mean, and the mean of the smallest and the largest numbers. The arithmetic mean is the most usual and the most satisfactory of all these averages.

The Variance and the Standard Deviation. In addition to a measure of central tendency, we need a number to tell us something about how widely spread a set of data is. After calculating the mean blood pressure in the set of 50 blood pressures, $\overline{Y} = 113.06$, we wish to know whether all the blood pressures tend to have values near 113.06 or whether they vary widely around 113.06.

Again there are numerous measures of variability that could be used. Usually, however, a statistic called the *variance* is calculated from the data and then its square root, the *standard deviation*, is used.

The formula for the variance (denoted by s^2) is

$$s^2 = \frac{(Y_1 - \bar{Y})^2 + (Y_2 - \bar{Y})^2 + \cdots + (Y_n - \bar{Y})^2}{n-1},$$

or, with summation notation,

$$s^2 = \frac{\sum_{i=1}^{n} (Y_i - \bar{Y})^2}{n-1}. \tag{1.2}$$

From its formula, the variance of a set of data is seen to be an average of squared deviations from the mean. In Table 1.3, the variance has been

TABLE 1.3 EXAMPLE OF COMPUTATION OF STANDARD DEVIATION FOR SYSTOLIC BLOOD PRESSURE

	Y_i	$Y_i - \bar{Y}$	$(Y_i - \bar{Y})^2$
1	114	0.86	0.74
2	90	-23.14	535.46
3	88	-25.14	632.02
4	96	-17.14	293.78
.	.	.	.
.	.	.	.
.	.	.	.
.	.	.	.
47	118	4.86	23.62
48	130	16.86	284.26
49	117	3.86	14.90
50	135	21.86	477.86
	$\sum_{i=1}^{n} Y_i = 5657$	$\sum_{i=1}^{n} (Y_i - \bar{Y}) = 0$	$\sum_{i=1}^{n} (Y_i - \bar{Y})^2 = 11760.01$

$\bar{Y} = \frac{5657}{50} = 113.1$

$s^2 = 11760.01/49$ or

$s^2 = 240$

$s = 15.5$

calculated for the 50 blood pressures. The standard deviation s is simply the square root of the variance and is in the same units as are the observations Y_i. Thus

$$s=\left[\sum_{i=1}^{n}\frac{\left(Y_i-\overline{Y}\right)^2}{n-1}\right]^{1/2} \qquad \text{or} \qquad s=15.49. \tag{1.3}$$

Several remarks should be made concerning the variance. It is apparent first of all that if a set of data has a very small spread, all the deviations $Y_i-\overline{Y}$ will be small in magnitude, and s^2 (and s) will be small. If the set of data is widely spread, some deviations from the mean will be large in magnitude, and s^2 (and s) will be large.

At first it seems surprising that in obtaining s^2 we average the squared deviations by dividing by $n-1$ rather than by n. The use of $n-1$ is discussed in Chapter 3; here we simply accept the formula as it stands.

The Covariance and the Correlation Coefficient. When two variables such as blood pressure and weight have been measured on the same set of individuals, a statistic measuring the degree of association between the two variables is often desirable. The usual statistic for this purpose is called the *correlation coefficient*. Before defining it, we define another statistic, *the covariance*.

Let Y and X denote two variables (e.g., blood pressure and weight); let Y_i and X_i be measurements of the two variables on the ith individual, for $i=1,\ldots,n$. The covariance of X and Y for this set of data, designated $\widehat{\text{Cov}}\,X, Y$, is defined as follows:

$$\widehat{\text{Cov}}\,X,Y=\frac{\left(X_1-\overline{X}\right)\left(Y_1-\overline{Y}\right)+\left(X_2-\overline{X}\right)\left(Y_2-\overline{Y}\right)+\cdots+\left(X_n-\overline{X}\right)\left(Y_n-\overline{Y}\right)}{n-1}$$

or

$$\widehat{\text{Cov}}\,X, Y=\frac{\sum_{i=1}^{n}\left(X_i-\overline{X}\right)\left(Y_i-\overline{Y}\right)}{n-1}, \tag{1.4}$$

where $\overline{Y}$ and $\overline{X}$ denote the means of the two variables.

The covariance is seen to be an average of the products of deviations from the mean. If the ith individual is heavy and has high blood pressure, $X_i-\overline{X}$ and $Y_i-\overline{Y}$ are both positive and their product is positive. Similarly, a light individual with very low blood pressure contributes a positive term to the covariance; for him, $X_i-\overline{X}$ and $Y_i-\overline{Y}$ are both negative and their product is positive. Thus if high weight tends to be associated with high blood pressure, the covariance of weight and blood pressure is positive. On the other hand, a tendency for high weights to appear with low blood pressures would be reflected in a negative covariance.

We can interpret the sign of the covariance but not its size, for its size depends on the units in which X and Y are measured. If the weights X_i were measured in ounces, the covariance of blood pressures and weights would be 16 times as large as it would be if the weights were measured in pounds. A measure of association that is unaffected by the units of measurement is needed. Such a statistic is obtained by simply dividing the covariance of X and Y by the product of the two standard deviations. This gives the *correlation coefficient* between X and Y, usually denoted by r. Of the many equivalent formulae for r, we use

$$r=\frac{\sum_{i=1}^{n}\left(X_i-\bar{X}\right)\left(Y_i-\bar{Y}\right)}{\left[\sum_{i=1}^{n}\left(X_i-\bar{X}\right)^2\sum_{i=1}^{n}\left(Y_i-\bar{Y}\right)^2\right]^{1/2}}. \tag{1.5}$$

Properties of the correlation coefficient as a measure of association are discussed at length in Chapter 10. Like the covariance, a negative correlation coefficient indicates a tendency for large X's to occur with small Y's and small X's with large Y's; a positive correlation coefficient means that large X's tend to go with large Y's and small X's with small Y's. The correlation coefficient is never less than -1 or greater than $+1$.

Alternate Formulae. The variance, covariance, and correlation coefficient can be calculated from (1.2), (1.4), and (1.5), but they are often calculated from algebraically equivalent expressions which are especially suitable in certain situations. Some formulae are well adapted for work done on a desk calculator. A multiplicity of formulae occurs throughout statistics, and because confusion often results from too many formulae for the same statistic, we attempt to use in the text one basic formula for each statistic. These basic formulae are appropriate for hand calculations and also for calculations done on large computers. Various formulae for the statistics introduced in this chapter are summarized in Table 1.4.

PROBLEMS IN PRACTICE

The graphical and numerical methods given in this chapter often suffice to answer the question the investigator is asking. For example, let us suppose that heights and weights of children have been measured in a grade school

TABLE 1.4. NOTATION AND COMPUTATIONS

	Statistic	Formula	Definition of Symbols	Type of Formula
1.	$\overline{Y}$	$\sum_{i=1}^{n} Y_i/n$	Y_i= ith observation	Basic formula
2.	s^2	$\sum_{i=1}^{n} (Y_i-\overline{Y})^2/(n-1)$	n= number of observations	Basic formula
3.	s^2	$[\sum_{i=1}^{n} Y_i^2 - (\sum_{i=1}^{n} Y_i)^2/n]/(n-1)$		Desk calculators
4.	s^2	$(\sum_{i=1}^{n} Y_i^2 - n\overline{Y}^2)/(n-1)$		Desk calculators
5.	s^2	$(n\sum_{i=1}^{n} Y_i^2 - (\sum_{i=1}^{n} Y_i)^2)/n(n-1)$		Desk calculators
6.	$\overline{Y}$	$\sum_{i=1}^{k} f_iY_i/n$	k= number of class intervals; f_i= frequency in ith class, i=1,...,k	Calculations from frequency table
7.	s^2	$[\sum_{i=1}^{k} f_iY_i^2 - (\sum_{i=1}^{k} f_iY_i)^2/n]/(n-1)$	Y_i= midpoint of ith interval, i=1,...,k	Calculations from frequency table
8.	s^2	$(\sum_{i=1}^{k} f_iY_i^2 - n\overline{Y}^2)/(n-1)$	$n=\sum_{i=1}^{k} f_i$	Calculations from frequency table
9.	$\widehat{\text{Cov}}$ X,Y	$\sum_{i=1}^{n} (X_i-\overline{X})(Y_i-\overline{Y})/(n-1)$	X_i,Y_i= ith observation	Basic formula
10.	$\widehat{\text{Cov}}$ X,Y	$(\sum_{i=1}^{n} X_iY_i-n\overline{X}\overline{Y})/(n-1)$	$\overline{X},\overline{Y}$= means	Desk calculators
11.	r	$\sum_{i=1}^{n} (X_i-\overline{X})(Y_i-\overline{Y})\bigg/\sqrt{\sum_{i=1}^{n} (X_i-\overline{X})^2 \sum_{i=1}^{n} (Y_i-\overline{Y})^2}$		Basic formula
12.	r	$(\text{Cov } X,Y)/s_xs_y$	s_x= standard deviation of X's; s_y= standard deviation of Y's	Basic formula
13.	r	$(\sum_{i=1}^{n} X_iY_i-n\overline{X}\overline{Y})\bigg/\sqrt{(\sum_{i=1}^{n} X_i^2-n\overline{X}^2)(\sum_{i=1}^{n} Y_i^2-n\overline{Y})^2}$		Desk calculators

*In using these formulae one always performs operations within the innermost parentheses before using quantities computed from a set of parentheses. For example in the formula for the variance in line 3, $\sum_{i=1}^{n} Y_i$ is computed first and then squared to obtain $(\sum_{i=1}^{n} Y_i)^2$; next $\sum_{i=1}^{n} Y_i^2 - (\sum_{i=1}^{n} Y_i)^2/n$ is computed. Usually with a desk calculator it is simpler to cpmpute the numerator and denominator, and then perform the division as a last step.

located in a poor neighborhood, and that the investigator wishes to know whether they differ from heights and weights of grade school children in the United States as a whole. He can compare histograms of his data with histograms of data available from the National Center for Health Statistics. If his histogram of weights from the grade school looks very much like that taken from a nationwide sample, the investigator can quickly conclude that there is no appreciable difference between weights in his school and weights of typical children in the United States.

Investigators frequently perform statistical analyses that could be avoided by graphical display of their data and computation of means and standard deviations. Prewritten computer programs (BMD, EPID, etc.) require very little computer sophistication to use; they perform the computations and provide graphical output. Such programs are particularly valuable for handling a large amount of data. With only 30 or 40 cases, an hour or two of work with a desk calculator, pencil, and paper can provide a great deal of information.

SUMMARY

In this chapter some methods of describing data have been presented. The first group includes frequency tables, histograms, frequency polygons, and scatter diagrams. The second group consists of summary statistics. The mean is a measure of central tendency; the variance or standard deviation describes the variability of the data; and the correlation coefficient measures association between two variables.

REFERENCES

APPLIED STATISTICS

1.1. Dunn, O. J., *Basic Statistics: A Primer for the Biomedical Sciences*, Wiley, New York, 1964.

1.2. Dixon, W. J., and F. J. Massey, Jr., *Introduction to Statistical Analysis*, 3rd ed., McGraw-Hill, New York, 1969.

1.3. Hoel, P. G., *Elementary Statistics*, Wiley, New York, 1960.

1.4. Smart, J. V., *Elements of Medical Statistics*, Staples Press, London, 1963.

1.5. Steel, R. G. D., and J. H. Torrie, *Principles and Procedures of Statistics*, McGraw-Hill, New York, 1960.

NATURE OF MEASUREMENT

1.6. Churchman, C. W., and P. Ratoosh, (Ed.), *Measurement: Definition and Theories*, Wiley, New York, 1959.

1.7. Stevens, S. S., (Ed.), *Handbook of Experimental Psychology*, Wiley, New York, 1951.

1.8. Torgerson, W. S., *Theory and Methods of Scaling*, Wiley, New York, 1958.

COMPUTER PROGRAMS

1.9. BMD, W. J. Dixon (Ed.), University of California Press, Los Angeles, 1970.

1.10. EPID, Epidemiology Division, School of Public Health, UCLA, 1973.

PROBLEMS

1.1. Using the data on IQ in Table 1.1:
- (*a*) Form a frequency table and draw a frequency histogram, a percentage histogram, and a polygon.
- (*b*) Calculate the arithmetic mean $\overline{Y}$ from the original data and from the frequency table. Compare the two answers.
- (*c*) Calculate the variance s^2 and the standard deviation s from the original data and from the frequency table. Compare the answers.
- (*d*) The IQ test was developed to have a mean of 100. Do you consider this group of girls to be typical?

1.2. Using the IQ and score on algebra examination data in Table 1.1:
- (*a*) Draw a scatter diagram.
- (*b*) Calculate the correlation coefficient between IQ and score on algebra examination.
- (*c*) Compare the results of (*a*) and (*b*).

1.3. Using the following data on the concentration of chlorides in the sweat of a group of normal individuals, as measured in milliequivalents per liter:

37.1 31.7 16.2 23.5 30.1 9.6 39.3 22.2 41.6 15.4

- (*a*) Calculate the mean.
- (*b*) Calculate the variance using both the basic formula and the desk calculator formula.
- (*c*) Calculate the standard deviation.

1.4. Using a standard statistical computer program, the investigator obtains means and standard deviations for all the variables in Table 1.1. For which variables is this output inappropriate?

CHAPTER 2

STATISTICAL INFERENCE: POPULATIONS AND SAMPLES

In research investigators are not always content with mere description of a set of data, however important such description may be. Usually they wish to learn something about a larger group of individuals. For example, the 50 systolic blood pressures in Table 1.1 were measured on 50 high school girls. They may have been measured, however, in order to gain information about blood pressures of high school girls in general and not because of interest in these 50 specific measurements. In this case, we can talk about an entire *population* of high school girls; the 50 girls are referred to as a *sample* of size 50 from the population.

A population can be defined as the set of individuals that we wish to study. A sample is a subset of the population; it is the set of individuals that are actually studied and about whom the data are obtained. Here an individual may be a person, an animal, a plant, an object, or a measurement of any attribute. Thus we can refer to a population of the girls of the entire county or to a population of girls' weights for the girls in the entire county. Similarly, by sample we can refer to the 50 girls or to the 50 weights.

It is usual to study a sample rather than an entire population. The population is often so large that it would be impractical to work with it in its entirety. Often the population does not actually exist but is nevertheless of importance. For example, in studying a new dye for cotton, we might try the new dye with just 10 one-yard pieces of material and make measurements on the resistance to fading. Here the sample consists of the 10 pieces of cotton treated with the new dye. The population in which we are interested, however, consists of all possible pieces of cotton of a certain type that might be treated with the new dye. This population does not really exist. We can imagine it, however, and we study the 10 pieces of cotton in order to make inferences concerning it.

DISTRIBUTIONS FOR POPULATIONS AND SAMPLES

The measurements of an attribute obtained from a sample form a set of data, and they can be described by making a frequency table and by drawing a histogram or frequency polygon as in Chapter 1. If we had measurements on the same attribute for the entire population, we could also construct a frequency table and draw a histogram and a frequency polygon. With a large population, the class intervals could be made very fine. With many, very short intervals, the frequency polygon would become very much like a curve. In taking many class intervals, we could adjust the vertical scale in such a way that the total area under the frequency polygon became one.

Populations are often so large that they can be considered to be of infinite size. Here we follow the practice of considering populations to be infinite unless otherwise specified. For such an infinite population, there exists a *frequency distribution* or *density function*, which corresponds to the frequency polygon for the sample. If we had blood pressures for the entire population of interest, we could draw the density function. With blood pressure measurements on a sample of only 50 girls, we can draw the frequency polygon for the sample and can only imagine the frequency distribution for the population.

PARAMETERS AND STATISTICS

In Chapter 1 we defined several statistics that can be calculated to describe a set of data — the mean, variance, standard deviation, covariance, and correlation coefficient. These numbers are calculated from the sample and are termed *sample statistics*. Corresponding to the statistics, there are *population parameters*—numbers that describe the population. For a large but not infinite population, we could calculate the population parameters if we had measurements for all the individuals in the population. In practice, however, this is rarely possible. We would like to know a certain population parameter; we have data only on a sample, and so we calculate the corresponding sample statistic.

Before discussing individual parameters, let us introduce a convenient notation, the expected value.

THE EXPECTED VALUE

The *expected value* of any quantity denotes its average value, averaged over the entire population. Thus for a large but finite population of blood pressure measurements, the expected value of Y (where Y denotes blood pressure) would be obtained by taking the arithmetic mean of all blood pressures. The expected value of Y^2 would be obtained by squaring every blood pressure measurement and then taking the mean of all these values of Y^2.

For a population of infinite size, the expected value of any quantity is still its average value over the whole population. Now the averaging must be done by methods of calculus (i.e., by a limiting process), but the method used is not of concern to us here.

The symbol E denotes expected value; EY^2 means the expected value of Y^2.

IMPORTANT POPULATION PARAMETERS

With the expected value notation, the parameters corresponding to the sample statistics introduced in Chapter 1 can be quickly defined. In general, Greek letters or roman script letters are used for population parameters.

The population mean for a variable Y is defined as the expected value of Y. It is usually denoted by μ (mu); its definition is

$$\mu = EY.$$

The population mean of the Y's is sometimes denoted by μ_y; when there is no ambiguity, however, the y subscript is omitted. For a finite population, we could compute EY by using

$$\mu = \frac{\sum_{i=1}^{\mathfrak{n}} Y_i}{\mathfrak{n}}$$

where $\mathfrak{n}$ is the number of observations in the population.

The *population variance* is defined as the expected value of the quantity $(Y-\mu)^2$. It is usually denoted by σ_y^2 or by σ^2 (sigma squared). In symbols, we have

$$\sigma^2 = E(Y-\mu)^2.$$

Thus the population variance is the mean over the entire population of the squares of deviations from the population mean. For a finite population, we can write

$$\sigma^2 = \frac{\sum_{i=1}^{\mathfrak{n}} (Y-\mu)^2}{\mathfrak{n}}.$$

The population variance is a measure of the variability of the population; a large value of σ^2 occurs when the population is widely spread about the mean.

The *population standard deviation* is the square root of the population variance and is denoted by σ (read "sigma") or sometimes by σ_y.

The *population covariance* is defined as the expected value of the quantity $(X-\mu_x)(Y-\mu_y)$. We are considering a population of individuals with two variables X and Y (say, blood pressure and weight) for each individual of the population. For each individual there is a product of the deviation of his blood pressure from the population mean blood pressure times the deviation of his weight from the population mean weight. The average over the entire population is the covariance of blood pressure and weight. In symbols,

$$\operatorname{Cov} X, Y = E(X-\mu_x)(Y-\mu_y).$$

The population covariance is positive when large values of X tend to correspond to large values of Y and small values correspond to small values; it is negative when large values of one variable correspond with small values of the other variable.

The *population correlation coefficient* between X and Y is defined as the covariance of X and Y divided by the product of their standard deviations. It is denoted by ρ or, if advisable for clarity, by ρ_{xy}. In symbols,

$$\rho = \frac{\operatorname{Cov} X, Y}{\sigma_x \sigma_y}.$$

The parameter ρ is similar to the covariance; it has the advantage of being independent of the units of measurement.

We have on the one hand the population, on the other hand the sample. There are parameters we wish to know but cannot know exactly because we do not have measurements for the entire population, and there are the statistics that we actually calculate from our data. One purpose of statistical inference is to make meaningful statements about the population parameters when all that is available is the sample statistics. The distinction between population and sample, parameter and statistics should always be kept in mind. Table 2.1 shows them side by side.

THE SIMPLE RANDOM SAMPLE

We have not yet discussed how to obtain a sample of individuals from the population. In order to draw inferences from the sample to the population,

TABLE 2.1. LIST OF POPULATION PARAMETERS AND SAMPLE STATISTICS

	Population Parameter	Sample Statistic
Mean	$\mu = E\,Y$	$\bar{Y} = \sum_{i=1}^{n} Y_i/n$
Variance	$\sigma^2 = E(Y-\mu)^2$	$s^2 = \sum_{i=1}^{n} (Y_i-\bar{Y})^2/(n-1)$
Standard Deviation	$\sigma = \sqrt{E(Y-\mu)^2}$	$s = \sqrt{\sum_{i=1}^{n} (Y_i-\bar{Y})^2/(n-1)}$
Covariance	$\text{Cov } X,Y = E(X-\mu_x)(Y-\mu_y)$	$\widehat{\text{Cov}}\ X,Y = \sum_{i=1}^{n} (X_i-\bar{X})(Y_i-\bar{Y})/(n-1)$
Correlation Coefficient	$\rho = (\text{Cov } X,Y)/\sigma_x\sigma_y$	$r = \sum_{i=1}^{n} (X_i-\bar{X})(Y_i-\bar{Y})/\sqrt{\sum_{i=1}^{n} (X_i-\bar{X})^2 \sum_{i=1}^{n} (Y_i-\bar{Y})^2}$

we wish our sample to be a random sample. The random sample as we define it here is sometimes called a *simple random sample*.

By a random sample of size n, we mean a subset of size n chosen from the population in a manner ensuring that every size n subset of the population is equally likely to be chosen.

It follows from this definition of random sample that we must know how a sample is obtained in order to know whether it is a random sample. We cannot know with certainty whether the sample is random by simply looking at it. For example, the dealer in a bridge game deals 13 cards to the player called North. These 13 cards form a sample of size 13 from a population of size 52. If the dealer has shuffled the cards properly and dealt them in proper order, North's hand is a simple random sample. This is true regardless of whether the hand is an ordinary one or consists of 13 spades. If the dealer cheated by trying to deal North a particularly good hand, the hand is not a random sample.

We wish to work with random samples because only then can we use the laws of probability to draw inferences from the sample to the population. The practical aspects of drawing random samples from large populations, such as the people of the United States or the fish in a lake, form a subject in themselves and are not treated here. For a discussion of various types of sample and for problems in sampling, see the references at the end of this chapter.

STATISTICAL INDEPENDENCE

We often say that one variable is statistically independent of another variable. By this we mean that knowledge of either variable tells us nothing about the other variable. For example, if blood pressure and height are statistically independent, knowledge of an individual's height will not help us in trying to guess his blood pressure. It is of interest that the covariance of statistically independent variables is zero. The converse—that variables with a zero covariance must be statistically independent—does not hold in general. However, if the variables have a particular distribution called normal, which is discussed later in this chapter, zero covariance *does* imply independence.

When we draw a random sample from an infinite population, each observation in the sample is statistically independent of all other observations in the sample. Knowledge that the first individual in the sample has a blood pressure that is unusually high for this population gives us no help in trying to guess whether the next individual in the sample has a particularly high or low blood pressure. Observations in a random sample are statistically independent only if the population is infinite. This may be seen by considering the sample of 13 cards from a 52-card deck. Looking at some of the cards in the hand gives us information concerning the rest of the hand.

LINEAR COMBINATIONS OF VARIABLES AND THEIR PARAMETERS

Often it is desirable to form new variables by combining several variables. In particular, linear combinations of variables occur frequently. The population means, variances, and covariances of such linear combinations are obtained easily in terms of the parameters of the original variables. Here we give the rules for calculating such means, variances, and covariances. The rules are given first in words, then illustrated, and finally given in symbols.

THE MEAN

The expected value of a linear combination of several variables is obtained by writing down the same linear combination of the expected values of the variables.

As an example, suppose that X_1 denotes a score on a midsemester examination in a mathematics course, that X_2 denotes the score on the final examination in the same course, and that in a large population

$EX_1 = 70$ and $EX_2 = 50$. As a measure of performance in the course, we use a weighted average, say $U = 0.4X_1 + 0.6X_2$. The mean of U's is $EU = 0.4EX_1 + 0.6EX_2 = 0.4(70) + 0.6(50) = 28 + 30 = 58$.

Another example is the mean of a sample of size n from a population. The formula for the sample mean is

$$\overline{Y} = \frac{Y_1 + Y_2 + \cdots + Y_n}{n}.$$

We see immediately that $\overline{Y}$ is simply a linear combination of the observations, with weights all equal to $1/n$. Therefore,

$$E\overline{Y} = \frac{1}{n}EY_1 + \frac{1}{n}EY_2 + \cdots + \frac{1}{n}EY_n.$$

If the sampling is done from a population whose mean is μ, then in repeated sampling the mean of the first observation in the sample is μ. In other words, $EY_1 = \mu$, and similarly $EY_2 = \mu, \ldots, EY_n = \mu$. Thus we can write

$$\begin{aligned} E\overline{Y} &= \frac{1}{n}\mu + \frac{1}{n}\mu + \cdots + \frac{1}{n}\mu \\ &= n\frac{1}{n}\mu \\ &= \mu. \end{aligned}$$

THE VARIANCE

The variance of a linear combination of variables is obtained by first squaring the linear combination; then in the expression for the square of the linear combination, we replace the square of a variable by that variable's variance, and we replace the product of two variables by their covariance.

In the example $U = 0.4X_1 + 0.6X_2$, let us assume that the population variances for the scores on the midsemester examination and the final examination are, respectively, 100 and 150, and that the covariance of the two scores is 80. That is, $\operatorname{Var} X_1 = 100$, $\operatorname{Var} X_2 = 150$, $\operatorname{Cov} X_1, X_2 = 80$. First squaring U, we have

$$\begin{aligned} U^2 = 0.16X_1^2 &+ 0.24X_1X_2 \\ &+ 0.24X_2X_1 + 0.36X_2^2. \end{aligned}$$

We now replace X_1^2 by $\operatorname{Var} X_1 = 100$, X_1X_2 by $\operatorname{Cov} X_1X_2 = 80$, and X_2^2 by $\operatorname{Var} X_2 = 150$, and have

$$\begin{aligned}\operatorname{Var} U &= 0.16(100) + 0.24(80) \\ &\quad + 0.24(80) + 0.36(150) \\ &= 16 + 19.2 + 19.2 + 54 \\ &= 108.4.^{*}\end{aligned}$$

As another example, we find the variance of the sample mean $\overline{Y}$:

$$\begin{aligned}\overline{Y}^2 = (&Y_1^2 + Y_1Y_2 + \cdots + Y_1Y_n \\ &+ Y_2Y_1 + Y_2^2 + \cdots + Y_2Y_n \\ &+ \cdots \\ &+ Y_nY_1 + Y_nY_2 + \cdots + Y_n^2)/n^2.\end{aligned}$$

The variance of Y_i, the ith observation in the sample, is simply the population variance σ^2. Furthermore, any two observations in a sample drawn from an entire population are statistically independent, and therefore their covariance is zero. Thus when we substitute variances and covariances in the expression for $\overline{Y}^2$ we have

$$\begin{aligned}\operatorname{Var} \overline{Y} = (&\sigma^2 + 0 + \cdots + 0 \\ &+ 0 + \sigma^2 + \cdots + 0 \\ &+ \cdots \\ &+ 0 + 0 + \cdots + \sigma^2)/n^2.\end{aligned}$$

---

* The formula for the variance of U can be derived as follows:

$$\begin{aligned}U &= 0.4X_1 + 0.6X_2 \\ EU &= 0.4EX_1 + 0.6EX_2 \\ U - EU &= 0.4(X_1 - EX_1) + 0.6(X_2 - EX_2)\end{aligned}$$

$$(U - EU)^2 = (0.4)^2(X_1 - EX_1)^2 + 2(0.4)(0.6)(X_1 - EX_1)(X_2 - EX_2) + (0.6)^2(X_2 - EX_2)^2.$$

Thus

$$\operatorname{Var} U = (0.4)^2 \operatorname{Var} X_1 + 2(0.4)(0.6) \operatorname{Cov} X_1, X_2 + (0.6)^2 \operatorname{Var} X_2.$$

In the numerator there are n variances so that

$$\text{Var}\,\overline{Y} = n\frac{\sigma^2}{n^2}$$

$$= \frac{\sigma^2}{n}.$$

This formula for the variance $\overline{Y}$ is extremely useful.

THE COVARIANCE

The covariance of two linear combinations of variables is obtained by first multiplying the two linear combinations; in the expression for the product, we replace the square of a variable by that variable's variance and the product of two variables by their covariance.

As an example, let us use as one variate $U = 0.4X_1 + 0.6X_2$ as before. As a second variate, we use $V = 0.3Y_1 + 0.3Y_2 + 0.4Y_3$, where Y_1, Y_2, and Y_3 are scores from three examinations in a statistics course. The variance of V is calculated as in the case of U from the variances and covariances of the three Y_i's and is found to be 98.9. The covariances between the X's and the Y's are

$$\text{Cov}\,X_1, Y_1 = 40, \qquad \text{Cov}\,X_1, Y_2 = 20, \qquad \text{Cov}\,X_1, Y_3 = 60,$$

$$\text{Cov}\,X_2, Y_1 = 50, \qquad \text{Cov}\,X_2, Y_2 = 10, \qquad \text{Cov}\,X_2, Y_3 = 50.$$

Multiplying the expressions for U and V, we have

$$UV = 0.12X_1Y_1 + 0.12X_1Y_2 + 0.16X_1Y_3 + 0.18X_2Y_1 + 0.18X_2Y_2 + 0.24X_2Y_3.$$

Substitution of the appropriate covariances gives

$$\begin{aligned} \text{Cov}\,U, V &= 0.12(40) + 0.12(20) + 0.16(60) \\ &\quad + 0.18(50) + 0.18(10) + 0.24(50) \\ &= 39.60. \end{aligned}$$

The two variables U and V happen to contain no variates in common, but it is perfectly possible to have two linear combinations containing some common variables.

To find the correlation coefficient of U and V, we now calculate their standard deviations. The standard deviation of U equals $\sqrt{108.4}$ or 10.4; the standard deviation of V equals $\sqrt{98.9}$ or 9.94. The correlation coefficient between U and V is then $39.60/(10.4)(9.94) = 0.383$. Thus individuals who perform well on the mathematics tests have some tendency to do well on the statistics tests.

The rules for finding the means, variances, and covariances of linear combinations of variables are simpler when written in symbols instead of words. They are

$$E\left(\sum_{i=1}^{m} a_i Y_i\right) = \sum_{i=1}^{m} a_i E Y_i. \tag{2.1}$$

$$\operatorname{Var}\left(\sum_{i=1}^{m} a_i Y_i\right) = \sum_{i=1}^{m} a_i^2 \operatorname{Var} Y_i + \sum_{\substack{i=1\\i\neq j}}^{m} \sum_{j=1}^{m} a_i a_j \operatorname{Cov} Y_i, Y_j. \tag{2.2}$$

$$\operatorname{Cov}\left(\sum_{i=1}^{m} a_i Y_i, \sum_{j=1}^{m} b_j Y_j\right) = \sum_{i=1}^{m} a_i b_i \operatorname{Var} Y_i + \sum_{\substack{i=1\\i\neq j}}^{m} \sum_{j=1}^{m} a_i b_j \operatorname{Cov} Y_i, Y_j. \tag{2.3}$$

In these expressions, $Y_1, Y_2, \ldots, Y_m$ denote any m variates, and $a_1, \ldots, a_m$, $b_1, \ldots, b_m$ denote any constants. Since some of the constants may be zero, the two linear combinations may or may not contain the same variables; they also may or may not contain the same number of variables. Often the Y_i and Y_j are independent. In this case, $\operatorname{Cov} Y_i, Y_j$ equals zero and the formulae simplify to

$$\operatorname{Var}\left(\sum_{i=1}^{m} a_i Y_i\right) = \sum_{i=1}^{m} a_i^2 \operatorname{Var} Y_i. \tag{2.4}$$

$$\operatorname{Cov}\left(\sum_{i=1}^{m} a_i Y_i, \sum_{i=1}^{m} b_i, Y_i\right) = \sum_{i=1}^{m} a_i b_i \operatorname{Var} Y_i. \tag{2.5}$$

These rules are used throughout the text to obtain needed expected values, variances, and covariances.

THE NORMAL DISTRIBUTION

Sometimes population frequency distributions can be approximated by mathematical curves. One advantage of such theoretical frequency distributions is that various areas or proportions under the curves can be

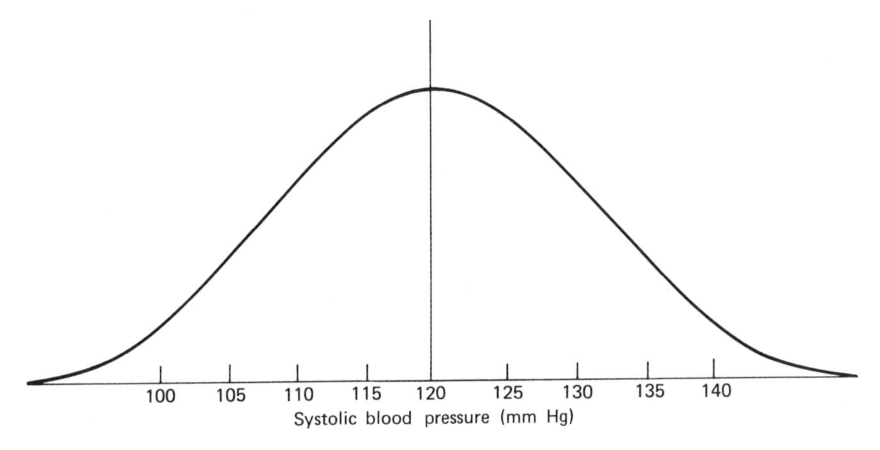

Figure 2.1. Normal distribution where $\mu = 120$ and $\sigma = 11$.

obtained directly from tables. One of the most important theoretical frequency distributions in statistics is called the normal distribution.*

If we say that a variable such as blood pressure of college girls has a normal distribution, we mean that an infinite population of blood pressures has as its distribution a curve that is symmetric and bell-shaped. The curve extends an infinite distance in both directions but comes very close to the horizontal axis. Thus, if the population of blood pressures followed a normal distribution, a very small proportion of blood pressures would be above any large number such as 400. This obviously would not occur, and of course it is also out of the question for a small proportion of the population to have blood pressures below zero. Thus it is clearly impossible for an infinite population of blood pressures to be exactly normally distributed. For practical purposes, however, blood pressure for girls might be considered to be normally distributed. That is, the areas under the normal curve might closely approximate the proportions of the population with the corresponding blood pressures.

A normal distribution has two parameters that describe it completely—the mean and the variance (or its square root, the standard deviation). That is, if we know that blood pressures are (approximately) normally

* The mathematical formula for the ordinate of a normal distribution whose mean is μ and variance is σ^2 is

$$f(y) = \frac{1}{\sqrt{2\pi}\,\sigma} e^{-(y-\mu)^2/2\sigma^2},$$

where $\pi = 3.1416$ and $e = 2.7183$.

distributed, that the mean blood pressure for the population is 120 mm Hg, and that the standard deviation of the population of blood pressures is 11 mm Hg, we can plot the distribution curve for blood pressures. The mean tells us where the highest point of the distribution lies. The standard deviation tells us how spread out the distribution is. Figure 2.1 shows a distribution of blood pressures for normal individuals with mean systolic blood pressure $\mu = 120$ mm Hg and standard deviation $\sigma = 11$ mm Hg.

AREAS UNDER THE NORMAL CURVE

From Figure 2.1, we can read roughly by eye that about 50% of normal individuals have blood pressures between 110 and 130. Similarly, about 90% of the area seems to lie between 100 and 140; therefore, about 90% of blood pressures should be between 100 and 140.

In order to measure areas under normal curves more precisely, we use tables that have been developed for obtaining areas under the standard normal distribution. The standard normal distribution is the normal distribution whose mean equals zero and whose variance equals one. Table A.1 of the Appendix gives areas under the standard normal curve for various points on the horizontal axis. The areas in the table (denoted by λ) extend from minus infinity to $z[\lambda]$ and are areas between the curve and the horizontal axis.

To obtain areas under the normal curve with mean $\mu = 120$ and $\sigma = 11$ mm Hg, we use the transformation $z = (X - 120)/11$ or, more generally, $z = (X - \mu)/\sigma$. For details on the use of the normal tables see Refs. 1 and 2.

We write the statement that blood pressures Y are normally distributed with mean $\mu = 120$ and variance $\sigma^2 = 121 (\sigma^2 = 11^2 = 121)$ as

$$Y \text{ND}(120, 121) \qquad \text{or} \qquad Y \text{ND}(\mu, \sigma^2)$$

where ND denotes normal distribution; the first number inside the parentheses is the mean, the second is the variance.

THE DISTRIBUTION OF THE SAMPLE MEAN

The normal distribution in statistics is important partly because many large populations have distributions that are approximately normally distributed. It is also important because many statistics have distributions that are approximately normal. We introduce here the idea of the distribution of a statistic, using the sample mean as an illustration.

For a population of finite size, we consider all possible samples of size n which could be formed from the population; each sample has a sample mean $\bar{Y}$. For a population of even moderate size, it is impractical to form all possible samples of size n, for there are too many of them and the process would take too long. For a very small population, we can actually write down all the samples of a given size and calculate all their sample means. All these values of $\bar{Y}$ form a population, which is called the population of sample means.

As an example of a population of sample means, consider the very small population consisting of the four numbers: 1, 3, 3, and 9. The population mean is $EY=(1+3+3+9)/4=4$. The population variance is

$$\sigma^2=E(Y-4)^2=\frac{(1-4)^2+(3-4)^2+(3-4)^2+(9-4)^2}{4}$$

$$=9.$$

Now let us write down all possible samples of size 2 from this population and calculate the sample means. They are:

Sample	Sample Mean
(1,3)	2
(1,3)	2
(1,9)	5
(3,3)	3
(3,9)	6
(3,9)	6

It should be clear that the two samples listed as (1,3) are actually two different samples, for one contains the first 3 of the population, the other the second 3. Taking samples of size 2 ($n=2$) and calculating all possible sample means, we have generated a population of $\bar{Y}$'s.

We now calculate the population mean for this population of sample means; it is

$$E\bar{Y}=\frac{2+2+5+3+6+6}{6}$$

$$=4.$$

We notice that the mean of the populations of $\overline{Y}$'s is the same as the mean of the original population of Y's; that is, both populations have means of 4. The mean of the population of $\overline{Y}$'s for samples of a given size will be denoted by $E\overline{Y}$. This is an illustration of an important fact concerning the distribution of sample means: $E\overline{Y}=\mu$.

From the rule given for calculating expected values of linear combinations, we showed earlier that $E\overline{Y}=\mu$. This is true regardless of the type of population from which we sample. The statistic $\overline{Y}$ is called an *unbiased* estimate of μ.

If the population is infinite in size, we can no longer write down all possible samples of a given size and calculate all possible sample means. Nevertheless, we can conceive of such a population of sample means. We can think of drawing one sample after another in a random fashion and calculating the mean of each sample. If we drew a very large number of such samples, we would not have the entire population of $\overline{Y}$'s, but we would have a set of $\overline{Y}$'s whose distribution should approximate very closely the distribution of $\overline{Y}$'s. The mean of all the $\overline{Y}$'s that we calculate should be very close to μ.

Another fact can be shown mathematically concerning the distribution of the sample means. We know from the section on variances of linear combinations that the formula for the variance of this distribution is

$$\operatorname{Var} \overline{Y}=\frac{\sigma^2}{n}, \tag{2.6}$$

where n is the sample size. Formula 2.6 holds for any population whatsoever, provided the distribution has a finite mean and variance and the population is of infinite size.

When the population is finite, as in the example of the small population consisting of the four numbers, 1, 3, 3, and 9, (2.6) no longer holds. The formula for Var $\overline{Y}$ when the population is of finite size is

$$\operatorname{Var} \overline{Y}=\frac{\sigma^2}{n}\left(1-\frac{n-1}{N-1}\right). \tag{2.7}$$

In almost all sampling problems we sample from a large but finite population, and the size of the sample is small compared with the size of the population. We therefore neglect the $(n-1)/(N-1)$ in (2.7) and thus use (2.6). When used in this manner, (2.6) overestimates the variance of the distribution of $\overline{Y}$'s; its use is conservative. Unless otherwise specified, we assume that our sample size is a small fraction of the population size, and we use Var $\overline{Y}=\sigma^2/n$.

A third fact concerning the distribution of the population of $\bar{Y}$'s can be shown mathematically. Under very general conditions it can be proved that, regardless of the kind of population from which we draw samples of size n, the distribution of the sample means is approximately normal, provided n is sufficiently large. This statement is intentionally vague—we have not specified how close to normality "approximately normal" means, nor have we defined "sufficiently large." For a given population of $\bar{Y}$'s, the sample size necessary to have the distribution of $\bar{Y}$'s approximately normal depends on how close to normality we wish the $\bar{Y}$ distribution to be. Furthermore, after deciding how close we want the approximation to normality to be, the necessary sample size depends on the distribution of the population from which the samples are drawn. For example, if the population of Y's has a normal distribution, the distribution of sample means is *exactly* normal, even with a sample size as small as $n=1$ or 2. On the other hand, in sampling from a distribution that is far from normal, we need a larger sample size in order to have the distribution of $\bar{Y}$'s nearly normal.

The foregoing facts concerning the distribution of $\bar{Y}$ have merely been stated and illustrated, but they can be proved mathematically. The importance of knowing these properties of the sampling distribution of $\bar{Y}$ has not been considered; this can be seen in the next chapter.

DISCUSSION

Distributions of observations such as heights and weights often tend to be slightly skewed to the right rather than being normally distributed. For example, since we find some extremely heavy persons and no one who weighs less than zero, the distribution of weights tends to have a long right-hand tail. In most problems concerning the mean of such a distribution, this slight skewness causes no difficulty and we can proceed as if we had normality. Extreme skewness can be handled by making a transformation on the data that lowers larger values more than it lowers smaller values. Statements are then made on the transformed data. Further discussion on skewness is found in Chapter 14. Good procedure includes making histograms of the data to check the form of their distribution before proceeding with the confidence intervals and tests discussed in the remainder of this book. Such checking can be done using prepared computer programs.

As mentioned previously, even though the population is far from normal, sample means tend to be normally distributed. Many students find it difficult to accept this idea. For them, simulation studies can be helpful. For example, random samples of size 25 can be drawn from the uniform distribution (a distribution in which all numbers between zero and one are equally likely to occur). Means of these random samples can be computed and their histograms displayed; such histograms tend to look convincingly normal. These simulations can be performed using readily available subroutines. Examples can be found in Dixon and Massey [2.2]. It is also possible to obtain samples from distributions that are even farther from normality than the uniform distribution.

SUMMARY

In this chapter we have discussed the problem of making inferences from samples to populations. Important population parameters were defined both verbally and in formulae. The formulae for linear combinations of variables were stated. The normal distribution was defined and its role as a limiting distribution for sample means discussed.

REFERENCES

APPLIED STATISTICS

2.1. Dunn, O. J., *Basic Statistics: A Primer for the Biomedical Sciences*, Wiley, New York, 1964.

2.2. Dixon, W. J., and F. J. Massey, *Introduction to Statistical Analysis*, 3rd ed., McGraw-Hill, New York, 1969.

2.3. Smart, J. V., *Elements of Medical Statistics*, Staples Press, London, 1963.

2.4. Steel, R. G. D., and J. H. Torrie, *Principles and Procedures of Statistics*, McGraw-Hill, New York, 1960.

MATHEMATICAL STATISTICS

2.5. Feller, W., *An Introduction to Probability Theory and Its Application*, Vol. 1, 2nd ed., Wiley, New York, 1957.

2.6. Hoel, P. G., *Introduction to Mathematical Statistics*, 3rd ed., Wiley, New York, 1962.

2.7. Mood, A., and F. Graybill, *Introduction to the Theory of Statistics*, McGraw-Hill, New York, 1963.

SAMPLING

2.8. Hanson, M. H., W. N. Hurwitz, and W. G. Madow, *Sampling Survey Methods and Theory*, Vols. I and II, Wiley, New York, 1953.

2.9. Kish, L., *Survey Sampling*, Wiley, New York, 1965.

PROBLEMS

2.1. Given a population of four elements whose values are 2, 5, 1, 7, draw all possible random samples of three elements each.
(*a*) How many possible samples are there?
(*b*) What is $E\overline{Y}$, the mean of all the sample means?
(*c*) What is Var $\overline{Y}$, the variance of all the sample means?
(*d*) What is the variance of the population (σ^2)?
(*e*) By what number do we divide σ^2 to obtain Var $\overline{Y}$?

2.2. If $Y = 1.8X + 32$, where X is temperature (°C) and Y is temperature (°F), find the expected value and the variance of the temperature in °F, given that $EX = 20$ and Var $X = 5$.

2.3. Given $EY_1 = 10$, $EY_2 = 12$, Var $Y_1 = 11$, Var $Y_2 = 12$, and Cov $Y_1, Y_2 = 2$, find the expected value and the variance for $V = Y_1 + 3Y_2 + 10$.

2.4. For the normal population with $\mu = 10$ and $\sigma^2 = 4$, find the proportion falling in the regions: 10 to 12, 10 to 14, below 14, above 14.

2.5. If in a normal population the average temperature is 98.6°F with $\sigma = 0.2$°F, what proportion of persons would you expect to have temperatures between 98.4 and 100°F?

CHAPTER 3

INFERENCE FROM A SINGLE SAMPLE

In this chapter we consider the simplest problem in statistical inference; we have a single random sample and a single variable Y measured on each individual of the sample. As an example, we use a sample of 8 pieces of a certain aluminum die casting. The strength (in 1000 psi) was measured for each of the pieces. The data are:

i	1	2	3	4	5	6	7	8
Y_i	29.4	30.8	30.6	31.5	32.1	31.7	30.3	30.8

First, we may wish to learn something about the population mean μ, the mean strength of all similar pieces of aluminum die casting. The first statistic to obtain is $\overline{Y}$, and we calculate

$$\overline{Y} = \frac{\sum_{i=1}^{8} Y_i}{8} = \frac{247.2}{8} = 30.90.$$

The sample mean 30.90 is our point estimate of μ, the population mean. The point estimate of μ is the single number to be used as an estimate for the population mean.

We need something besides $\overline{Y}$, however. We wonder how far this estimate is from the true population mean. We cannot make any progress in this direction unless we know something about the variance of the distribution of sample means. If the variance is large, our value of $\overline{Y} = 30.90$ may easily be far from the population mean μ; if the variance is small, our value of $\overline{Y} = 30.90$ is probably close to μ.

We now apply the facts discussed in Chapter 2 to samples of size 8. If our sample were large, we would expect the mean $\overline{Y}$ to be approximately normally distributed. Here our sample is too small. Let us assume, however, that histograms drawn from previous sets of similar data have led us to believe that strengths of die castings are approximately normal. If the data themselves are from a normal distribution, the population of sample means is also normally distributed, with mean μ and variance $\sigma^2/8$. We also know from Chapter 2 that for any variable having a normal distribution we can make a change of scale to a standard normal variable by subtracting the mean and dividing the difference by the standard deviation. Here the normally distributed variable of interest is $\overline{Y}$, and we have

$$z=\frac{\overline{Y}-\mu}{\sqrt{\sigma^2/n}}=\frac{\overline{Y}-\mu}{\sqrt{\sigma^2/8}}. \tag{3.1}$$

We should like to learn something about μ by using the standard normal distribution, and we can do this if we know the population variance σ^2. Occasionally the population variance is known and the normal tables can be used, but usually σ^2 is unknown. Because an unknown population variance is the rule rather than the exception, we shall consider that situation first.

We estimate the unknown σ^2 by calculating s^2. From the basic formula (1.2) we have for our example

$$s^2=\frac{(29.4-30.9)^2+(30.8-30.9)^2+\cdots+(30.8-30.9)^2}{7}$$

$$=0.737.$$

The statistic $s^2=0.737$ is our point estimate for σ^2, the variance of the population of measurements of strength of the die castings. Because σ^2/n is the variance of the $\overline{Y}$ population, it is estimated by

$$\frac{s^2}{n}=\frac{s^2}{8}=\frac{0.737}{8}=0.0921.$$

Instead of using $z=(\overline{Y}-\mu)/\sqrt{\sigma^2/n}$, we use a corresponding expression with s^2 substituted for σ^2; this we denote by t, where

$$t=\frac{\overline{Y}-\mu}{\sqrt{s^2/n}}. \tag{3.2}$$

The data were assumed to be from a normal distribution. With this assumption for any particular population mean μ, the quantity t has a distribution very similar to the standard normal distribution. Like the normal distribution function, it is bell-shaped and symmetric about the mean value zero. The difference between the standard normal distribution and the t distribution is that somewhat less area is concentrated near zero and somewhat more area lies in the tails of the t distribution. The exact shape of the t distribution depends on the sample size n. We can easily imagine that the distribution of t for large samples would be very similar to the standard normal distribution, whereas for small samples the distribution would have a wider spread; this is indeed true.

The distribution of t was first obtained by Gosset. Since he published under the name of Student, the distribution is referred to as Student's t distribution.* Table A.2 gives some percentage points in the distribution of t. To use the table, we must know the appropriate *degrees of freedom* (df). The degrees of freedom in our example are $n-1=7$, which is the number used in the denominator of the estimate of the variance σ^2; that is, in $s^2=\sum_{i=1}^{n}(Y_i-\bar{Y})^2/(n-1)$. The degrees of freedom are denoted by either df or ν.

We need to know the degrees of freedom whenever we do not know σ^2. If we knew the population mean μ, our estimate of σ^2 would be $\sum_{i=1}^{n}(Y_i-\mu)^2/n$, and the degrees of freedom would be n. When the population mean is unknown, it is estimated by the sample mean $\bar{Y}$. One degree of freedom is subtracted; thus the degrees of freedom for s^2 is $n-1$. In estimating μ by $\bar{Y}$, we are limiting ourselves to samples that have $\bar{Y}$ as their sample mean; with such samples, if the first $n-1$ observations were chosen, the last observation could be determined in such a way that the mean of the n observations would be $\bar{Y}$. Thus we say that there are only $n-1$ degrees of freedom in estimating s^2.

In general, the degrees of freedom for an estimate of a variance are the number of observations used minus the number of parameters which must be estimated from the data in order to calculate the estimated variance.

The percentages given in the body of Table A.2 are numbers below which a certain proportion of the distribution lies. That is, under .75, we find, for degrees of freedom 7 that 75% of the distribution lies between $-\infty$ and 0.711. We denote percentiles of the t distribution by $t[\lambda;\nu]$, where λ is the percentage and ν is the degrees of freedom.

* The density function for the t distribution with degrees of freedom $=\nu$ is

$$f(t)=\frac{[(\nu-1)/2]!(1+t^2/\nu)^{-(1/2)(\nu+1)}}{\sqrt{\pi\nu}\,[(\nu-2)/2]!}.$$

A CONFIDENCE INTERVAL FOR μ

We now use the t distribution to obtain what is called a 95% *confidence interval* for μ, the population mean. We are looking for an interval within which we have reason to believe that the true population mean may lie. First we obtain the confidence interval for the strength of aluminum die castings and then we explain its meaning.

We know that $\bar{Y}$ is $ND(\mu, \sigma^2/8)$, and this tells us that the variable $t = (\bar{Y} - \mu)/\sqrt{s^2/8}$ has a t distribution. From Table A.2 under $\nu = 7$ we find the 97.5% point of the t distribution, $t[.975; 7] = 2.365$. That is, in repeated sampling with samples of size 8 from a normal population with mean μ and variance σ^2, if t is calculated from each sample, 97.5% of the t's calculated will be less than $+2.365$. By symmetry, 2.5% of calculated t's will be less than -2.365. Therefore, 95% of all t's calculated from repeated samples of size 8 will lie between -2.365 and $+2.365$ (Figure 3.1). The formula for the 95% confidence interval for μ is then

$$\bar{Y} \pm t[.975; n-1]\sqrt{s^2/n}\,. \tag{3.3}$$

This formula is justified in the next section.

In the example, the interval for μ is

$$30.90 \pm 2.365\sqrt{0.737/8}\,,$$

$$30.90 \pm 0.72, \qquad \text{or} \qquad 30.18 < \mu < 31.62.$$

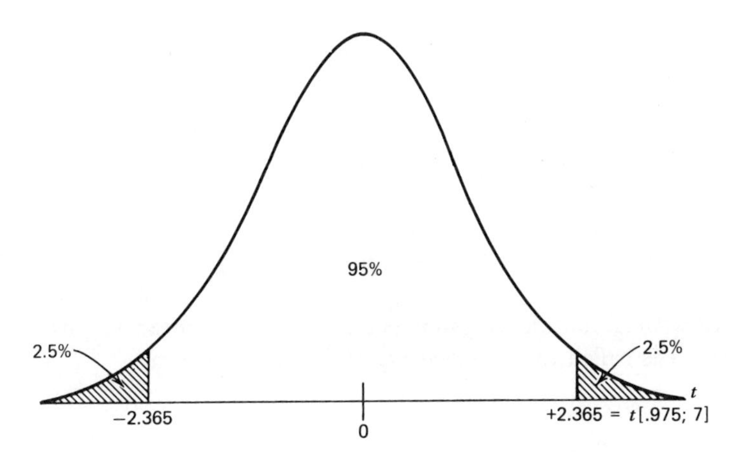

Figure 3.1. Area under a t distribution for $\nu = 7$.

MEANING AND JUSTIFICATION OF THE CONFIDENCE INTERVAL

The meaning of the 95% confidence interval just obtained for the population mean is as follows. We know that our point estimate of the population mean was based on a single sample of 8 observations; the confidence interval also was based on this single sample. If we started over and drew another sample of 8 pieces of aluminum casting, we could calculate another sample mean and another 95% confidence interval, again using (3.3). These values would undoubtedly differ from those calculated from the first sample. If we continued to draw many samples of size 8 and to calculate a 95% confidence interval from each, in the long run 95% of such intervals would cover the population mean μ. We think that the population mean probably lies between 30.18 and 31.62, because, using the same method of obtaining a confidence interval for each sample, 95% of the intervals obtained in repeated sampling cover the population mean. Five percent of the time we would be unlucky and draw a sample yielding a confidence interval that does not cover the population mean.

That 95% of all such intervals cover μ can be seen as follows. From Figure 3.1 we know that the area under the t distribution between -2.365 and $+2.365$ is 95%. In other words, the proportion of t's between -2.365 and $+2.365$ is 95%. We can express the same idea by saying that the probability is .95 that a single t drawn at random lies between -2.365 and $+2.365$. In symbols, we write

$$\text{Prob}[-2.365 < t < +2.365] = .95$$

or

$$\text{Prob}\left[-2.365 < \frac{\bar{Y}-\mu}{\sqrt{s^2/n}} < +2.365\right] = .95.$$

Algebraic manipulation of the expression in the brackets yields the equivalent statement

$$\text{Prob}\left[\bar{Y} - 2.365\sqrt{s^2/n} < \mu < \bar{Y} + 2.365\sqrt{s^2/n}\right] = .95. \tag{3.4}$$

If we wish a confidence interval that covers the mean $(1-\alpha)\%$ of the time, $1-\alpha$ is called the *confidence level*. We then have

$$\text{Prob}\left\{\bar{Y} - t[1-\alpha/2; n-1]\sqrt{s^2/n} < \mu < \bar{Y} + t[1-\alpha/2; n-1]\sqrt{s^2/n}\right\} = 1-\alpha. \tag{3.5}$$

Expression 3.5 tells us that the $(1-\alpha)\%$ confidence interval for μ is

$$\bar{Y} \pm t[1-\alpha/2; n-1]\sqrt{s^2/n}\ . \tag{3.6}$$

DISCUSSION OF THE CONFIDENCE INTERVAL

In obtaining a confidence interval for the population mean strength of aluminum castings, we have presented our first example of statistical inference. This represents a real achievement, for we now have reason to believe that the population mean lies between 30.18 and 31.62. Previously, we had only 30.90 as an estimate of the population mean and could merely guess that with a sample as small as 8, we should not necessarily expect the sample mean to lie close to μ. The confidence interval is invaluable in communicating results of research. If a research worker reports a sample mean of 30.90, we know nothing about the actual size of the population mean; if he gives a sample mean of 30.90 and a sample size of 8, we still do not know where it is reasonable to expect the population mean to lie. If he reports a sample mean of 30.90 and a 95% confidence interval of 30.18 to 31.62, however, we can be reasonably confident that the true mean falls between 30.18 and 31.62.

In connection with a 95% confidence interval, we speak of a 95% *confidence level*. Often confidence levels other than 95% are used; for example, 80, 90, or 99% may be preferred. To obtain an 80% confidence interval for the previous example, we use the 90th percentile of the t distribution (then 10% of the area is in each tail and 80% of the area is in the middle of the t distribution). The interval is

$$30.90 \pm 1.415(0.3035) \qquad \text{or} \qquad 30.47 < \mu < 31.33.$$

The 80% confidence interval is shorter than the 95% confidence interval, but we are not nearly so sure that it covers the population mean. If we wanted to be almost certain that the interval covers the mean, we might conceivably take a 99.999% interval. Here we would pay for our increased confidence in the interval by having a longer interval. The most frequently used confidence levels are 95 and 99%.

TESTS OF HYPOTHESES ON THE POPULATION MEAN

Another method of drawing inferences from the sample to the population is called the *statistical test*. We illustrate a statistical test of a population mean by using the data on die casting. To make a statistical test concerning a parameter, the experimenter should have some hypothesis in mind concerning the parameter and should then seek to find out whether the hypothesis is true. Such a hypothesis is called the *null hypothesis*, H_0. For example, before taking his measurements on the 8 pieces of aluminum, the experimenter might know that the population mean for an older type of aluminum die casting is 32.0. He then looks at the 8 castings in order to decide whether they are from a population with mean 32.0. In other words, he wishes to answer the question "Is the population mean strength of aluminum die casting equal to 32.0 or is it not?" His null hypothesis, then, is that the population mean for the newer type of die casting is 32.0; in symbols, $H_0: \mu = 32$.

We now attempt to prove or disprove the null hypothesis on the basis of the sample data. If μ is actually 32.0, then $t = (\bar{Y} - 32.0)/\sqrt{0.737/8}$ has a t distribution with 7 degrees of freedom. To make the test we calculate the value of t and see whether that value is very unusual when $\mu = 32.0$. We calculate

$$t = \frac{30.9 - 32.0}{.3035}$$

$$= \frac{-1.1}{.3035}$$

$$= -3.62.$$

We now look in Table A.2 under $\nu = 7$ and find that the proportion of t values larger than $+3.62$ is about .005; by symmetry, the proportion of t values smaller than -3.62 is about .005. Adding these two proportions, we obtain a quantity that we call the *P value* of the test. The P value is the probability under the null hypothesis (or the probability if the null hypothesis is true) of obtaining a value as unusual as or more unusual than the one we actually obtained from the sample. In our example, $P = .01$. Figure 3.2 shows the distribution of t; the shaded areas were added to obtain P. It is obvious that if the null hypothesis is true (i.e., if $\mu = 32.0$) we have obtained an unusual value of t with $P = .01$. We therefore conclude that the null hypothesis is probably not true and that the mean is not equal to 32.0. We *reject* the null hypothesis.

If the value obtained for P had been .47 instead of .01, we would have felt that our value of t was a usual one under the null hypothesis; we therefore would have had no reason to doubt the null hypothesis. We would accept (i.e., fail to reject) the null hypothesis and decide that μ may indeed be 32.0. With very small values of P, we reject H_0; with large values, we accept H_0. We then have the problem of deciding what is a large and what is a small value of P. As a dividing line between values of P

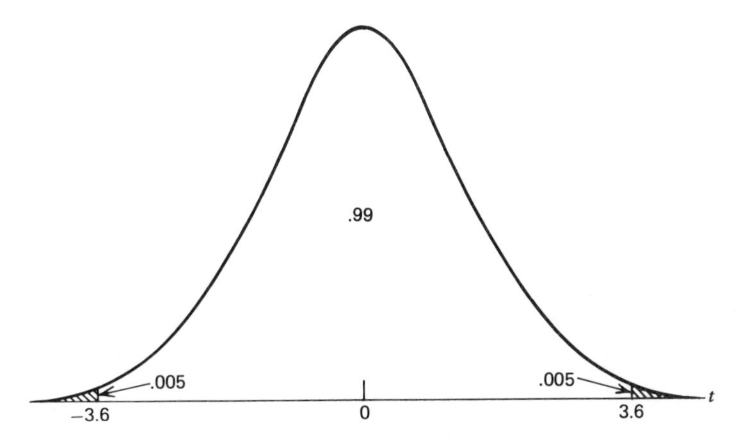

Figure 3.2. Example of t distribution for $\nu=7$.

that are small enough to cause us to reject the null hypothesis and those which will allow us to accept it, a *significance level* is usually chosen in advance for the test. The significance level is denoted by α. The significance level used most often is $\alpha=.05$; sometimes $\alpha=.01$ is used, and sometimes $\alpha=.10$. With a significance level of $\alpha=.05$, if P is smaller than .05, we reject the null hypothesis.

If the null hypothesis is actually true and $\mu=32.0$, repeated drawing of samples from the population and testing with $\alpha=.05$ will result for 5% of the samples in t values which are so far from zero that we reject the null hypothesis. Thus the level of significance α is equal to the proportion of the time that we would mistakenly reject the null hypothesis when it is true.

It is possible to make the level of significance extremely small in order to reduce the probability of rejecting the null hypothesis mistakenly. This approach creates difficulties when the null hypothesis is *not* true. If the null hypothesis is not true, we want to reject the null hypothesis and avoid the mistake of accepting H_0 when it is false. Two types of mistake can be made in the decision concerning H_0: if it is true, we may decide that it is false; if it is false, we may decide that it is true. The two types of error can be illustrated by considering a decision regarding whether it will rain before nightfall and the consequences thereof. If it rains but we have decided that it will not, we may be drenched because we left home without an umbrella; if it does not rain but we have decided that it will, we find ourselves carrying an unnecessary umbrella.

With any test, the probability of making the second type of error is called β. Thus we have as definitions of the two probabilities of error

$$\alpha = \text{probability of rejecting } H_0 \text{ when } H_0 \text{ is true;}$$

$$\beta = \text{probability of accepting } H_0 \text{ when } H_0 \text{ is false.}$$

It is customary to set the value of α in advance. The value of β then depends on the particular test, on the values of α and n, and on the actual value of the parameter. In the example, if $\mu = 33.0$ the value of β would be very different from what it would have been if μ were 50.0. For μ close to 32, β is large, because we usually fail to reject the null hypothesis. If μ is distant from 32.0, the null hypothesis is usually rejected, and the chance of accepting $H_0 : \mu = 32.0$ erroneously is small.

After making a statistical test, we decide either to reject or to accept the null hypothesis. If we reject it, we know that if the null hypothesis is true there was a probability of size α of rejecting it. With a small value of α, we can feel reasonably sure that the null hypothesis is not true. On the other hand, if we accept the null hypothesis, our decision should be that the null hypothesis *may be* true rather than that it *is* true. We have merely failed to disprove it, perhaps because our sample was small. If the test had resulted in acceptance of the null hypothesis in the aluminum casting example, we would not be convinced that the mean is really 32.0; we would merely have failed to establish that it differs from 32.0.

In the foregoing discussion of hypothesis testing, we have restricted ourselves to alternative hypotheses of the form $H_A : \mu \neq \mu_0$. There exist other alternative hypotheses such as $H_A : \mu = \mu_A$ where μ_A is a particular mean value different from μ_0, $H_A : \mu > \mu_0$ and $H_A : \mu < \mu_0$. A treatment in depth of hypothesis testing that includes a variety of alternative hypotheses can be found in Ferguson [3.8].

SUMMARY OF TEST PROCEDURE

Because formal test procedures are used so often, we outline the procedure we have presented:

1. State the null hypothesis—$H_0 : \mu = \mu_0$.
2. Decide on alternative hypotheses—$H_A : \mu \neq \mu_0$.
3. Pick a test statistic—$t = (\overline{Y} - \mu_0)/\sqrt{s^2/n}$.
4. Decide on α and n.
5. Draw sample and calculate statistic.
6. Calculate P value from tables.
7. Decide whether to accept or reject H_0 by comparing 4 and 6.

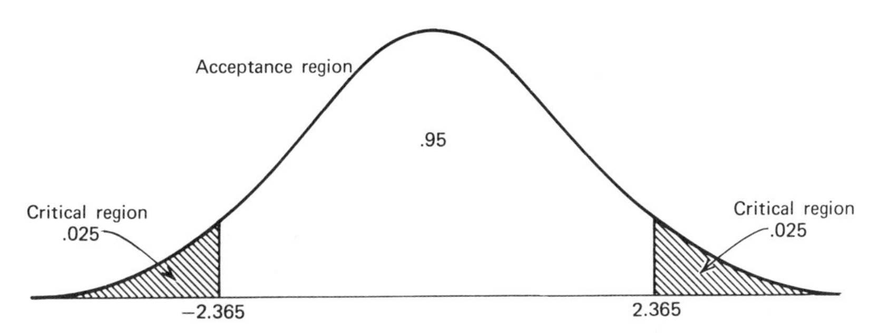

Figure 3.3. Acceptance and critical regions for t distributions for $\alpha = .05$ and $\nu = 7$.

AN ALTERNATIVE FORMULATION OF STATISTICAL TESTS

In making a statistical test, it is frequently impossible to obtain the exact P value because the available tables provide relatively few values. The test can still be made using an equivalent procedure. After stating H_0 and H_A, choosing a test statistic and a significance level α, we determine from the tables a *critical region* (sometimes called the *rejection region*). The critical region is the set of values of the calculated test statistic that will cause the null hypothesis to be rejected. Similarly, the set of values of the calculated test statistic that will cause the null hypothesis to be accepted is called the *acceptance region.*

In our example, we had $\alpha = .05$, $H_0 : \mu = 32.0$, $H_A : \mu_A \neq 32.0$, and we had decided on a test statistic $t = (\overline{Y} - \mu)/\sqrt{s^2/n}$, where $n = 8$. We choose as a critical region two equal tails of the the t distribution, each containing $\alpha/2 = 2.5\%$ of the entire distribution. From Table A.2, we find $t[.975, 7] = 2.365$. We know that 2.5% of the t values are larger than 2.365, and, by symmetry, 2.5% of the t values are less than -2.365. Therefore, we include in the critical region all values of $t > 2.365$ and $t < -2.365$, rejecting H_0 whenever the calculated t falls into the critical region. If H_0 is true, it will be rejected 5% of the time in repeated experimentation. Figure 3.3 indicates the critical region and acceptance region for the test. Since $t = -3.62$ in our example, H_0 is rejected.

INFERENCE CONCERNING THE MEAN WHEN THE POPULATION VARIANCE IS KNOWN

Occasionally a population variance is known. Then we use σ^2/n as the variance of the distribution of population means $\overline{Y}$. Instead of using the Student's t distribution, we use the normal tables. Thus the $1 - \alpha$ level confidence interval for μ becomes

$$\bar{Y} - z[1-\alpha/2]\sqrt{\sigma^2/n} < \mu < \bar{Y} + z[1-\alpha/2]\sqrt{\sigma^2/n}\,. \tag{3.7}$$

In the test of $H_0: \mu = \mu_0$ when the variance is known, we use $z = (\bar{Y} - \mu_0)/\sqrt{\sigma^2/n}$ as a test statistic.

INFERENCE CONCERNING THE POPULATION VARIANCE

With a single sample from a normal distribution, we have tables that enable us to state confidence intervals for σ^2, the population variance, or to test that it is equal to some specified number. These tests and intervals are highly dependent on the normality of the data. For example, they should not be used if the data are skewed.

It can be shown mathematically that provided the data are from a normal distribution, the quantity $(n-1)s^2/\sigma^2$ has a distribution that is called the χ^2 *distribution*.* As in the case of the t distribution, there are many χ^2 distributions, each characterized by the number of degrees of freedom ν. For the quantity $(n-1)s^2/\sigma^2$, $\nu = n-1$. The χ^2 distributions are skewed curves extending from zero to infinity. Figure 3.4 presents examples of the χ^2 distribution for $\nu = 1$, 2, and 6. Percentage points for the χ^2 distribution are in Table A.3.

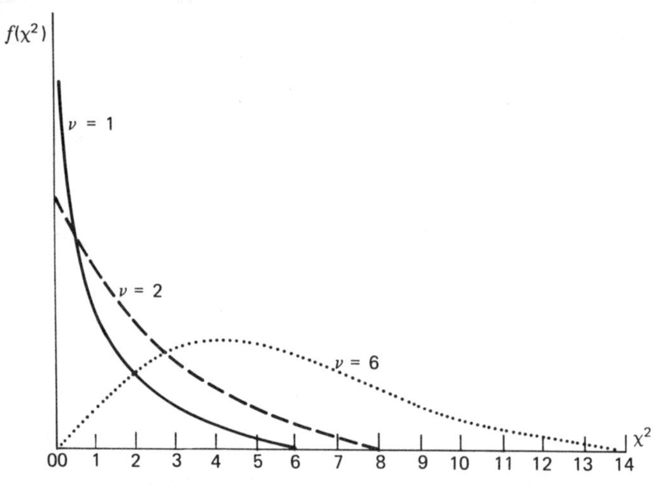

Figure 3.4. Example of several χ^2 distributions.

---

* The χ^2 distribution with ν degrees of freedom has the mathematical formula

$$f(\chi^2) = \frac{(\chi^2)^{\nu/2-1} e^{-\chi^2/2}}{2^{\nu/2}[(\nu-2)/2]!}$$

CONFIDENCE INTERVALS FOR σ^2

We now obtain a confidence interval for σ^2. For the sample of 8 aluminum castings we have already calculated $s^2=0.737$ as our point estimate of the population variance. We wish to obtain a 95% confidence interval for the population variance. In the last section we mentioned that the quantity $(n-1)s^2/\sigma^2$ has a χ^2 distribution with $n-1$ degrees of freedom. If $\chi^2[.025;n-1]$ denotes the point below which 2.5% of the distribution lies, and if 97.5% of the distribution lies above $\chi^2[.975;n-1]$, we have

$$\text{Prob}\left\{\chi^2[.025;n-1]<\frac{(n-1)s^2}{\sigma^2}<\chi^2[.975;n-1]\right\}=.95.$$

As in the case of the confidence interval for the population mean, this equation can be manipulated algebraically to the form

$$\text{Prob}\left\{\frac{(n-1)s^2}{\chi^2[.975;n-1]}<\sigma^2<\frac{(n-1)s^2}{\chi^2[.025;n-1]}\right\}=.95.$$

In other words, in repeated sampling, the intervals

$$\frac{(n-1)s^2}{\chi^2[.975;n-1]} \quad \text{to} \quad \frac{(n-1)s^2}{\chi^2[.025;n-1]}$$

cover σ^2 95% of the time. In our example, $\chi^2[.025;7]=1.690$ and $\chi^2[.975;7]=16.013$, and the 95% confidence interval is

$$0.322<\sigma^2<3.053.$$

In general, the $1-\alpha$ level confidence interval for σ^2 is

$$\frac{(n-1)s^2}{\chi^2[1-\alpha/2;n-1]}<\sigma^2<\frac{(n-1)s^2}{\chi^2[\alpha/2;n-1]}. \tag{3.8}$$

A $1-\alpha$ level confidence interval for σ can be obtained by taking the square root of the expressions in (3.8).

CONFIDENCE INTERVAL VERSUS TEST

Whenever it is equally convenient to make a confidence interval and a test, it is advisable to calculate the confidence interval. It is more informative to know that the population mean probably lies between 30.18 and 31.62 than to make a test and decide only that the population mean is not 32.0. As a matter of fact, after obtaining the 95% confidence interval for the population mean, we can use it as a test of $H_0: \mu = 32.0$ (or of any other value). Because 32.0 does not lie in the confidence interval, we reject the null hypothesis. Such a two-sided test has a level of significance of .05; for, if the null hypothesis is true, 5% of the time the confidence interval fails to contain 32.0 and we mistakenly reject the null hypothesis.

For certain problems, tables exist for making tests, but there are no tables for making intervals. Whenever possible, we use confidence intervals.

ILLUSTRATIVE EXAMPLE

It has been suggested that entering students at a certain college are brighter than students were in the old days. For a period of several years in the 1940s the college gave an IQ test to entering freshmen; the average score was 115 and the standard deviation was 10. A random sample of 49 students is drawn from the current entering class and the same examination is given to the 49 students.

The hypothesis that $\mu = 115$ can be tested directly, or confidence limits for the present-day mean can be examined to see whether they include the former mean, 115. We might also wish to see if IQs have the same variability today as formerly; then we might test $H_0: \sigma^2 = 10^2$ by calculating a confidence interval for σ^2 and observing whether 100 lies within the interval. The confidence interval for the population variance can be useful both as a test of $\sigma^2 = 10^2$ and to note the length of the interval. Confidence intervals for variances are often disappointingly long, although when converted into limits for the standard deviation they may not look too long.

SUMMARY OF CONFIDENCE INTERVALS AND TESTS FOR THE SINGLE SAMPLE

In this chapter, we introduced the concepts of statistical inference. The rationales for confidence intervals and for tests of hypotheses were given for the single sample case. In Table 3.1, which summarizes the confidence intervals and tests presented in this chapter, the test procedure is outlined in seven steps, listed in the order of their performance.

TABLE 3.1. SUMMARY OF CONFIDENCE INTERVALS AND TEST PROCEDURES FOR A SINGLE SAMPLE

Parameter	μ With σ^2 Unknown	μ With σ^2 Known	σ^2
Assumptions	1. Random Sample 2. Normal Population	1. Random Sample 2. Normal Population and/or n large	1. Random Sample 2. Normal Population
Tables Used	Student's t	Normal	Chi square
Degrees of Freedom	n-1		n-1
Confidence Interval	$\overline{Y} \pm t[1-\alpha/2;n-1]\sqrt{s^2/n}$	$\overline{Y} \pm z[1-\alpha/2]\sqrt{\sigma^2/n}$	$(n-1)s^2/\chi^2[1-\alpha/2;n-1]$ to $(n-1)s^2/\chi^2[\alpha/2;n-1]$
Test of Hypothesis			
1. State H_o	$\mu = \mu_o$	$\mu = \mu_o$	$\sigma^2 = \sigma_o^2$
2. State Alternative H_A	$\mu \neq \mu_o$	$\mu \neq \mu_o$	$\sigma^2 \neq \sigma_o^2$
3. Choose Test Statistic	$t = (\overline{Y}-\mu_o)/\sqrt{s^2/n}$	$z = (\overline{Y}-\mu_o)/\sqrt{\sigma^2/n}$	$\chi^2 = (n-1)s^2/\sigma_o^2$
4. Choose α,n	Usually α = .10, .05 or .01	Usually α = .10, .05 or .01	Usually α = .10, .05 or .01
5. Calculate Test Statistic	Use sample data to find $\overline{Y}$ and s^2	Use sample data for $\overline{Y}$ but not σ^2	Use sample data for s^2
6. Find P From Tables	Student's t Tables	Normal Tables	Chi Square Tables
7. Make Decision by Comparing 4 and 6	Reject if $\alpha > P$	Reject if $\alpha > P$	Reject if $\alpha > P$

REFERENCES

APPLIED STATISTICS

3.1. Dixon, W. J., and F. J. Massey, Jr., *Introduction to Statistical Analysis*, 3rd ed., McGraw-Hill, New York, 1969.

3.2. Dunn, O. J., *Basic Statistics: A Primer for Biomedical Sciences*, Wiley, New York, 1964.

3.3. Ostle, B., *Statistics in Research*, Iowa State College Press, Ames, Iowa, 1963.

3.4. Snedecor, G. W., and W. G. Cochran, *Statistical Methods*, 6th ed., Iowa State College Press, Ames, Iowa, 1967.

3.5. Smart, J. V., *Elements of Medical Statistics*, Staples Press, London, 1963.

3.6. Steel, R. G. D., and J. H. Torrie, *Principles and Procedures of Statistics*, McGraw-Hill, New York, 1960.

MATHEMATICAL STATISTICS

3.7. Brownlee, K. A., *Statistical Theory and Methodology in Science and Engineering*, 2nd ed., Wiley, New York, 1965.

3.8. Ferguson, T. S., *Mathematical Statistics: A Decision Theoretic Approach*, Academic Press, New York, 1967.

3.9. Hoel, P. G., *Introduction to Mathematical Statistics*, 3rd ed., Wiley, New York, 1966.

3.10. Mood, A. and F. Graybill, *Introduction to the Theory of Statistics*, McGraw-Hill, New York, 1963

PROBLEMS

3.1. The right-hand grip strength in pounds was measured for a group of adult men using a dynamometer. The data were as follows:

86 88 80 83 93 87 71 91 76 82 88 100 92 97 78.

(*a*) Calculate a point estimate of the right-hand grip strength mean.
(*b*) Calculate 99% confidence intervals for the mean and variance.
(*c*) What is the P value for rejecting the hypothesis that the population mean is equal to 90 lb?
(*d*) Find the acceptance region for testing the hypothesis in (*c*) with $\alpha = .01$.
(*e*) For $\alpha = .01$, do you accept or reject the hypothesis $\mu = 90$? State your decision in terms of the problem.

3.2. In the process of staining thick blood films for a malariometric survey, the following losses in area (measured in screen squares) of films were recorded for 11 blood smears:

11 0 25 12 10 29 11 12 1 14 13.

(*a*) Calculate a point estimate of the mean loss in area of the blood films and its standard deviation.
(*b*) Calculate 95% confidence intervals for the mean and variance.

(*c*) Interpret the intervals in (*b*).

3.3. Income (in thousands of dollars per year) was measured for a homogeneous group of people. The results are as follows:

7.0 7.4 5.1 4.6 7.4 2.4 4.1 3.7 8.6 8.1.

(*a*) Give point estimates of the population mean and variance.
(*b*) Give a 95% confidence interval for the population mean.
(*c*) Test the hypothesis that the population mean is equal to 7 with a level of significance of .01.
(*d*) Repeat (*c*) using $\alpha = .20$.
(*e*) Compare (*c*) with (*d*).

CHAPTER 4

SAMPLES FROM TWO POPULATIONS

Often the researcher has available two samples, one from each of two populations in which he is interested. For example, he may wish to study a population of individuals with a certain disease who receive the standard treatment and a population of individuals with the same disease who receive a new treatment. This could be done in several ways. The researcher might make a survey, or he might perform an experiment. If in a certain hospital the patients of one physician are given the new treatment while other physicians continue to use the standard treatment, he may try to compare the success of the two treatments. Such a study is called a *survey*. In contrast to the survey, the researcher may perform an *experiment*, randomly assigning patients to the two treatments without regard to their physician. By random assignment of patients to the two treatments, he hopes to avoid large differences between the two treatment groups in such factors as age, economic status, and condition. The responses of individuals in two treatment groups are no doubt affected by individual differences in many factors. However, with random assignment such individual differences should tend to average out and the differences in response between the two treatment groups should reflect treatment differences.

In this book we consider a study to be an experiment when treatments are randomly assigned to the individuals (here an individual may be a person or an object); otherwise we call it a survey. The data from surveys and from experiments are analyzed in the same way. Inferences must be made much more cautiously in the survey than in the experiment, however. For example, if the standard treatment is used in one hospital and the new treatment in another, differences in response may be due to differences in hospital procedures as well as to treatment differences.

As an example of the two-sample situation, we use two sets of IQ scores. Here we have IQ scores for 10 girls whose parents are from a high economic level and for 9 girls whose parents are from a low economic level. Certainly the girls are not randomly assigned to parents, and we simply assume that we have a random sample of girls for each of these two populations. Thus we have surveyed girls from two populations. We use Y_1 to denote IQ measurements of girls of the high economic level and Y_2 for IQ measurements of girls of the low economic level. The data are as follows:

	Observation Number										$\sum_{j=1}^{n_i} Y_{ij}$	$\bar{Y}_{i.}$
	1	2	3	4	5	6	7	8	9	10		
Y_1	124	114	115	106	84	96	106	126	124	116	1111	111.1
Y_2	113	97	108	95	105	69	113	98	118		916	101.78

A second subscript is now used to denote the number of the observation in the sample. Thus Y_{ij} denotes the jth measurement in the ith sample; $Y_{12}=114$, $Y_{25}=105$. For the mean of the ith sample, we use $\bar{Y}_{i.}$; the dot in the second place indicates that we have averaged over the second subscript. Preliminary calculations for the sample means are

$$\bar{Y}_{1.} = \frac{\sum_{j=1}^{n_1} Y_{1j}}{n_1} = \frac{1111}{10} = 111.1$$

$$\bar{Y}_{2.} = \frac{\sum_{j=1}^{n_2} Y_{2j}}{n_2} = \frac{916}{9} = 101.78.$$

For the two sample variances, we have

$$s_1^2 = \frac{\sum_{j=1}^{n_1} \left(Y_{1j} - \bar{Y}_{1.}\right)^2}{n_1 - 1} = 179.66$$

$$s_2^2 = \frac{\sum_{j=1}^{n_2} \left(Y_{2j} - \bar{Y}_{2.}\right)^2}{n_2 - 1} = 215.19.$$

THE POOLED ESTIMATE OF THE VARIANCE

In working with two samples it is usual to pool the estimates of the variances, assuming that the populations have equal variances. Moreover, we meet many situations later in which the experimenter has several samples from populations whose variances he believes are at least approximately equal. Because the population means may be unequal, he does not wish to combine his samples into one large set of data from which a single variance is calculated. Instead, he calculates a sample variance from each sample and then obtains a weighted average of these sample variances. He calls this average the *pooled estimate of the variance* s_e^2. The formula for s_e^2 is

$$s_e^2 = \frac{\sum_{i=1}^{a} (n_i - 1) s_i^2}{N - a}. \tag{4.1}$$

Here we have a samples with sample sizes $n_1, \ldots, n_a$ and $N = \sum_{i=1}^{a} n_i$; the sample variances of the a samples, calculated with the usual formula, are $s_1^2, \ldots, s_a^2$. The pooled estimate of the variance is simply a weighted mean of the sample variances, with weights proportional to the degrees of freedom of the s_i^2.

In our problem, $a = 2$, and the pooled estimate of the variance is

$$s_e^2 = \frac{9(179.66) + 8(215.19)}{19 - 2}$$

$$= \frac{1616.94 + 1721.52}{17}$$

$$= 196.38.$$

CONFIDENCE INTERVAL FOR THE DIFFERENCE BETWEEN TWO POPULATION MEANS

We now wish to make inferences concerning the means of the two populations; that is, for the population mean IQ for girls whose parents are from a high economic level and for those whose parents are from a low economic level. In making inferences about the means, we usually assume that the two population variances are equal.

We designate the two population means μ_1 and μ_2 and the variance for the two populations σ_e^2. The two sample mean IQs $\overline{Y}_{1.}$ and $\overline{Y}_{2.}$ are assumed to be approximately normally distributed with means μ_1 and μ_2, respectively, and variances σ_e^2/n_1, σ_e^2/n_2, respectively. In other words,

$$\overline{Y}_{i.}\text{ND}(\mu_i, \sigma_e^2/n_i) \qquad \text{for} \quad i = 1,2.$$

To estimate $\mu_1 - \mu_2$, the difference between the population means, we use the statistic $\overline{Y}_{1.} - \overline{Y}_{2.}$; we wish to know the distribution of this statistic.

The quantity $\overline{Y}_{1.} - \overline{Y}_{2.}$ is simply a linear combination of two normally distributed variables. It can be shown mathematically that such linear combinations are always normally distributed. Furthermore, we know from Chapter 2 that its expected value is the same linear combination of the expected values; that is, $E(\overline{Y}_{1.} - \overline{Y}_{2.}) = \mu_1 - \mu_2$.

We also need an expression for the population variance of $\overline{Y}_{1.} - \overline{Y}_{2.}$. To obtain this we must square the expression and then substitute variances and covariances as in Chapter 2. Thus we write

$$\left(\overline{Y}_{1.} - \overline{Y}_{2.}\right)^2 = \overline{Y}_{1.}^2 + \overline{Y}_{2.}^2 - 2\overline{Y}_{1.}\overline{Y}_{2.}$$

and

$$\text{Var}\left(\overline{Y}_{1.} - \overline{Y}_{2.}\right) = \text{Var}\,\overline{Y}_{1.} + \text{Var}\,\overline{Y}_{2.} - 2\,\text{Cov}\,\overline{Y}_{1.}, \overline{Y}_{2.}. \tag{4.2}$$

Because we have two independent samples, $\text{Cov}\,\overline{Y}_{1.}, \overline{Y}_{2.}$ is zero, and we have

$$\text{Var}\left(\overline{Y}_{1.} - \overline{Y}_{2.}\right) = \frac{\sigma_e^2}{n_1} + \frac{\sigma_e^2}{n_2} = \sigma_e^2\left(\frac{1}{n_1} + \frac{1}{n_2}\right)$$

or if $n_1 = n_2 = n$, $2\sigma_e^2/n$.

Thus for two independent samples of size n_i, $i = 1, 2$, from approximately normally distributed populations with means μ_1 and μ_2 and with both variances equal to σ_e^2, we can write

$$\left(\overline{Y}_{1.} - \overline{Y}_{2.}\right)\text{ND}\left[\mu_1 - \mu_2, \sigma_e^2\left(\frac{1}{n_1} + \frac{1}{n_2}\right)\right]. \tag{4.3}$$

This statement is of the same form as the one we made when working with a single sample: $\overline{Y}\text{ND}(\mu, \sigma^2/n)$. We can form a standard normal variate in the same way as before:

$$z=\frac{(\overline{Y}_{1.}-\overline{Y}_{2.})-(\mu_1-\mu_2)}{\sqrt{\sigma_e^2(1/n_1+1/n_2)}}. \tag{4.4}$$

Again, we can replace σ_e^2 by its estimate s_e^2; when we do this, we have a statistic that has a Student's t distribution with n_1+n_2-2 degrees of freedom:

$$t=\frac{(\overline{Y}_{1.}-\overline{Y}_{2.})-(\mu_1-\mu_2)}{\sqrt{s_e^2(1/n_1+1/n_2)}}. \tag{4.5}$$

The $1-\alpha$ confidence interval for $\mu_1-\mu_2$ is then constructed in the same way as the interval for μ was constructed in Chapter 3. It is

$$(\overline{Y}_{1.}-\overline{Y}_{2.})\pm t[1-\alpha/2;n_1+n_2-2]\sqrt{s_e^2(1/n_1+1/n_2)}\;, \tag{4.6}$$

where $t[1-\alpha/2;\mathrm{n}_1+\mathrm{n}_2-2]$ is the $1-\alpha/2$ point of the t distribution with n_1+n_2-2 degrees of freedom. In the example, the 95% confidence interval uses the 97.5% point of the t distribution with 17 degrees of freedom; it is

$$9.32\pm 2.110\sqrt{41.458} \qquad \text{or} \qquad -4.27<\mu_1-\mu_2<22.91.$$

The interpretation of this confidence interval is the same as in the previous chapter. In repeated sampling, with two independent samples of size 10 and 9 drawn repeatedly from the two populations, 95% of the confidence intervals formed in this way will cover $\mu_1-\mu_2$. We can conclude that there may be no difference between the two population means, for the interval contains both negative and positive values.

TESTING THE EQUALITY OF TWO MEANS

If the research worker is not interested in the magnitude of the difference between two population means but merely wants to know whether a difference exists, he may wish to test the null hypothesis that $\mu_1=\mu_2$ or, equivalently, $\mu_1-\mu_2=0$. He then chooses as his null hypothesis $H_0:\mu_1=\mu_2$ and as his alternate hypothesis $H_A:\mu_1\neq\mu_2$. For a significance level he may choose $\alpha=.05$. If, then, he rejects the null hypothesis, he decides that children whose parents are of high economic status have IQs different from those whose parents are of low economic status. If he accepts the null hypothesis, he decides that there may be no difference in the population mean IQs.

The test statistic is the t statistic in (4.5) with $\mu_1-\mu_2=0$ under the null hypothesis; the degrees of freedom are 17. From the sample, we have

$$t=\frac{9.32-0}{\sqrt{41.458}}=1.45.$$

Looking for 1.45 in the t table, we find that $t[.90;17]=1.333$ and $t[.95;17]=1.740$. Therefore, $P=2\,\text{Prob}(t>1.45)$ is between $2(1-.90)$ and $2(1-.95)$, that is between .10 and .20. Because the computed P is greater than .05, he cannot reject the hypothesis of equal means. The mean IQs of the two populations of girls may be equal.

CONFIDENCE INTERVALS AND TESTS WITH KNOWN VARIANCES

Occasionally we know the two population variances σ_1^2 and σ_2^2, although such knowledge is infrequent. The two variances can, of course, be unequal and the variance of $\bar{Y}_1-\bar{Y}_2$ is $\sigma_1^2/n_1+\sigma_2^2/n_2$ from (4.2). The standard normal variable z is then

$$z=\frac{\left(\bar{Y}_{1.}-\bar{Y}_{2.}\right)-(\mu_1-\mu_2)}{\sqrt{\sigma_1^2/n_1+\sigma_2^2/n_2}}.$$

The $1-\alpha$ level confidence interval for $\mu_1-\mu_2$ is

$$\bar{Y}_{1.}-\bar{Y}_{2.}\pm z[1-\alpha/2]\sqrt{\sigma_1^2/n_1+\sigma_2^2/n_2}\,. \tag{4.7}$$

The α level test of $H_0:\mu_1=\mu_2$ is made by calculating the statistic $z=(\bar{Y}_{1.}-\bar{Y}_{2.})/\sqrt{\sigma_1^2/n_1+\sigma_1^2/n_2}$ and comparing its P value with .05 or, alternately, by comparing the calculated z with the $1-\alpha/2$ point of the normal distribution.

A CONFIDENCE INTERVAL FOR THE VARIANCE

We can form a confidence interval for the population variance σ_e^2 by using the pooled estimate s_e^2. It can be proved mathematically that the quantity $(N-a)s_e^2/\sigma_e^2$ has a χ^2 distribution and that the degrees of freedom are $N-a$. Therefore, we can use s_e^2 to form a confidence interval for σ_e^2 in exactly the way we used s^2 from a single sample. The $1-\alpha$ confidence interval is

$$\frac{(N-a)s_e^2}{\chi^2[1-\alpha/2; N-a]} < \sigma_e^2 < \frac{(N-a)s_e^2}{\chi^2[\alpha/2; N-a]}, \tag{4.8}$$

where $\chi^2[1-\alpha/2; N-a]$ and $\chi^2[\alpha/2; N-a]$ denote the $1-\alpha/2$ point and the $\alpha/2$ point of the χ^2 distribution with $N-a$ degrees of freedom. In the example, a 95% confidence interval is

$$\frac{17(196.38)}{30.191} \quad \text{to} \quad \frac{17(196.38)}{7.564}, \quad \text{or} \quad 110.6 < \sigma_e^2 < 441.40.$$

We can also test $H_0: \sigma_e^2 = \sigma_0^2$. Here the statistic to be calculated is $(N-a)s_e^2/\sigma_0^2$, and the degrees of freedom to be used in entering the table are $N-a$.

TESTING THE EQUALITY OF VARIANCES

Sometimes the research worker wishes to test whether two population variances are equal. In the IQ example, with samples from two populations of IQs, he may wish to test whether the two population variances σ_1^2 and σ_2^2 are equal before pooling the two sample variances s_1^2 and s_2^2. The null hypothesis $H_0: \sigma_1^2 = \sigma_2^2$ can be tested, provided the two populations are normally distributed. To make the test, we use Table A.4, which gives percentage points of a distribution called the *F distribution*.* The F distribution is related to the χ^2 distribution. We know from Chapter 3 that if we have a sample variance s_1^2 calculated from a size n_1 sample from a normal distribution with mean μ_1 and with variance σ_1^2, the quantity $(n_1-1)s_1^2/\sigma_1^2$ has a χ^2 distribution with n_1-1 degrees of freedom. With a second sample variance s_2^2 calculated from a second normally distributed population with mean μ_2 and variance σ_2^2, we have a second χ^2 variate $(n_2-1)s_2^2/\sigma_2^2$. These two χ^2 variates are statistically independent of each other because they have been calculated from two independent random samples. A new variate can be obtained which has the F distribution by taking the ratio, not of the two χ^2 variates, but of the χ^2 variates when each has been divided by its own degrees of freedom. Here the degrees of freedom of the two independent χ^2 variates are n_1-1 and n_2-1, and the F variate is $(s_1^2/\sigma_1^2)/(s_2^2/\sigma_2^2)$. To use the tables of the F distribution, we must

* $f(F) = \dfrac{\nu_1^{\nu_1/2}\nu_2^{\nu_2/2}[(\nu_1+\nu_2)/2]!}{[(\nu_1-2)/2]![(\nu_2-2)/2]!} F^{(\nu_1-2)/2}(\nu_2+\nu_1 F)^{(\nu_1+\nu_2)/2}.$

know the two degrees of freedom of the two χ^2 variables that were used in forming the F variable.

The body of Table A.4 contains various percentiles of the F distribution; these are denoted by $F[\lambda;\nu_1,\nu_2]$, where ν_1 and ν_2 are the degrees of freedom for the numerator and denominator, respectively, of the F statistic, and where λ percent of the distribution lies between zero and $F[\lambda;\nu_1,\nu_2]$. It is important to note that we always write the numerator degrees of freedom before the denominator degrees of freedom.

We now wish to make a test of $H_0:\sigma_1^2=\sigma_2^2$ with alternate hypothesis H_A: $\sigma_1^2\neq\sigma_2^2$ using Table A.4. Regardless of whether the null hypothesis is true, the quantity $(s_1^2/\sigma_1^2)/(s_2^2/\sigma_2^2)$ has an F distribution; but with unknown parameters in the expression, we could not use the tables. However, if the two variances are equal, the parameters cancel and we are able to calculate

$$F=\frac{s_1^2}{s_2^2}.$$

In order to have a level of significance of .05, we use as a critical region $0<F<F[.025;9,8]$ and $F>F[.975;9,8]$. From Table A.4 we find that $F[.025;9,8]$ equals 0.244 and $F[.975;9,8]$ equals 4.36; our critical region consists of two parts, F between 0 and 0.244 and F greater than 4.36. This critical region seems sensible, for we reason that under the null hypothesis s_1^2 and s_2^2 both estimate the same variance. If their ratio is either very small $(F<0.244)$ or very large $(F>4.36)$, we decide that the null hypothesis is incorrect and that $\sigma_1^2\neq\sigma_2^2$.

We calculate $F=179.66/215.19=0.835$. Because 0.835 lies between 0.244 and 4.36, we decide that the two population variances may be equal.

The exact shape of the F distribution depends on the two degrees of freedom ν_1 and ν_2. Figure 4.1 illustrates several typical F distributions.

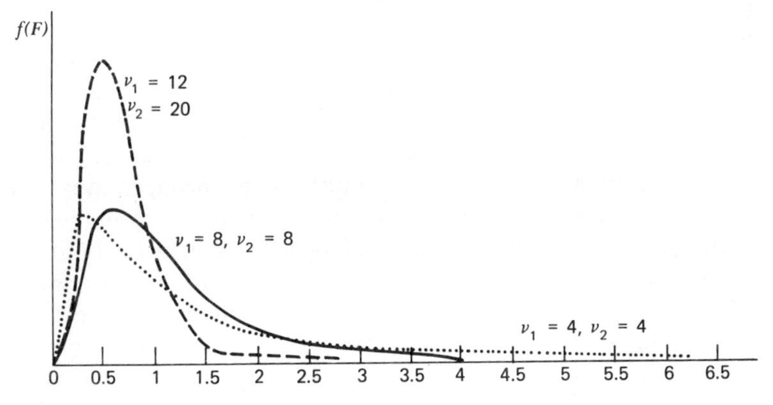

Figure 4.1. Three F distributions.

DISTRIBUTIONS RELATED TO THE NORMAL DISTRIBUTION

We have been introduced to all the most commonly used distributions, the standard normal distribution, the Student's t distribution, the χ^2 distribution, and the F distribution. We know that from two independent χ^2 variables we can form a variable that has an F distribution. In a similar way it can be shown mathematically that if we have a standard normal variable z and an independent χ^2 variable, which we shall call u, we can form a variable having a t distribution; this variable is $t = z/\sqrt{u/\nu}$, where ν is the degrees of freedom of u and the t distribution also has ν degrees of freedom. The relationships between the distributions are summarized in Table 4.1. The various tests of hypotheses and confidence intervals are summarized in Table 4.2.

ILLUSTRATIVE EXAMPLE

As an example in which methods of this chapter can be applied, let us consider the following situation. We wish to measure the effects of remedial training given to students with poor high school records during the summer preceding college entry. We randomly assign half the students to the remedial program; the remaining half receives no additional training. As a response variate, we measure freshman college grade point average.

Our analysis may consist of testing the hypothesis that the population mean grade point average for students receiving remedial training equals the mean for students with no training. If we reject this hypothesis, we can state that a difference exists between grade point average under the two treatments. Confidence limits for the difference in the population means should also be of value, for if the difference in grade point averages is very small, the program may not be worthwhile.

Let us consider some of the problems that may arise in this experiment. The random assignment of students to the two treatments may cause difficulties; some students may be irritated at not being allowed to attend with friends, and so on. In general, random assignment is easier when the experimental units are rats or plants, rather than uncooperative human beings. Another problem arises if some students assigned to the remedial treatment drop out, perhaps through lack of interest. It is important that students who drop the remedial program still be treated as if they were in that program, just as they were placed by the random assignment. Such students provide evidence concerning the success of the training program. Data on attendance should also be reported; the information may be helpful in evaluating the program.

TABLE 4.1. RELATIONSHIPS AMONG z, t, χ^2 AND F DISTRIBUTIONS

	From:	Obtain	From:	Test Statistic
z	$Y\ ND(\mu,\sigma^2)$	$z = (Y-\mu)/\sigma$	$\overline{Y}\ ND(\mu,\sigma^2/n)$	$z = (\overline{Y}-\mu)/\sqrt{\sigma^2/n}$
			$\overline{Y}_1-\overline{Y}_2\ ND(\mu_1-\mu_2,\sigma_e^2(\frac{1}{n_1}+\frac{1}{n_2}))$	$z = [(\overline{Y}_1-\overline{Y}_2)-(\mu_1-\mu_2)]/\sqrt{\sigma_e^2(\frac{1}{n_1}+\frac{1}{n_2})}$
t	$z\ ND(0,1)$	$t(\nu) = z/\sqrt{u/\nu}$	$z = (\overline{Y}-\mu)/\sqrt{\sigma^2/n}$	$t(n-1) = (\overline{Y}-\mu)/\sqrt{s^2/n}$
	u: $\chi^2(\nu)$		$u = (n-1)s^2/\sigma^2$	
	z,u indep.			
			$z = [(\overline{Y}_{1.}-\overline{Y}_{2.})-(\mu_1-\mu_2)]/\sqrt{\sigma_e^2(\frac{1}{n_1}+\frac{1}{n_2})}$	$t(n_1+n_2-2) =$
			$u = (n_1+n_2-2)s_e^2/\sigma^2$	$[(\overline{Y}_{1.}-\overline{Y}_{2.})-(\mu_1-\mu_2)]/\sqrt{s_e^2(\frac{1}{n_1}+\frac{1}{n_2})}$
F	u_1: $\chi^2(\nu_1)$	$F(\nu_1,\nu_2) =$	$u_1 = (n_1-1)s_1^2/\sigma_1^2$	$F(n_1-1,n_2-1)=(s_1^2/\sigma_1^2)/(s_2^2/\sigma_2^2)$
	u_2: $\chi^2(\nu_2)$	$(u_1/\nu_1)/(u_2/\nu_2)$	$u_2 = (n_2-1)s_2^2/\sigma_2^2$	$= \frac{s_1^2}{s_2^2}$ if $\sigma_1^2 = \sigma_2^2$
	u_1,u_2 indep.			

TABLE 4.2. SUMMARY OF CONFIDENCE INTERVALS AND TESTS

Parameter	Number of Samples	Sample Size	Variance	Degrees of Freedom	1-α Confidence Interval	H_o	Test Statistic
μ	1	n	Unknown	n-1	$\overline{Y}\pm t[1-\alpha/2;n-1]\sqrt{s^2/n}$	$\mu = \mu_o$	$t = (\overline{Y}-\mu_o)/\sqrt{s^2/n}$
μ	1	n	Known	None	$\overline{Y}\pm z[1-\alpha/2]\sqrt{\sigma^2/n}$	$\mu = \mu_o$	$z = (\overline{Y}-\mu_o)/\sqrt{\sigma^2/n}$
$\mu_1-\mu_2$	2	n_1,n_2	Unknown Equal	n_1+n_2-2	$(\overline{Y}_{1.}-\overline{Y}_{2.})\pm t[1-\alpha/2;n_1+n_2-2]\sqrt{s_e^2(\frac{1}{n_1}+\frac{1}{n_2})}$	$\mu_1 = \mu_2$	$t = [(\overline{Y}_{1.}-\overline{Y}_{2.})-0]/\sqrt{s_e^2(\frac{1}{n_1}+\frac{1}{n_2})}$
$\mu_1-\mu_2$	2	n_1,n_2	Known	None	$(\overline{Y}_{1.}-\overline{Y}_{2.})\pm z[1-\alpha/2]\sqrt{(\frac{\sigma_1^2}{n_1}+\frac{\sigma_2^2}{n_2})}$	$\mu_1 = \mu_2$	$z = [(\overline{Y}_{1.}-\overline{Y}_{2.})-0]/\sqrt{(\frac{\sigma_1^2}{n_1}+\frac{\sigma_2^2}{n_2})}$
σ^2	1	n	-	n-1	$\frac{(n-1)s^2}{\chi^2[1-\alpha/2;n-1]}$ to $\frac{(n-1)s^2}{\chi^2[\alpha/2;n-1]}$	$\sigma^2 = \sigma_o^2$	$\chi^2 = (n-1)s^2/\sigma_o^2$
σ_2^2	2	n_1,n_2	-	n_1-1,n_2-1		$\sigma_1^2 = \sigma_2^2$	$F = s_1^2/s_2^2$
σ_e^2	a	$\sum_{i=1}^{a} n_i=N$	-	N-a	$\frac{(N-a)s_e^2}{\chi^2[1-\alpha/2;N-a]}$ to $\frac{(N-a)s_e^2}{\chi^2[\alpha/2:N-a]}$	$\sigma_e^2 = \sigma_o^2$	$\chi^2 = (N-a)s_1^2/\sigma_o^2$

Another consideration is that the distribution of grade point averages may not be very close to a normal distribution. The use of the t statistic as given in this chapter is *not* highly sensitive to moderate departures from normality. It can safely be used unless the departures are quite noticeable or the sample sizes in the two groups quite different. On the other hand, the F test for equality of variances is more sensitive to departures from normality; thus the experimenter should not bother to make a test of $H_0: \sigma_1^2 = \sigma_2^2$ but should proceed directly to the use of the t statistic. Fortunately, the t statistic is quite robust with respect to inequality of variances, provided the two sample sizes are nearly equal, as they are in this experiment.

If the size of the two groups is small, it is advisable to look at other characteristics of the students, such as age and sex. With large samples, the students in the two treatment groups are probably comparable in these other variables, but there may be important differences in smaller samples. If some variable such as age differs widely between the two groups, the relation of age to grade point average should be assessed by making a scatter diagram of grade point average on the vertical axis and age on the horizontal axis. If, unfortunately, age seems to be related to grade point average, methods of analysis such as covariance (Chapter 13) should be used. Even with small samples, however, the random assignment of students often results in the two treatment groups being essentially alike.

SUMMARY

This chapter has introduced methods for making confidence intervals for the difference between means and tests on the equality of two means on the basis of two independent random samples from normal distributions having equal variances. Concepts of the experiment and survey were discussed. The pooled estimate of the variance was defined. A confidence interval for the population variance was given, and a test for equality of two variances. Relationships among the χ^2, F, t, and normal distributions were mentioned.

REFERENCES

APPLIED STATISTICS

4.1. Dixon, W. J., and F. J. Massey, Jr., *Introduction to Statistical Analysis*, 3rd ed., McGraw-Hill, New York, 1969.

4.2. Dunn, O. J., *Basic Statistics: A Primer for the Biomedical Sciences*, Wiley, New York, 1964.

4.3. Ostle, B., *Statistics in Research*, Iowa State College Press, Ames, Iowa, 1963.

4.4. Snedecor, G. W., and W. G. Cochran, *Statistical Methods*, 6th ed., Iowa State College Press, Ames, Iowa, 1967.

4.5. Smart, J. V., *Elements of Medical Statistics*, Staples Press, London, 1963.

4.6. Steel, R. G. D., and J. H. Torrie, *Principles and Procedures of Statistics*, McGraw-Hill, New York, 1960.

MATHEMATICAL STATISTICS

4.7. Brownlee, K. A., *Statistical Theory and Methodology in Science and Engineering*, 2nd ed., Wiley, New York, 1965.

4.8. Hoel, P. G., *Introduction to Mathematical Statistics*, 3rd ed., Wiley, New York, 1966.

4.9. Mood, A., and F. Graybill, *Introduction to the Theory of Statistics*, McGraw-Hill, New York, 1963.

PROBLEMS

4.1. To compare language teaching methods for school children, methods A and B were randomly assigned to two groups of children of the same age. The following are the results of the language test performed on each child.

A	B
65	90
79	98
90	73
75	88
61	83
85	90
98	98
80	81
97	84
75	79

(*a*) Test the hypothesis that $\mu_1 = \mu_2$ at $\alpha = .01$ level.
(*b*) Find a 99% confidence interval for $\mu_1 - \mu_2$.
(*c*) Does the interval in (*b*) cover zero?
(*d*) If you reject the hypothesis, would you expect the confidence interval to cover zero? Why?

4.2. The following are the results of a malariometric survey conducted in seven localities in two ecologically different areas A and B:

Locality Number	% Positive Slides A	B
1	5	13
2	12	18
3	13	11
4	2	17
5	4	14
6	3	11
7	8	21

(*a*) Test for equality of the variance in the two populations using as significance level $\alpha = .05$.
(*b*) Test for equality of two means at significance level $\alpha = .05$.
(*c*) Find a 95% confidence interval for the difference of means of areas *A* and *B*.
(*d*) Find a 95% confidence interval for the variance.

4.3. In an agricultural experiment, yields of two varieties of corn were measured:

Yield (bushels/acre)	
Variety I	Variety II
68.5	81.5
83.0	85.2
83.0	87.1
66.5	69.3
58.1	73.5
82.4	65.5
	73.4
	56.1

(*a*) Test the hypothesis of equal means at significance level $\alpha = .01$.
(*b*) Test the hypothesis $\sigma_e^2 = 100$ at significance level $\alpha = .05$.
(*c*) Assuming that the variances equal 100, find a 99% confidence interval for $\mu_1 - \mu_2$.

CHAPTER 5

ONE-WAY ANALYSIS OF VARIANCE

Frequently an investigator wishes to compare three or more treatments in a single experiment. In a survey, too, he may wish to study several populations; for example, he may be interested in IQ scores from a standard test for students at five schools. Such comparisons could be accomplished using the results of Chapter 4 by looking at the samples two at a time and comparing the means. Although feasible, this is an inefficient method of comparison for more than two populations.

One reason for its inefficiency is that the standard deviation for the difference between the two sample means is not calculated from observations from all the samples but instead uses samples only from the two populations under immediate consideration. Second, we feel intuitively that we shall almost always find a significant difference between at least one pair of means (the extreme ones, e.g.) if we consider enough identical populations. We can no longer trust our level of significance.

Therefore, instead of using two samples at a time, we wish to make a single test to find out whether the students from the five schools are from five populations having the same population mean. The null hypothesis we wish to test is $H_0: \mu_1 = \mu_2 = \mu_3 = \mu_4 = \mu_5$. Our reason for making such a test is not that we think the five population means may be equal. They probably are unequal. However, if a preliminary test fails to show that the means are unequal, we may feel that the differences are rather small and do not warrant further investigation.

In this chapter we are concerned principally with estimation and tests for population means. To study the means, it is necessary to "analyze the variance."

Let us consider a particular example. An investigator wished to study the effect of fertilizers on the yield of corn. He divided a field into 24 rectangular plots of the same size and shape. His four treatments consisted of (1) no fertilizer, (2) $K_2O + N$, (3) $K_2O + P_2O_5$, and (4) $N + P_2O_5$. He

TABLE 5.1. YIELDS OF CORN UNDER FOUR TREATMENTS

Observation		1	2	3	4	5	6	$\sum_{j=1}^{6} Y_{ij}$	$\bar{Y}_{i.}$
Treatment									
1.	Control	99	40	61	72	76	84	432	72
2.	$K_2O + N$	96	84	82	104	99	105	570	95
3.	$K_2O + P_2O_5$	63	57	81	59	64	72	396	66
4.	$N + P_2O_5$	79	92	91	87	78	71	498	83

$$\sum_{i=1}^{4} \sum_{j=1}^{6} Y_{ij} = 1{,}896\,, \quad \bar{Y}_{..} = 79$$

assigned each treatment at random to six of the 24 plots. His yields, in bushels per acre, are presented in Table 5.1. Here we again use i to denote the number of the treatment and j to refer to the number of the observation. The sample means are again designated by $\bar{Y}_{i.}$, where the dot in the second subscript position indicates that we have averaged over the second subscript j; in other words $\bar{Y}_{i.}$ is the mean of all the observations of the ith sample. The overall mean is denoted by $\bar{Y}_{..}$; the two dots indicate that the mean is obtained by summing over both subscripts and then dividing by the total number.

MODEL I

The investigator's purpose is usually to learn something about the populations from which the samples are drawn. To accomplish this, he needs an underlying model. Here we describe what is commonly called Model I. Model I is the simplest of three models that we designate Models I, II, and III. By and large we feel that it is the most useful of the models, but there are many experiments in which Models II and III fit the situation better. For completeness, we present Models II and III in Chapter 9, a starred chapter. For an introduction to analysis of variance, we suggest that such starred material be omitted during a first reading of the portion of the book devoted to the analysis of variance. It can be studied later.

For Model I we assume that our four samples, each consisting of six corn yields, are independent random samples from four populations, that each of the four populations has a normal distribution, and, finally, that the variances of the four populations are equal. The investigator should consider these assumptions carefully; for some discussion on departures from the assumptions, see Chapter 14.

The four population means may be designated μ_1, μ_2, μ_3, and μ_4. We arbitrarily divide each of these four means into two parts. The first part is the mean of the four population means, which we call the "overall mean," and the second part is the difference between the mean of each population and the overall mean. In symbols, the means are written as $\mu_1 = \mu + \alpha_1, \ldots,$ $\mu_4 = \mu + \alpha_4$, where μ denotes the overall mean and α_i is the difference $\mu_i - \mu$. The overall mean μ has been chosen in such a way that $\sum_{i=1}^{a} \alpha_i = 0$, where a is the number of treatments (in this case, 4). If in our example the four population means were $\mu_1 = 70$, $\mu_2 = 100$, $\mu_3 = 70$, $\mu_4 = 80$, we would have $\mu = (70 + 100 + 70 + 80)/4 = 80$. The population means then could be written

$$\mu_1 = \mu + \alpha_1 = 80 + (-10)$$

$$\mu_2 = \mu + \alpha_2 = 80 + (+20)$$

$$\mu_3 = \mu + \alpha_3 = 80 + (-10)$$

$$\mu_4 = \mu + \alpha_4 = 80 + (+0).$$

The difference $\alpha_i = \mu_i - \mu$ is often called the *effect* of the particular treatment. It should not be confused with α, the probability of rejecting a null hypothesis which is true. In the example, the α_i sum to zero, and this is always the case.

We think, then, of a population mean as the sum of two parts: an overall mean μ (which, as an average of the four population means, may be of little interest to us) and the part that we attribute to the particular treatment. Using this notation, we can summarize the model just described by saying that each observation Y_{ij} is an independent observation from a normally distributed population whose mean is $\mu + \alpha_i$ and whose variance is denoted by σ_e^2. This can be written

$$Y_{ij}\,\text{IND}(\mu + \alpha_i, \sigma_e^2), \quad \sum_{i=1}^{a} \alpha_i = 0, \qquad i = 1, \ldots, a; \qquad j = 1, \ldots, n, \qquad (5.1)$$

where a is the number of treatments, n is the number of observations on each treatment, and IND is read as "independently normally distributed." In Model I, we are studying only these particular a populations.

An equivalent way of writing down the model which is often convenient is

$$Y_{ij} = \mu + \alpha_i + \varepsilon_{ij} \qquad (5.2)$$

$$\sum_{i=1}^{a} \alpha_i = 0; \qquad \varepsilon_{ij}\,\text{IND}(0, \sigma_e^2); \qquad i = 1, \ldots, a; \qquad j = 1, \ldots, n.$$

Here we think of a particular corn yield Y_{ij} as made up of its population mean $\mu+\alpha_i$, plus whatever is left over, which we call ε_{ij}. In our data, for example, $Y_{23}=82=80+20+(-18)$; therefore, if the mean yield for the second population is 100, $\varepsilon_{23}=-18$. These 24 deviations form a random sample of 24 ε_{ij}, all from a normal population with zero mean and variance σ_e^2. Note that because in practice we do not know the population means, we cannot know the values of the 24 ε_{ij}'s. They can, however, be estimated from the sample data.

THE ASSUMPTIONS

The Expression 5.1 (or, alternatively, 5.2) is a shorthand statement of our model. Let us emphasize here the assumptions implied by the model:

1. The four populations of corn yields are normally distributed.
2. The variances of the four populations are equal.
3. The 24 observations are independent.

These assumptions concerning the populations are the only ones implied by the model given in (5.1) or (5.2). It should be noted that in expressing the means of the population as $\mu+\alpha_i$, $i=1,\ldots,a$, where $\sum_{i=1}^{a}\alpha_i=0$, we are not making an additional assumption; any a means can be expressed in this way. It is also true that inferences are to be made about only these four particular populations. The assumptions are the same as those assumed in Chapter 4, where we dealt with just two populations. In our example of corn yields, the assumption of equal variances seems reasonable; it is unlikely that the variances of the four populations differ appreciably.

ESTIMATES OF PARAMETERS

We can estimate the values of μ, α_i, and ε from our data. As estimates we use unbiased statistics (i.e., statistics whose expected values equal the parameters being estimated). Among unbiased statistics, we choose those having the smallest possible variances.

The first calculations performed by the experimenter with the data in Table 5.1 are the sample means $\overline{Y}_{i.}$ and the overall mean $\overline{Y}_{..}$. We could guess which parameters these statistics should be used to estimate. Instead, we calculate the expected values of $\overline{Y}_{i.}$ and $\overline{Y}_{..}$ and use each statistic to estimate its expected value. The mean of the ith sample is

$$\overline{Y}_{i.}=\frac{Y_{i1}+\cdots+Y_{in}}{n}; \tag{5.3}$$

this is a linear combination of the Y_{ij}. From Chapter 2, we find the expected value of $\overline{Y}_{i.}$ by taking the same linear combination of the expected values of the Y_{ij}. From $EY_{ij}=\mu+\alpha_i$, we can write

$$E\overline{Y}_{i.}=\frac{(\mu+\alpha_i)+\cdots+(\mu+\alpha_i)}{n}$$

$$=\frac{n\mu+n\alpha_i}{n}$$

$$=\mu+\alpha_i. \tag{5.4}$$

This is not new to us, for it is merely the expected value of a mean. Thus we use $\overline{Y}_{i.}$ to estimate $\mu+\alpha_i$.
Similarly, $\overline{Y}_{..}$ can be expressed as follows:

$$\overline{Y}_{..}=\frac{(Y_{11}+\cdots+Y_{1n})+\cdots+(Y_{a1}+\cdots+Y_{an})}{an}. \tag{5.5}$$

Its expected value, therefore, is

$$E\overline{Y}_{..}=\frac{(\mu+\alpha_1)+\cdots+(\mu+\alpha_1)+\cdots+(\mu+\alpha_a)+\cdots+(\mu+\alpha_a)}{an}$$

$$=\frac{an\mu+n(\alpha_1+\cdots+\alpha_a)}{an}$$

$$=\mu, \tag{5.6}$$

because $\alpha_1+\cdots+\alpha_a=0$. Therefore, not unexpectedly, $\overline{Y}_{..}$ can be used to estimate μ. With $\overline{Y}_{i.}$ as an estimate of $\mu+\alpha_i$ and $\overline{Y}_{..}$ as an estimate of μ, we have $\overline{Y}_{i.}-\overline{Y}_{..}$ as our estimate of α_i.

The population variance σ_e^2 is estimated by s_e^2, the pooled estimate of the variance. The list of the parameters of the problem and their point estimates is as follows:

Parameter	Estimate
μ	$\overline{Y}_{..}$
α_i	$\overline{Y}_{i.}-\overline{Y}_{..}$
$\mu+\alpha_i$	$\overline{Y}_{i.}$
σ_e^2	s_e^2

DIVISION OF THE BASIC SUM OF SQUARES

When the unknown parameters in $Y_{ij}=\mu+\alpha_i+\varepsilon_{ij}$ are replaced by their estimates from the data, we have

$$Y_{ij}=\overline{Y}_{..}+\left(\overline{Y}_{i.}-\overline{Y}_{..}\right)+\left(Y_{ij}-\overline{Y}_{i.}\right), \tag{5.7}$$

an algebraic identity. (To be an identity, the equation must hold for any values of Y_{ij}. Note that if we cancel out the two $\overline{Y}_{i.}$ and $\overline{Y}_{..}$ we have the same thing on the right side of the equation as on the left.)

If we knew the true values of the parameters, say $\mu=80$ and $\alpha_2=20$, we would express Y_{23} as $82=80+20+(-18)$; inasmuch as we do *not* know the parameters, we express it as $82=79+(95-79)+(82-95)=79+16+(-13)$, using the sample estimates obtained in Table 5.1. This could be done for all 24 observations. Instead, however, we usually write

$$Y_{ij}-\overline{Y}_{..}=\left(\overline{Y}_{i.}-\overline{Y}_{..}\right)+\left(Y_{ij}-\overline{Y}_{i.}\right), \tag{5.8}$$

subtracting $Y_{..}$ from both sides of the identity (5.7). For Y_{23}, this equation is $3=16+(-13)$. In other words, we look at the deviation of this particular corn yield from the overall mean and break that deviation up into two terms. The first term represents the deviation of the average yield for the second treatment from the average for all treatments; the second term represents the difference between the yield for this particular plot and the average yield for plots using the second treatment.

If we square the deviation $3=16+(-13)$, we can use the formula $(a+b)^2=a^2+b^2+2ab$ to obtain

$$3^2=16^2+(-13)^2+2(16)(-13), \qquad \text{or} \qquad 9=256+169-416.$$

This can be done for all 24 observations. Table 5.2 shows this computation and also gives sums for each of the four columns in the table. The sum of the right-hand column is seen to be zero. This always happens; it can be proved algebraically that the cross-product terms always add to zero when summed over an entire set of data.

Thus the sum of the squared deviations of observations from the overall mean always equals the sum of squared deviations of the individual means from the overall mean, plus the sum of squared deviations of the observations from the individual means. In symbols we have

$$\sum_{i=1}^{a}\sum_{j=1}^{n}\left(Y_{ij}-\overline{Y}_{..}\right)^2=\sum_{i=1}^{a}\sum_{j=1}^{n}\left(\overline{Y}_{i.}-\overline{Y}_{..}\right)^2+\sum_{i=1}^{a}\sum_{j=1}^{n}\left(Y_{ij}-\overline{Y}_{i.}\right)^2 \tag{5.9}$$

TABLE 5.2. VALUES OF THE INDIVIDUAL TERMS OF THE SUM OF SQUARES FOR THE DATA FROM TABLE 5.1

Observations	$(Y_{ij}-\bar{Y}_{..})^2$	$(\bar{Y}_{i.}-\bar{Y}_{..})^2$	$(Y_{ij}-\bar{Y}_{i.})^2$	$2(\bar{Y}_{i.}-\bar{Y}_{..})(Y_{ij}-\bar{Y}_{i.})$
Y_{11}	400	49	729	-378
Y_{12}	1521	49	1024	448
Y_{13}	324	49	121	154
Y_{14}	49	49'	0	0
Y_{15}	9	49	16	- 56
Y_{16}	25	49	144	-168
Y_{21}	289	256	1	32
Y_{22}	25	256	121	-352
Y_{23}	9	256	169	-416
Y_{24}	625	256	81	288
Y_{25}	400	256	16	128
Y_{26}	676	256	100	320
Y_{31}	256	169	9	78
Y_{32}	484	169	81	234
Y_{33}	4	169	225	-390
Y_{34}	400	169	49	182
Y_{35}	225	169	4	52
Y_{36}	49	169	36	-156
Y_{41}	0	16	16	- 32
Y_{42}	169	16	81	72
Y_{43}	144	16	64	64
Y_{44}	64	16	16	32
Y_{45}	1	16	25	- 40
Y_{46}	64	16	144	- 96
Sum	6212	2940	3272	0

TABLE 5.3. ANALYSIS OF VARIANCE TABLE FOR ONE-WAY CLASSIFICATION, EQUAL NUMBERS

Source of Variation (1)	Sum of Squares (2)	d f (3)	Mean Square (4)	EMS (5)	Computed F (6)	Tabled F (7)
Due to Treatment	$SS_a = n \sum_{i=1}^{a} (\overline{Y}_{i.} - \overline{Y}_{..})^2$	a-1	$MS_a = SS_a/(a-1)$	$\sigma_e^2+n\sigma_a^2$	MS_a/s_e^2	$F[1-\alpha;a-1,a(n-1)]$
Residual	$SS_r = \sum_{i=1}^{a} \sum_{j=1}^{n} (Y_{ij} - \overline{Y}_{i.})^2$	a(n-1)	$s_e^2 = SS_r/a(n-1)$	σ_e^2		
Total	$SS_t = \sum_{i=1}^{a} \sum_{j=1}^{n} (Y_{ij} - \overline{Y}_{..})^2$	an-1				

where a denotes the number of groups or treatments. It can be shown that (5.9) is an algebraic identity. The third sum, the cross-product term, is not included because it equals zero. In (5.9), the sum of squared deviations from the overall mean has been broken into two parts—the first, $\sum_{i=1}^{a}\sum_{j=1}^{n}(\overline{Y}_{i.}-\overline{Y}_{..})^2$, is said to be "due to treatment"; the second, $\sum_{i=1}^{a}\sum_{j=1}^{n}(Y_{ij}-\overline{Y}_{i.})^2$, is said to be "within treatment." The second term represents the variation of the individual observations about their own sample means; it is sometimes called the residual sum of squares.

In the expression $\sum_{i=1}^{a}\sum_{j=1}^{n}(\overline{Y}_{i.}-\overline{Y}_{..})^2$ there is no subscript j within the summation sign; thus over j we add the same expression n times. Therefore, $\sum_{i=1}^{a}\sum_{j=1}^{n}(\overline{Y}_{i.}-\overline{Y}_{..})^2$ can be written $n\sum_{i=1}^{a}(\overline{Y}_{i.}-\overline{Y}_{..})^2$.

ANALYSIS OF VARIANCE TABLE

The division of the basic sum of squared deviations is usually summarized in an analysis of variance table. The table displays in a particular order the sums of squares and certain other quantities involved in the computations. Such a display simplifies the arithmetic and algebraic processes, which tend to become complicated in more involved designs. Table 5.3 shows the general form of the analysis of variance table for the Model I one-way design with equal n. In Table 5.4 an analysis of variance table is provided for the corn yield problem.

In the first column of the analysis of variance table we write the "source of variation"; that is, a title to tell which sum of squares is listed in each row of the table. The second column of the table contains the sums of squares, which we denote by SS. The third column contains the degrees of freedom for each sum of squares. The "degrees of freedom" are the numbers by which we divide the sum of squares in order to obtain an

TABLE 5.4. ANALYSIS OF VARIANCE TABLE FOR CORN YIELDS

Source of Variation (1)	SS (2)	df (3)	MS (4)	EMS (5)	Computed F (6)	Tabled F (7)
Due Treatment	2940	3	980	$\sigma_e^2+6\sigma_a^2$	5.99	3.10
Residual	3272	20	163.6	σ_e^2		
Total	6212	23				

unbiased estimate of the population parameter found in column 5. The fourth column, the mean square (MS) column, is obtained by dividing each sum of squares by the degrees of freedom in the same row. The values in the mean square column are used as estimates of the parameters in column 5. In column 5 we list the expected mean squares (EMS). These are the mathematical expected values of the mean squares. Each entry in the EMS column is the parameter (or sum of parameters) that can be estimated by the corresponding statistic in the mean square column. Columns 6 and 7 are used for making F tests; these are discussed in the next section.

In Table 5.3 we have introduced several new symbols, some of them self-explanatory. We use the subscripts a, r, and t on SS to denote the "due to treatment," "residual," and "total," sums of squares, respectively. We denote the "due to treatment" mean square by MS_a. For the "residual" mean square, we use the symbol s_e^2; this estimates the parameter σ_e^2. In this design, s_e^2 is simply the pooled estimate of the variance. The symbol σ_a^2, which is defined as $\sigma_a^2 = \sum_{i=1}^{a}\alpha_i^2/(a-1)$, is discussed later.

In Table 5.4, the sum of squares column contains values calculated by the formulae in Table 5.3. Because there are four populations in this problem, the "due to treatment" degrees of freedom is $a-1=4-1=3$. The "residual" degrees of freedom is that of the pooled estimate of the variance, $24-4=20$ [or $a(n-1)$]. The "total" degrees of freedom is of course $24-1=23$ (or $an-1$). The degrees of freedom for the residual and the due to treatment (in many books denoted by "among" or "between" treatments) sums of squares always add to the degrees of freedom for the total sum of squares. The usual estimate of the variance σ_e^2 is simply s_e^2, obtained from the residual sum of squares. When the four population means are equal, it can be shown that the "due to treatment" sum of squares $\mathrm{SS}_a = n\sum_{i=1}^{a}(\overline{Y}_{i.}-\overline{Y}_{..})^2$, when divided by the proper degrees of freedom, also provides an unbiased estimate of the variance σ_e^2. Let us consider what happens to these two estimates of σ_e^2 if the population means differ widely from one another. The four sample means $\overline{Y}_{i.}$ probably vary considerably around the overall mean $\overline{Y}_{..}$; thus $\mathrm{MS}_a = n\sum_{i=1}^{4}(\overline{Y}_{i.}-\overline{Y}_{..})^2/3$ tends to be large. On the other hand, $s_e^2 = \sum_{i=1}^{4}\sum_{j=1}^{6}(Y_{ij}-\overline{Y}_{i.})^2/20$ has no tendency to be large because of large differences among the population means. In repeated sampling, it still estimates σ_e^2.

In repeated sampling when the population means are unequal, MS_a estimates something larger than σ_e^2. It can be shown mathematically that the expected value for the due to treatment mean square equals $\sigma_e^2 + n\sum_{i=1}^{a}\alpha_i^2/(a-1)$, or as given in column 5 of Table 5.3, $\sigma_e^2 + n\sigma_a^2$. For convenience, and only for convenience, the quantity $\sum\alpha_i^2/(a-1)$ has been denoted by σ_a^2. This quantity is *not* the variance of a sample of size a from the populations of α's. In Model I we are concerned with only a α's, based on a populations chosen in some purposeful fashion; we are not interested in any other populations. Whenever all population means are equal, all the α_i are zero and $\sigma_a^2 = \sum_{i=1}^{a}\alpha_i^2/(a-1)$ is also zero. Then the expected value of the due to treatment mean square is σ_e^2.

We proceed finally to an F test of the null hypothesis that all four population means are the same. This null hypothesis can now be formulated in several ways:

$$H_0: \alpha_i = 0, \qquad i = 1, \ldots, 4$$

$$H_0: \sigma_a^2 = 0$$

$$H_0: EY_{ij} = \mu$$

$$H_0: \alpha_1 = \alpha_2 = \alpha_3 = \alpha_4$$

or

$$H_0: \mu_1 = \mu_2 = \mu_3 = \mu_4.$$

Here we usually write $H_0: \alpha_i = 0$, $i = 1, \ldots, a$.

The two statistics MS_a and s_e^2 can be shown to be statistically independent of each other. In other words, the value of s_e^2, whether particularly large or small, tells us nothing about whether the value of MS_a is particularly large or small. When H_0 is true, the quantity $(a-1)\text{MS}_a/\sigma_e^2$ has a χ^2 distribution with $a-1$ degrees of freedom; for under H_0, MS_a is simply a variance calculated from a sample of size a to estimate σ_e^2. We know already that the quantity $a(n-1)s_e^2/\sigma_e^2$ has a χ^2 distribution with $a(n-1)$ degrees of freedom.

Because the two χ^2 variables are independent, after dividing each by its degrees of freedom, their ratio has an F distribution with degrees of freedom $\nu_1 = a - 1 = 3$ and $\nu_2 = a(n-1) = 20$. The ratio is

$$F = \frac{\text{MS}_a/\sigma_e^2}{s_e^2/\sigma_e^2} = \frac{\text{MS}_a}{s_e^2}. \tag{5.10}$$

The unknown σ_e^2 has again canceled out of the ratio, and F can be written down directly from the analysis of variance table. In the example, we have

$$F = \frac{980}{163.6} = 5.99. \tag{5.11}$$

This calculated value of F is written under computed F in the sixth column of Table 5.4.

If we wish to make a test of H_0 with level of significance .05, we write under tabled F the 95th percentage point of the F distribution with 3 and 20 degrees of freedom; that is, $F[.95;3,20]=3.10$. The test is usually made as a one-sided test with the values of F larger than $F[1-\alpha;a-1,a(n-1)]$ as the critical region. For if the null hypothesis is not true, the numerator of F is an estimate of $\sigma_e^2+n\sigma_a^2$, which is certainly larger than σ_e^2; the denominator of F estimates σ_e^2. Thus the F value calculated tends to be larger than it would be if the numerator and denominator both estimated the same parameter. In our example, $F=5.9902$ is larger than $F[.95;3,20]=3.10$; therefore, we decide that the four population means are not the same.

The problem of what to do with extremely small values of the computed F is discussed in Chapter 14.

ESTIMATION OF PARAMETERS

We have now accomplished the purpose of making an overall test of whether differences exist among several population means. It should be emphasized that preparation of the analysis of variance table and performance of the F test by no means complete the analysis. The experimenter's primary purpose was surely not to decide whether all four means are equal, but rather to estimate various parameters or linear combinations of parameters. He may be interested in estimating the population yields under the four different treatments (the $\mu+\alpha_i$). He may wish to estimate the difference in yield between each fertilizer treatment and control treatment $[(\mu+\alpha_i)-(\mu+\alpha_1)=(\alpha_i-\alpha_1)]$. He may wish to estimate the difference between two fertilizers, $\alpha_3-\alpha_4$, for example. He may wonder whether fertilizers are better than no fertilizers at all; he then estimates $(\mu+\alpha_1)-[(\mu+\alpha_2)+(\mu+\alpha_3)+(\mu+\alpha_4)]/3=\alpha_1-(\alpha_2+\alpha_3+\alpha_4)/3$. He may wish to estimate any α_i, the difference between the mean for the ith treatment and the overall mean. The parameters μ and α_i are probably less interesting to him than the population means and the various comparisons among the population means that have been mentioned. Finally, he may want to estimate σ_e^2, the variation in yield from one plot to another when both receive the same fertilizer.

We have already presented the best point estimates that can be obtained from the data. It is now time to obtain the corresponding confidence intervals.

CONFIDENCE INTERVALS FOR PARAMETERS

The Variance. The $1-\alpha$ level confidence interval for σ_e^2 is (from 4.8)

$$\frac{a(n-1)s_e^2}{\chi^2[1-\alpha/2;a(n-1)]} \quad \text{to} \quad \frac{a(n-1)s_e^2}{\chi^2[\alpha/2;a(n-1)]}. \tag{5.12}$$

In the data example, for $\alpha=.05$ we have

$$\frac{4(5)(163.6)}{34.170} \quad \text{to} \quad \frac{4(5)(163.6)}{9.591}$$

or

$$96<\sigma_e^2<341.$$

Means and Linear Combinations of Means. Earlier, when we considered a single sample, we stated that the sample mean for a sample of size n was normally distributed with mean μ and with variance equal to σ^2/n; in symbols this was written

$$\overline{Y}\text{ND}(\mu,\sigma^2/n). \tag{5.13}$$

We also used s^2 an estimate of σ^2 and learned that $(n-1)s^2/\sigma^2$ has a χ^2 distribution with $n-1$ degrees of freedom. Furthermore, we learned that $\overline{Y}$ and s^2 are statistically independent. From these three statements, we were able to formulate a $1-\alpha$ level confidence interval for μ:

$$\overline{Y}\pm t[1-\alpha/2;n-1]\sqrt{s^2/n}\,. \tag{5.14}$$

Similarly in Chapter 4, in working with the difference between two means, we learned that

$$\overline{Y}_{1.}-\overline{Y}_{2.}\text{ND}\left[\mu_1-\mu_2,\sigma^2\left(\frac{1}{n_1}+\frac{1}{n_2}\right)\right], \tag{5.15}$$

that s_e^2 is an estimate of σ^2, and that $(n_1+n_2-2)s_e^2/\sigma^2$ is an *independent* χ^2 variable with n_1+n_2-2 degrees of freedom. From this, the confidence interval with level $1-\alpha$ for $\mu_1-\mu_2$ was

$$\left(\overline{Y}_{1.}-\overline{Y}_{2.}\right)\pm t[1-\alpha/2;n_1+n_2-2]\sqrt{s_e^2(1/n_1+1/n_2)}\,. \tag{5.16}$$

For any parameter or linear combination of parameters we care to estimate (μ, α_i, $\mu+\alpha_i$, $\alpha_i-\alpha_{i'}$, where $i \neq i'$, etc.), we can obtain a confidence interval with confidence level $1-\alpha$ in exactly the same fashion.

First we find the point estimate. If, for example, we wish to estimate $\alpha_1-\alpha_3$, $\bar{Y}_{1.}-\bar{Y}_{3.}$ estimates $\alpha_1-\alpha_3$. We know that the expected value of $\bar{Y}_{1.}-\bar{Y}_{3.}$ is $\alpha_1-\alpha_3$. We now need the expression for the variance of $\bar{Y}_{1.}-\bar{Y}_{3.}$. We know that $\bar{Y}_{1.}$ and $\bar{Y}_{3.}$ each has a variance σ_e^2/n; furthermore, $\bar{Y}_{1.}$ and $\bar{Y}_{3.}$ are statistically independent because our data consist of a independent random samples. Therefore, the covariance of $\bar{Y}_{1.}$ and $\bar{Y}_{3.}$ is zero; thus the variance of $\bar{Y}_{1.}-\bar{Y}_{3.}$ is $\sigma_e^2/n+\sigma_e^2/n$ or $2\sigma_e^2/n$. We have

$$\bar{Y}_{1.}-\bar{Y}_{3.}\,\text{ND}\left(\alpha_1-\alpha_3, \frac{2\sigma_e^2}{n}\right), \tag{5.17}$$

and the $1-\alpha$ confidence interval for $\alpha_1-\alpha_3$ is

$$(\bar{Y}_{1.}-\bar{Y}_{3.}) \pm t[1-\alpha/2; a(n-1)]\sqrt{2s_e^2/n}\,. \tag{5.18}$$

In the example of the four fertilizers for corn, the 95% interval for $\alpha_1-\alpha_3$ is

$$(72-66) \pm 2.086\sqrt{2(163.6)/6} \qquad \text{or} \qquad -9.40<\alpha_1-\alpha_3<21.40. \tag{5.19}$$

As a second example, $\alpha_1-(\alpha_2+\alpha_3+\alpha_4)/3$ is estimated by $\bar{Y}_{1.}-(\bar{Y}_{2.}+\bar{Y}_{3.}+\bar{Y}_{4.})/3$. Because of the independence of the four different samples, the variance of $\bar{Y}_{1.}-(\bar{Y}_{2.}+\bar{Y}_{3.}+\bar{Y}_{4.})/3$ is simply the sum of the variances of $\bar{Y}_{1.}$, $\bar{Y}_{2.}/3$, $\bar{Y}_{3.}/3$, and $\bar{Y}_{4.}/3$:

$$\begin{aligned}\text{Var}\left(\bar{Y}_{1.}-\frac{\bar{Y}_{2.}+\bar{Y}_{3.}+\bar{Y}_{4.}}{3}\right) &= \frac{\sigma_e^2}{n}+\frac{1}{9}\frac{\sigma_e^2}{n}+\frac{1}{9}\frac{\sigma_e^2}{n}+\frac{1}{9}\frac{\sigma_e^2}{n} \\ &= \frac{12}{9n}\sigma_e^2. \end{aligned} \tag{5.20}$$

The $1-\alpha$ level confidence interval is then

$$\left(\bar{Y}_{1.}-\frac{\bar{Y}_{2.}+\bar{Y}_{3.}+\bar{Y}_{4.}}{3}\right) \pm t[1-\alpha/2; a(n-1)]\sqrt{(12/9n)s_e^2}\,. \tag{5.21}$$

Here again the degrees of freedom are $a(n-1)$. Note that s_e^2 and the degrees of freedom are easily read from the completed analysis of variance table.

In the corn yield problem, the interval for $\alpha_1-(\alpha_2+\alpha_3+\alpha_4)/3$ is

$$\left(72-\frac{95+66+83}{3}\right)\pm 2.086\sqrt{12(163.6)/54} \tag{5.22}$$

or

$$-21.91<\alpha_1-\frac{\alpha_2+\alpha_3+\alpha_4}{3}<3.244.$$

A third example involves slightly more work in determining the variance. If we wish a confidence interval for α_1, we find that its point estimate is $\overline{Y}_{1.}-\overline{Y}_{..}$. By our usual rule, the variance $\overline{Y}_{1.}-\overline{Y}_{..}$ is the variance of $\overline{Y}_{1.}$ plus the variance of $\overline{Y}_{..}$ minus twice their covariance. However, we know that $\overline{Y}_{1.}$ and $\overline{Y}_{..}$ are *not* statistically independent, since some of the observations in $\overline{Y}_{..}$ are also in $\overline{Y}_{1.}$. Therefore, we do not expect their covariance to be zero. Rather than find the covariance for $\overline{Y}_{1.}$ and $\overline{Y}_{..}$, we express $\overline{Y}_{1.}-\overline{Y}_{..}$ as a linear combination of the treatment means, which are independent; then we find the variance from that expression. Thus we write

$$\begin{aligned}\overline{Y}_{1.}-\overline{Y}_{..}&=\overline{Y}_{1.}-\frac{\overline{Y}_{1.}+\overline{Y}_{2.}+\overline{Y}_{3.}+\cdots+\overline{Y}_{a.}}{a}\\&=\frac{a-1}{a}\overline{Y}_{1.}-\frac{\overline{Y}_{2.}+\cdots+\overline{Y}_{a.}}{a}\end{aligned} \tag{5.23}$$

$$\begin{aligned}\operatorname{Var}\left(\overline{Y}_{1.}-\overline{Y}_{..}\right)&=\left(\frac{a-1}{a}\right)^2\frac{\sigma_e^2}{n}+\frac{a-1}{a^2}\frac{\sigma_e^2}{n}\\&=\frac{\sigma_e^2}{n}\frac{a-1}{a}\frac{a-1+1}{a}\\&=\frac{\sigma_e^2}{n}\frac{a-1}{a}.\end{aligned}$$

Then we have

$$\overline{Y}_{1.}-\overline{Y}_{..}\text{ND}\left(\alpha_1,\frac{a-1}{an}\sigma_e^2\right). \tag{5.24}$$

The $1-\alpha$ level confidence interval is

$$\overline{Y}_{1.}-\overline{Y}_{..}\pm t[1-\alpha/2;a(n-1)]\sqrt{[(a-1)/an]s_e^2}\,. \qquad (5.25)$$

In the corn yield problem, the confidence interval for α_1 is

$$(72-79)\pm 2.086\sqrt{3(163.6)/24} \qquad \text{or} \qquad -16.43<\alpha_1<2.43. \quad (5.26)$$

Often the linear combinations that are most interesting to estimate are the ones called linear contrasts (sometimes, linear comparisons) among the means. We have had two examples of these. When we wanted to compare the yield of corn under the control treatment with that of corn under treatment 3, we estimated the difference $(\mu+\alpha_1)-(\mu+\alpha_3)=\alpha_1-\alpha_3$. Similarly, to compare the control treatment with the average of the three fertilizers, we estimated $(\mu+\alpha_1)-[(\mu+\alpha_2)+(\mu+\alpha_3)+(\mu+\alpha_4)]/3=\alpha_1-(\alpha_2+\alpha_3+\alpha_4)/3$. These are both linear contrasts among the means; and in both the coefficients of the population means add to zero $(1-\frac{1}{3}-\frac{1}{3}-\frac{1}{3}=0)$. *When the coefficients of the population means add to zero, a linear combination is called a linear contrast.* In linear contrasts, the μ's always cancel out, and we are left with the linear contrast among the α_i's to estimate; this is estimated by the same linear contrast among the sample means.

Another example of a linear contrast among means in the corn yield problem would be a comparison between the fertilizers containing K_2O with the fertilizer that does not contain it. This would be

$$\frac{(\mu+\alpha_2)+(\mu+\alpha_3)}{2}-(\mu+\alpha_4)=\frac{\alpha_2+\alpha_3}{2}-\alpha_4,$$

and it would be estimated by $(\overline{Y}_{2.}+\overline{Y}_{3.})/2-\overline{Y}_{4.}$.

For a general expression for a linear contrast among the means, we write $\sum_{i=1}^{a}h_i\alpha_i$, where $\sum_{i=1}^{a}h_i=0$; this is estimated by $\sum_{i=1}^{a}h_i\overline{Y}_{i.}$. In the contrast $(\alpha_2+\alpha_3)/2-\alpha_4$, $h_1=0$, $h_2=\frac{1}{2}$, $h_3=\frac{1}{2}$, and $h_4=-1$.

The variance of the linear contrast $\sum_{i=1}^{a}h_i\overline{Y}_{i.}$ is

$$\operatorname{Var}\sum_{i=1}^{a}h_i\overline{Y}_{i.}=\frac{\sum h_i^2}{n}\sigma_e^2. \qquad (5.27)$$

The confidence interval for $\sum_{i=1}^{a}h_i\alpha_i$ is

$$\sum_{i=1}^{a}h_i\overline{Y}_{i.}\pm t[1-\alpha/2;a(n-1)]\sqrt{(\sum h_i^2/n)s_e^2}\,. \qquad (5.28)$$

TABLE 5.5. ANALYSIS OF VARIANCE TABLE FOR ONE-WAY CLASSIFICATION, MODEL I, UNEQUAL NUMBERS

Source of Variation (1)	Sum of Squares (2)	d f (3)	Mean Square (4)	EMS (5)	Computed F (6)	Tabled F (7)
Due Treatment	$SS_a = \sum_{i=1}^{a} n_i(\overline{Y}_{i.} - \overline{Y}_{..})^2$	a-1	$MS_a = SS_a/(a-1)$	$\sigma_e^2 + \sum_{i=1}^{a} n_i\alpha_i^2/(a-1)$	MS_a/s_e^2	$F[1-\alpha;a-1,N-a]$
Residual	$SS_r = \sum_{i=1}^{a} \sum_{j=1}^{n_i} (Y_{ij} - \overline{Y}_{i.})^2$	N-a	$s_e^2 = SS_r/(N-a)$	σ_e^2		
Total	$SS_t = \sum_{i=1}^{a} \sum_{j=1}^{n_i} (Y_{ij} - \overline{Y}_{..})^2$	N-1				

UNEQUAL SAMPLE SIZES

Other things being equal, it is desirable to have equal numbers of observations on each treatment. Sometimes this is not possible, however. In the one-way analysis of variance, only minor changes need be made in the formulae to account for unequal sample sizes; the analysis is essentially the same. To avoid repetition, we merely mention the necessary changes in notation and give a summary table.

We let n_i be the number of observations in the ith sample (replacing n). We let $N = n_1 + \cdots + n_a$ be the total number of observations (replacing an). We define the overall mean μ in such a way that $\sum_{i=1}^{a} n_i\alpha_i = 0$ (instead of $\sum_{i=1}^{a} \alpha_i = 0$). With this notation, the analysis of variance table is as given in Table 5.5.

Table 5.6 gives the variances of the various statistics. From these formulae we can construct confidence intervals for both equal and unequal sample sizes.

TABLE 5.6. VARIANCES OF SAMPLE STATISTICS, MODEL I

Parameter Being Estimated	Point Estimate of Parameter	Variance
μ	$\overline{Y}_{..}$	σ_e^2/N
$\mu+\alpha_i$	$\overline{Y}_{i.}$	σ_e^2/n_i
α_i	$\overline{Y}_{i.} - \overline{Y}_{..}$	$\sigma_e^2(N-n_i)/(n_iN)$
$\alpha_i-\alpha_{i'}$	$\overline{Y}_{i.} - \overline{Y}_{i'.}$	$\sigma_e^2(\frac{1}{n_i} + \frac{1}{n_{i'}})$
$\sum_{i=1}^{a} h_i\alpha_i$	$\sum_{i=1}^{a} h_i\overline{Y}_{i.}$	$\sigma_e^2 \sum_{i=1}^{a} (h_i^2/n_i)$

where

$$\sum_{i=1}^{a} h_i = 0$$

CONFIDENCE INTERVALS WITH AN OVERALL CONFIDENCE LEVEL

Constructing numerous confidence intervals from a single set of data leads to the same uncomfortable feeling that we have when we make many tests. We know that each interval covers its parameter $1-\alpha$ percent of the time in repeated sampling. Nevertheless, we feel intuitively that among many intervals, one or more may fail to cover their parameters. This intuitive feeling can be shown mathematically to be correct.

An answer to this difficulty lies in making confidence intervals having an overall confidence level $1-\alpha$. If a set of intervals is constructed so that at least $1-\alpha$ percent of the time every interval calculated in the set covers its parameter, the intervals are called multiple confidence intervals with confidence level $1-\alpha$. We give here the two methods we consider most generally advantageous—multiple-t intervals and Tukey's intervals.

MULTIPLE-t CONFIDENCE INTERVALS

We can construct m intervals for m linear combinations of population means using the same t statistic as in the last section and simply adjust the level of the percentile that is read from the t distribution. If, in place of the $1-\alpha/2$ point of the t distribution, we use the $1-\alpha/2m$ point, we obtain a set of m confidence intervals such that, in repeated experimentation, the proportion of such sets of m intervals which cover all m linear combinations of means is greater than or equal to $1-\alpha$.

In our example of corn yields, if the experimenter decided (before looking at his data) that the parameters of interest to him were μ, $\mu+\alpha_1$, $\mu+\alpha_2$, $\mu+\alpha_3$, $\mu+\alpha_4$, $\alpha_1-\alpha_2$, $\alpha_1-\alpha_3$, $\alpha_1-\alpha_4$, $\alpha_1-(\alpha_2+\alpha_3+\alpha_4)/3$, and $(\alpha_2+\alpha_3)/2-\alpha_4$, he wishes to make 10 confidence intervals, and $m=10$. If he wishes an overall confidence level of .95, then $1-.05/(2)(10)=1-.0025=.9975$. As before, the degrees of freedom for the t distribution is $a(n-1)=20$, and from Table A.2 we find that $t[.9975;20]=3.153$. The experimenter then calculates his 10 intervals; for example, the first two intervals are

$$\bar{Y}_{..}\pm t[.9975;\ 20]\sqrt{s_e^2/an} \quad \text{or} \quad 70.77<\mu<87.23 \tag{5.29}$$

$$\bar{Y}_{1.}\pm t[.9975;20]\sqrt{s_e^2/n} \quad \text{or} \quad 55.54<\mu+\alpha_1<88.46.$$

The general formulation for $\sum_{i=1}^{a}h_i(\mu+\alpha_i)$ or ($\sum_{i=1}^{a}h_i\alpha_i$ if $\sum_{i=1}^{a}h_i=0$) is

$$\sum_{i=1}^{a} h_i \bar{Y}_{i.} \pm t[1-\alpha/2m; a(n-1)] \sqrt{\left(\sum_{i=1}^{a} \frac{h_i^2}{n}\right) s_e^2} \,. \tag{5.30}$$

For unequal sample sizes, the intervals are

$$\sum_{i=1}^{a} h_i \bar{Y}_{i.} \pm t[1-\alpha/2m; N-a] \sqrt{\left(\sum_{i=1}^{a} \frac{h_i^2}{n_i}\right) s_e^2} \tag{5.31}$$

As mentioned earlier, the experimenter decides what is appropriate to estimate before looking at his data. The set of confidence intervals to be estimated should be completely planned in advance. Otherwise, the experimenter could look at the data, notice a contrast that seems interesting, and estimate it with a single confidence interval (i.e., with $m=1$). In this way he would obtain a shorter interval than if he had decided to form numerous intervals. The confidence level for such an interval becomes meaningless.

The experimenter pays for a multiple confidence level by having longer intervals. However, beyond very small values of m, $t[1-\alpha/2m;\nu]$ increases very slowly with m; therefore, it is just as well to plan to have a large number of confidence intervals with a multiple confidence level. Then, if examination of the data shows that some of the intervals are uninteresting, the calculations for all m intervals need not actually be made. However, $t[1-\alpha/2m;\nu]$ is used for any intervals that *are* calculated. If, after planning the analysis, examination of the data discloses an interesting contrast that the experimenter failed to anticipate, he notes this result. He does not, however, claim to have established it by means of a confidence interval. For those who have studied probability, proof that these intervals have a confidence level greater than or equal to $1-\alpha$ is based on the simplest Bonferroni inequality (see Dunn [5.16] or [5.17]).

If the interval for the contrast $\alpha_1-\alpha_2$ does *not* cover zero, the investigator should decide (as in Chapter 4 with $\mu_1-\mu_2$) that $\alpha_1 \neq \alpha_2$. Thus, by finding the various confidence intervals for contrasts, he may be able to decide which contrasts among the means differ from zero. Even if the F test from the analysis of variance table leads him to accept the null hypothesis of equal means, it is still advantageous to make confidence intervals for interesting contrasts. The location and length of an interval in which a contrast lies is of interest, even if it covers zero. For example, a very short interval containing zero tells us that the difference between these means is probably very small; an interval from a small negative number to a large positive one tells us that if the difference is negative, its magnitude is probably small. We even occasionally find intervals for contrasts that do not contain zero when the decision on the F test has been that all the means may be the same.

TUKEY'S INTERVALS FOR CONTRASTS

A method due to Tukey for obtaining confidence intervals for linear contrasts among means can be applied when the treatment groups are equal ($n_i = n$). The intervals are formed using tables of the distribution of the Studentized range. The range of a sample is the largest observation minus the smallest observation. The Studentized range, usually denoted by q, is the range divided by an independent estimate of the population standard deviation. For samples from normal populations, some percentage points for q are given in Table A.5. There are two parameters appearing in the table—ν, the degrees of freedom of the χ^2 estimate of the population variance, and k. For the case of a means, $k = a$ and $\nu = a(n-1)$.

Tukey's intervals for $\sum_{i=1}^{a} h_i \overline{Y}_{i.}$, where $\sum_{i=1}^{a} h_i = 0$ and the sum of the positive h_i's is 1 are given by:

$$\sum_{i=1}^{a} h_i \overline{Y}_{i.} \pm q[1-\alpha; a, a(n-1)]\sqrt{s_e^2/n}\ . \tag{5.32}$$

Table 5.7 shows confidence intervals for the example using the two methods. We note that Tukey's intervals are all of the same length and are advantageously short for contrasts involving just two means.

DISCUSSION OF MULTIPLE CONFIDENCE INTERVALS

From (5.30) and (5.32) we see that Tukey's intervals in a particular problem are always of the same length, whereas the multiple-t intervals differ in length from one interval to another. Tukey's intervals are most advantageous if it is desired to estimate only quantities like $\alpha_1 - \alpha_2$. In such cases, they are usually the shortest and should be used whenever possible. For intervals like $\alpha_1 - (\alpha_2 + \alpha_3)/2$, they are usually longer than the multiple-t intervals. Tukey's intervals are limited to the case of equal sample sizes, whereas the multiple-t can be used for unequal sample sizes.

There is another difference between the Tukey and the multiple-t intervals. Tukey's intervals are for all possible contrasts; in other words, we do not have to specify the contrasts in advance but can test as many as we wish after seeing the data. This sounds like a great advantage over using the t distribution, which necessitates a decision in advance of the contrasts

TABLE 5.7. CONFIDENCE INTERVALS WITH OVERALL LEVEL AT LEAST .95 IN CORN YIELD PROBLEM

Parameter	Multiple t, m = 10	Multiple t, m = 5	Tukey
μ	70.77 to 87.23	-	-
$\mu + \alpha_1$	55.54 to 88.46	-	-
$\mu + \alpha_2$	78.54 to 111.46	-	-
$\mu + \alpha_3$	49.54 to 82.46	-	-
$\mu + \alpha_4$	66.54 to 99.46	-	-
$\alpha_1 - \alpha_2$	-46.28 to 0.28	-44.01 to -1.99	-43.68 to -2.32
$\alpha_1 - \alpha_3$	-17.28 to 29.28	-15.01 to 27.01	-14.68 to 26.68
$\alpha_1 - \alpha_4$	-34.28 to 12.28	-32.01 to 10.01	-31.68 to 9.68
$\alpha_1 - (\alpha_2 + \alpha_3 + \alpha_4)/3$	-28.34 to 9.68	-26.49 to 7.82	-30.01 to 11.35
$(\alpha_2 + \alpha_3)/2 - \alpha_4$	-22.66 to 17.66	-20.69 to 15.69	-23.18 to 18.18

that are to be estimated. However, this disadvantage of the multiple-t method can be overcome by deciding beforehand to make intervals for all interesting quantities.

MULTIPLE TESTS

It can be argued that when several tests are made from the same data, they should be made with an overall significance level. The principle used in constructing multiple-t intervals is perfectly general and can be applied to tests. When we wish m tests with an overall level of α, each of the m tests is made using a single level of α/m. As an example, in succeeding chapters we make several F tests from the same set of data. Each calculated F statistic can then be compared with $F[1-\alpha/m;\nu_1,\nu_2]$, rather than with $F[1-\alpha;\nu_1,\nu_2]$.

ILLUSTRATIVE EXAMPLE

As an illustration of a one-way analysis of variance problem, we consider an experiment that has been performed before the data are brought for analysis to the statistician. Three diets have been given to rats: a standard diet, a standard diet plus saturated fat, and a standard diet plus unsaturated fat. Fortunately, since the experimenter randomly assigned the rats to the three treatment groups, the study can be considered an experiment. Care has been taken to treat all three groups as much alike as possible except for the diet. However, rats in the same treatment group were placed in adjacent cages, and cage location thus may be mixed with treatment effect.

The statistician decides first to set up the model ignoring the possible cage location complication and to carry out the analysis. Because there are three treatment groups, the model is

$$Y_{1j} = \mu + \alpha_1 + \varepsilon_{1j} \qquad \text{(standard)}$$

$$Y_{2j} = \mu + \alpha_2 + \varepsilon_{2j} \qquad \text{(standard + saturated fat)}$$

$$Y_{3j} = \mu + \alpha_3 + \varepsilon_{3j} \qquad \text{(standard + unsaturated fat),}$$

where $\sum n_i \alpha_i = 0$ because the treatments 1 and 2 have 20 rats, but treatment 3 has only 18 rats.

The variable observed is cholesterol level in milligram%. The statistician first checks visually to see whether cholesterol level is approximately normally distributed and to ensure that no observations are included that are obviously in error. He may use a histogram program such a BMD07D for this purpose, or he may plot by hand. When each sample of data consists of just 18 or 20 rats, only gross departures from normality can be detected visually. If he draws a histogram using all the residuals $Y_{ij} - \overline{Y}_{i.}$, he has 58 observations and can check normality reasonably well. Let us assume that the data appear to be normally distributed. If they are quite skewed, transformations as discussed in Chapter 14 should be considered.

The analysis of variance is performed either on a desk calculator or by computer (the BMD07D program also performs the analysis of variance), and the problem of interpretation begins. If the F test for testing $H_0: \alpha_i = 0$ results in acceptance of the hypothesis, the statistician reports this result and also confidence limits for the variance σ_e^2. He may also report confidence limits for differences among treatments, although he may not bother with these.

If the preliminary F test rejects the null hypothesis, the statistician should certainly report confidence intervals for individual differences. The high F value may be caused by one treatment mean being quite different from the others, or there may be differences among all the means.

To consider whether there may be differences in cholesterol due to cage location, the statistician may plot cholesterol levels versus cage location in various ways, keeping the three groups separate, because any differences among the groups as a whole can be due to either treatment or location. If smaller differences in location within the groups seem to have no effect on cholesterol level, perhaps the larger between-group differences have no effect. It is impossible to be sure, however, and unless the experimenter knows from experience that location does not matter, results must be reported cautiously.

SUMMARY

One-way analysis of variance, the simplest analysis of variance design, was introduced. The F test of the null hypothesis that no differences exist among treatment means was made, and confidence intervals for parameters were obtained. Multiple comparisons among means using a Student t method and Tukey's method were presented and discussed.

REFERENCES

APPLIED STATISTICS

5.1. Anderson, R. L., and T. A. Bancroft. *Statistical Theory in Research*, McGraw-Hill, New York, 1952.

5.2. Bennett, C. A., and N. L. Franklin, *Statistical Analysis in Chemistry and the Chemical Industry*, Wiley, New York, 1954.

5.3. Bliss, C. I., *Statistics in Biology*, McGraw-Hill, New York, 1967.

5.4. Brownlee, K. A., *Statistical Theory and Methodology in Science and Engineering*, 2nd ed., Wiley, New York, 1965.

5.5. Davies, O. L. (Ed.), *The Design and Analysis of Industrial Experiments*, 2nd ed., Oliver & Boyd, Edinburgh, 1956.

5.6. Dixon, W. J., and F. J. Massey, Jr., *Introduction to Statistical Analysis*, 3rd ed., McGraw-Hill, New York, 1969.

5.7. Fisher, R. A., and F. Yates, *Statistical Tables for Biological, Agricultural, and Medical Research*, 6th ed., Oliver & Boyd, London, 1963.

5.8. Fisher, R. A., *Statistical Methods for Research Workers*, Oliver & Boyd, Edinburgh, 1925.

5.9. Kempthorne, O., *The Design and Analysis of Experiments*, Wiley, New York, 1952.

5.10. Ostle, B., *Statistics in Research*, Iowa State University Press, Ames, Iowa, 1963.

5.11. Snedecor, G. W., and W. G. Cochran, *Statistical Methods*, 6th ed., Iowa State University Press, Ames, Iowa, 1967.

5.12. Winer, B. J., *Statistical Principles in Experimental Design*, McGraw-Hill, New York, 1962.

5.13. Peng, K. E., *The Design and Analysis of Scientific Experiments*, Addison-Wesley, Reading, Mass., 1967.

MATHEMATICAL STATISTICS

5.14. Mann, H. B., *Analysis and Design of Experiments*, Dover, New York, 1949.
5.15. Scheffé, H., *Analysis of Variance*, Wiley, New York, 1967.

COMPARISONS

5.16. Dunn, O. J., "Multiple Comparisons Among Means," *Journal of the American Statistical Association*, Vol. 56 (1961), pp. 52–64.
5.17. Dunn, O. J., "Estimation of the Means of Dependent Variables," *Annals of Mathematical Statistics*, Vol. 29 (1958), pp. 1095–1111.
5.18. O'Neill, R., and G. B. Wetherill, "The Present State of Multiple Comparison Methods," *Journal of the Royal Statistical Society, Series B*, Vol. 33 (1971), pp. 218–250.

PROBLEMS

5.1. Four chemicals were used to combat plant lice on sugar beets, each chemical being applied to one plot. Twenty-five leaves were picked from each plot, and the number of plant lice on each leaf was recorded. The data obtained were as follows:

Chemical	1	2	3	4	5	6	7	8	9	10	11	12	13	14	15	16	17	18	19	20	21	22	23	24	25
I	12	13	26	13	17	24	14	10	6	4	2	10	8	6	7	13	18	10	18	3	4	18	13	10	21
II	10	21	34	15	5	22	12	25	18	12	2	2	10	22	17	20	19	20	12	11	16	5	11	17	16
III	23	14	14	20	27	25	17	18	29	14	31	5	13	18	23	16	13	23	4	16	17	9	28	23	19
IV	32	26	24	16	32	18	33	16	34	18	9	19	29	30	18	25	20	21	27	31	25	33	24	16	24

Preliminary calculations gave the following

Chemical	$\sum_{j=1}^{25} Y_{ij}$	$\sum_{j=1}^{25} Y_{ij}^2$	$\left(\overline{Y}_{i.} - \overline{Y}_{..}\right)^2$
I	300	4600	28.409
II	374	6922	5.617
III	459	9603	1.061
IV	600	15490	44.489

$$\sum_{i=1}^{4} \sum_{j=1}^{25} \left(Y_{ij} - \overline{Y}_{..}\right)^2 = 6582.110$$

Assume that $Y_{ij} = \mu + \alpha_i + \varepsilon_{ij}$, where $\sum_{i=1}^{4} \alpha_i = 0$ and ε_{ij} IND$(0, \sigma_e^2)$.

(*a*) Fill in an analysis of variance table and test at a .05 level whether differences exist among the four chemicals.
(*b*) Write down point estimates of the following parameters:
 (i) The overall mean number of plant lice per leaf for the four chemicals.
 (ii) $\mu+\alpha_1$, the mean for the first chemical.
 (iii) $\alpha_1-\alpha_2$.
 (iv) $\alpha_1-(\alpha_2+\alpha_3+\alpha_4)/3$.
 (v) σ_e^2.
(*c*) Obtain confidence intervals with an overall 95% level for the quantities μ, $\mu+\alpha_1$, $\alpha_1-\alpha_2$, and $\alpha_1-(\alpha_2+\alpha_3+\alpha_4)/3$ using the multiple-*t* technique.
(*d*) Obtain confidence intervals for $\alpha_1-\alpha_2$, $\alpha_1-\alpha_3$, and $\alpha_1-\alpha_4$ using Tukey's technique.
(*e*) Obtain a 95% confidence interval for σ_e^2.

5.2. Serum cholesterol was measured on 24 men whose ages were in the range 65 to 74 years. Classified by annual income, the values are as follows:

	Observation									
Income	1	2	3	4	5	6	7	8	9	10
<\$5,000	229.5	228.3	209.3	240.4	214.1	240.1	210.5	218.8	223.4	217.6
\$5,000–10,000	235.1	244.9	243.7	232.9	253.1	251.9	241.0	224.9	—	—
>\$10,000	241.2	227.9	244.8	240.5	252.1	232.1	—	—	—	—

(*a*) State the model.
(*b*) List the parameters or linear combinations of parameters (contrasts) you might wish to estimate.
(*c*) Make an analysis of variance table.
(*d*) Test the hypothesis $H_0: \alpha_1=\alpha_2=\alpha_3$. Use $\alpha=.05$.
(*e*) Estimate the following parameters and contrasts by confidence intervals with a joint confidence level of 95%:
 (i) $\mu+\alpha_1$.
 (ii) α_1.
 (iii) $\alpha_3-\alpha_1$.
 (iv) $\alpha_1-(\alpha_2+\alpha_3)/2$.
(*f*) Compute the 95% confidence interval for $\alpha_3-\alpha_1$, assuming that it is the only interval you wish to obtain. Compare the length of the interval with that for $\alpha_3-\alpha_1$ found in (*e*).

5.3. IQ scores were recorded for 20 girls classified into two equal groups according to the economic status of their parents. There were 10 girls in each group, and the scores were as follows:

Parental Status	Observation Number 1	2	3	4	5	6	7	8	9	10
High	124	114	115	106	84	96	106	126	124	116
Low	113	97	108	95	105	69	113	98	118	70

Preliminary calculations gave the following:

	$\sum_{j=1}^{10} Y_{ij}$	$\sum_{j=1}^{10} Y_{ij}^2$	$(\bar{Y}_{i.}-\bar{Y}_{..})^2$
High	1111.	125049.	39.0625
Low	986.	99850.	39.0625

$$\sum_{i=1}^{2}\sum_{j=1}^{10}(Y_{ij}-\bar{Y}_{..})^2=5028.55$$

(*a*) State the model.
(*b*) Make an analysis of variance table.
(*c*) Make an F test of $H_0:\alpha_1=\alpha_2$ with a .01 significance level.
(*d*) List the parameters or linear combinations of parameters that might be estimated.
(*e*) Obtain point estimates of the parameters listed in (*d*).
(*f*) Find a 95% confidence interval for σ_e^2.
(*g*) Obtain confidence intervals with an overall 95% level for all parameters (other than σ_e^2) listed in (*d*).
(*h*) Use the methods of Chapter 4 to make a t test of $H_0:\mu_1=\mu_2$ (or $H_0:\alpha_1=\alpha_2$). Note that the t statistic obtained is the square root of the F statistic in (*d*).

5.4. Measurement of chlorides in sweat (m/liter) in groups of normals, individuals heterozygotic for cystic fibrosis of the pancreas, and individuals homozygotic for the same disease yielded the following values:

Category	Observation 1	2	3	4	5	6	7	8	9	10
Normal	37.1	31.7	16.2	23.5	30.1	9.6	39.3	22.2	41.6	15.4
Heterozygote	40.0	37.2	46.1	53.0	53.2	—	—	—	—	—
Homozygote	110.0	103.9	123.7	108.4	105.6	108.4	—	—	—	—

Preliminary calculations gave the following values:

Category	n_i	$\overline{Y}_{i.}$	$\sum_{j=1}^{n_i} Y_{ij}^2$	$(\overline{Y}_{i.}-\overline{Y}_{..})^2$
Normal	10	26.67	8199.21	805.830
Heterozygote	5	45.90	10748.29	83.853
Homozygote	6	110.00	72849.38	3018.716

$$\sum_{i=1}^{3}\sum_{j=1}^{n_i}\left(Y_{ij}-\overline{Y}_{..}\right)^2=28139.811$$

(*a*) State the model.
(*b*) List the parameters or linear combination of parameters (contrasts) you might wish to estimate.
(*c*) Make an analysis of variance table.
(*d*) Test the hypothesis $H_0: \alpha_1=\alpha_2=\alpha_3$, using $\alpha=.05$.
(*e*) Estimate σ_e^2 with a point estimate and with a 95% confidence interval.
(*f*) Estimate μ, $\mu+\alpha_1$, α_1, $\alpha_1-\alpha_3$, and $\alpha_2-(\alpha_1+\alpha_3)/2$ by point estimates and by confidence interval estimates having an overall .95 level.

CHAPTER 6

SINGLE FACTOR ANALYSIS OF VARIANCE—OTHER DESIGNS

There are many instances in which an investigator wishes to study the effect of a single factor such as economic level or type of fertilizer, but because of practical problems he needs a design more complicated than that given in Chapter 5. In this chapter, we explore three types of analysis of variance that may be appropriate under such circumstances. These are the randomized complete block design, hierarchical or nested designs, and the Latin square. The Latin square design, which has been placed in a section marked with asterisks, can be omitted in a first reading without loss of continuity. All three designs can be extended and used when the investigator is studying more than one set of factors. It is advantageous to introduce them for the one-way analysis of variance.

THE RANDOMIZED COMPLETE BLOCK DESIGN

In one-way classification experiments considered in Chapter 5, the treatments were assigned at random. Such designs are called *completely randomized designs*. For example, if the treatments are three drugs and there are 24 patients, eight patients are assigned at random to each of the three treatments.

The 24 patients may vary widely in initial condition, and their initial condition may affect their response to the drugs. In the completely randomized design, we try to take care of these differences among the patients by assigning them at random into groups of eight patients. Unfortunately, it is possible that all the patients receiving drug 1 may be comparatively healthy and all those receiving drug 2 may be comparatively unhealthy, even though the assignment was randomly made. By randomization, however, at least we have given each drug an equal chance with respect to the initial condition of the groups. Furthermore, we can expect that if the experiment is large enough, randomization will roughly equalize the initial condition of the three groups.

There are other possible ways of dealing with initial condition. We might, for instance, attempt to select 24 patients who are all very similar in initial condition. Besides initial condition, the experimenter may feel that other factors might influence the response to the drugs (e.g., age or weight). One way or another, the experimenter takes account of certain of these factors in his design. For example, covariance analysis (see Chapter 13) can sometimes be used for this purpose. Factors that are not eliminated by choice of design (and of whose existence the experimenter may even be unaware), are taken care of by randomization.

A block design is a much-used method for dealing with factors that are known to be important and which the investigator wishes to eliminate rather than to study.

In the randomized complete block design, still with three treatments and 24 patients, the patients are divided into eight blocks, each consisting of three patients. These blocks are formed so that each block is as homogeneous as possible. Each block consists of as many experimental units as there are treatments—three, in this case. The blocks might be easily formed on the basis of age, for example, with blocks 1 and 8 consisting of the three youngest and the three oldest patients, respectively. The individuals in a particular block are as alike as possible. On the other hand, there may be wide differences between the individuals for different blocks.

After the blocks are formed, the three drugs are assigned at random to the three patients within each block. If the blocking has been done on a factor such as initial condition, and if initial condition is important in determining the level of the response, the responses to the drugs will differ widely from block to block. However, because each drug is used exactly once in each block, the design is balanced and the mean treatment responses to the three drugs will be comparable. The differences observed among the drugs should be largely unaffected by initial condition. Below are the eight blocks with a possible treatment assignment:

Patient Number	Block Number 1	2	...	8
1	Drug 3	Drug 2	...	Drug 1
2	Drug 1	Drug 1	...	Drug 3
3	Drug 2	Drug 3	...	Drug 2

When the data are gathered, they are arranged in rows according to treatment (drug), and we have

Drug Number	Block Number 1	2	...	8
1	Y_{11}	Y_{12}	...	Y_{18}
2	Y_{21}	Y_{22}	...	Y_{28}
3	Y_{31}	Y_{32}	...	Y_{38}

Note that this is a balanced design—each treatment occurs once in each block; thus if we obtain the mean response for drug 1 over the eight blocks, it will be comparable to the mean response for drugs 2 or 3.

This type of design is widely used. For example, industrial material frequently arrives in batches that tend to be homogeneous; thus a batch may be used as a block. In laboratories, to take another case, results often differ from day to day, and therefore days frequently serve as blocks. A common practice is to block out technician effect. In agricultural experiments, the blocks are sometimes separate plots of land. The technique of randomized blocks is a very useful one for removing unwanted variation. Frequently the investigator can obtain significant differences among treatment effects using a smaller sample size with the randomized complete block design than with a completely randomized design.

In planning an experiment it is important to identify in advance the factors that may introduce unwanted variation in the response (i.e., variation not due to the treatment effect) and to block accordingly. If results differ from day to day, days become blocks, and the design should be balanced within days. Each treatment must occur exactly the same number of times within each day, and it must be assigned at random within the day.

THE MODEL

We assume that each observation can be described as follows:

$$Y_{ij} = \mu + \alpha_i + \beta_j + \varepsilon_{ij}, \qquad i = 1,\ldots,a; \qquad j = 1,\ldots,b, \tag{6.1}$$

with

$$\sum_{i=1}^{a} \alpha_i = 0, \qquad \sum_{j=1}^{b} \beta_j = 0, \qquad \text{and} \qquad \varepsilon_{ij}\text{IND}(0, \sigma_e^2).$$

Here α_i is the effect of the ith treatment and β_j is the effect of the jth block. Note that the treatments correspond to rows and the blocks to columns. Thus Y_{ij} with $i=2$ and $j=3$ denotes the second treatment and the third block.

From the model as just given, we can read four assumptions:

1. The response to the ith treatment in the jth block Y_{ij} is from a normal distribution. (There are ab distributions.)

2. The means of these ab normal distributions can be expressed in the form $\mu+\alpha_i+\beta_j$. This property is often called *additivity*, or alternatively, *no interaction*.

3. The variances of the ab populations are all equal. This property is known as *homoscedasticity*.

4. The ε_{ij} (deviations from the means) are statistically independent. If we know that ε_{11} is large, we have no reason to expect ε_{12} to be small (or large, for that matter).

ADDITIVITY

Of the four assumptions, the only unfamiliar one is that of additivity. From (6.1) we see that the population mean of the ith treatment in the jth block is of the form $\mu+\alpha_i+\beta_j$. This tells us that the effect of the ith treatment on the response is the same (α_i) regardless of the block in which the treatment is used. Similarly, the effect of the jth block is the same regardless of the treatment. If a particular treatment increases (or decreases) the mean response by the same amount regardless of the block, we say that the effects are additive.

A set of additive population means is as follows:

Treatment	Block 1	Block 2	Block 3
1	$20+1+2=23$	$20+1+5=26$	$20+1-7=14$
2	$20-1+2=21$	$20-1+5=24$	$20-1-7=12$

In this example, $\mu=20$, $\alpha_1=1$, $\alpha_2=-1$, $\beta_1=2$, $\beta_2=5$, $\beta_3=-7$, and

$$\alpha_1+\alpha_2=0 \qquad \text{and} \qquad \beta_1+\beta_2+\beta_3=0.$$

A set of population means that are *not* additive is the following:

	Block 1	Block 2	Block 3
Treatment 1	$20+1+2-4=19$	$20+1+5-2=24$	$20+1-7+6=20$
Treatment 2	$20-1+2+4=25$	$20-1+5+2=26$	$20-1-7-6=\ 6$

TABLE 6.1. DATA AND PRELIMINARY CALCULATIONS ON SQUARE ROOTS OF NUMBERS OF ZERO COUNTS IN FIVE EXPERIMENTS

	Block					
	1	2	3	4	5	$\bar{Y}_{i.}$
Control	3.74	4.58	4.58	4.47	4.79	4.432
Treatment 1	4.47	6.78	5.19	5.19	6.85	5.696
Treatment 2	5.65	7.00	6.08	5.74	7.55	6.404
$\bar{Y}_{.j}$	4.620	6.120	5.283	5.133	6.397	5.511 = $\bar{Y}_{..}$

In the second case, the EY_{ij} (the population means) are not explained by μ, α_i, and β_j. The block and treatment effects are not additive, and we need a fourth term to express the population means. We then say that there is interaction between the blocks and treatments.

AN EXAMPLE

As an example of a randomized complete block design we use the data in Table 6.1, from a series of five bacterial experiments. The data consist of the square roots of the numbers of zero counts obtained. Each of the five experiments forms a block; there were three treatments including a control treatment.

ANALYSIS OF VARIANCE TABLE AND F TESTS

As in Chapter 5, we proceed now to divide the basic sum of squares into meaningful parts and to make an analysis of variance table. Corresponding to

$$Y_{ij} - \mu = \alpha_i + \beta_j + \varepsilon_{ij}, \tag{6.2}$$

we can express the difference $Y_{ij} - \bar{Y}_{..}$ as the sum of three terms:

$$Y_{ij} - \bar{Y}_{..} = \left(\bar{Y}_{i.} - \bar{Y}_{..}\right) + \left(\bar{Y}_{.j} - \bar{Y}_{..}\right) + \left(Y_{ij} - \bar{Y}_{i.} - \bar{Y}_{.j} + \bar{Y}_{..}\right). \tag{6.3}$$

Note that (6.3) is an algebraic identity. Note also that $\bar{Y}_{i.} - \bar{Y}_{..}$ provides an estimate of α_i and that $\bar{Y}_{.j} - \bar{Y}_{..}$ provides an estimate of β_j. The third term on the right-hand side of (6.3) must estimate ε_{ij}.

As before, we now square both sides of (6.3). If we do this for every observation and sum over the entire set of data, the cross-product terms again sum to zero. The basic sum of squares is thus expressed as the sum of three terms:

$$\sum_{i=1}^{a}\sum_{j=1}^{b}\left(Y_{ij}-\bar{Y}_{..}\right)^2 = b\sum_{i=1}^{a}\left(\bar{Y}_{i.}-\bar{Y}_{..}\right)^2 + a\sum_{j=1}^{b}\left(\bar{Y}_{.j}-\bar{Y}_{..}\right)^2$$

$$+\sum_{i=1}^{a}\sum_{j=1}^{b}\left(Y_{ij}-\bar{Y}_{i.}-\bar{Y}_{.j}+\bar{Y}_{..}\right)^2. \qquad (6.4)$$

We recognize the first term on the right-hand side of (6.4) as the sum of squares among treatments; the second term of similar form is the sum of squares among blocks. The third term represents the residual variation, after variation due to blocks and to treatments has been removed. Under the assumption of no interaction, this term provides an estimate of σ_e^2.

Table 6.2 is the analysis of variance table. The symbols σ_a^2 and σ_b^2 are defined by

$$\sigma_a^2 = \sum_{i=1}^{a}\frac{\alpha_i^2}{a-1} \qquad \text{and} \qquad \sigma_b^2 = \sum_{j=1}^{b}\frac{\beta_j^2}{b-1},$$

respectively. The degrees of freedom for the "due block," "due treatment," and "total" sums of squares are what we expect from the earlier model; the degrees of freedom for the residual sum of squares can be obtained by subtraction; thus we have

$$(a-1)(b-1)=(ab-1)-(b-1)-(a-1).$$

The EMS column needs little explanation; the "among treatment" EMS (EMS_a) and the "among block" EMS (EMS_b) are exactly what we should expect. We might mention, however, that this EMS column is correct only when there is no interaction.

Table 6.3 presents the analysis of variance for the data in Table 6.1. There are two possible F tests. To test whether there are treatment differences, $H_0:\alpha_i=0,\ i=1,\ldots,a$, we compare the calculated statistic $F=\text{MS}_a/s_e^2$ with $F[1-\alpha;a-1,\ (a-1)(b-1)]$.

We can similarly test $H_0:\beta_j=0,\ j=1,\ldots,b$ by comparing $F=\text{MS}_b/s_e^2$ with $F[1-\alpha;\text{b}-1,\ (a-1)(b-1)]$. This test helps us decide whether there are actually response differences among blocks, and we might wish to make it to determine whether similar blocking should be done in future work. Because the investigator is primarily studying treatment effects and merely eliminating block effects, he often omits the second test.

TABLE 6.2. ANALYSIS OF VARIANCE TABLE, RANDOMIZED COMPLETE BLOCK DESIGN, MODEL I

Source of Variation (1)	SS (2)	d.f. (3)	MS (4)	EMS (5)	F Computed (6)	F Tabled (7)
Due Treatments	$SS_a = b \sum_{i=1}^{a} (\overline{Y}_{i.} - \overline{Y}_{..})^2$	$a - 1$	$MS_a = SS_a/(a-1)$	$\sigma_e^2 + b\sigma_a^2$	MS_a/s_e^2	$F[1-\alpha; a-1, (a-1)(b-1)]$
Due Blocks	$SS_b = a \sum_{j=1}^{b} (\overline{Y}_{.j} - \overline{Y}_{..})^2$	$b - 1$	$MS_b = SS_b/(b-1)$	$\sigma_e^2 + a\sigma_b^2$	MS_b/s_e^2	$F[1-\alpha; b-1, (a-1)(b-1)]$
Residual	$SS_r = \sum_{i=1}^{a} \sum_{j=1}^{b} (Y_{ij} - \overline{Y}_{i.} - \overline{Y}_{.j} + \overline{Y}_{..})^2$	$(a-1)(b-1)$	$s_e^2 = SS_r/(a-1)(b-1)$	σ_e^2		
Total	$SS_t = \sum_{i=1}^{a} \sum_{j=1}^{b} (Y_{ij} - \overline{Y}_{..})^2$	$ab - 1$				

TABLE 6.3. ANALYSIS OF VARIANCE TABLE FOR SQUARE ROOTS OF NUMBERS OF ZERO COUNTS IN FIVE EXPERIMENTS

Source of Variation (1)	SS (2)	df (3)	MS (4)	EMS (5)	F Computed (6)	F Tabled (7)
Due Treatments	9.980	2	4.990	$\sigma_e^2 + 5\,\sigma_a^2$	26.26	4.46
Due Blocks	6.434	4	1.609	$\sigma_e^2 + 3\,\sigma_b^2$	8.47	3.84
Residual	1.523	8	.190	σ_e^2		
Total	17.93	14				

For the example, the F test for treatment effects ($H_0:\alpha_i=0$, $i=1,2,3$) is

$$F=\frac{\mathrm{MS}_a}{s_e^2}=\frac{4.990}{0.190}=26.26.$$

When compared with $F[.95;2,8]=4.46$, this result causes us to reject the null hypothesis. The mean numbers of zero counts differ from treatment to treatment.

ESTIMATES OF PARAMETERS AND CONFIDENCE INTERVALS

The estimate of the variance σ_e^2 is seen from the analysis of variance table to be

$$s_e^2=\frac{\sum_{i=1}^{a}\sum_{j=1}^{b}\left(Y_{ij}-\bar{Y}_{i.}-\bar{Y}_{.j}+\bar{Y}_{..}\right)^2}{(a-1)(b-1)}. \tag{6.5}$$

The confidence interval for σ_e^2 is obtained (see Table 6.4) exactly as before, using a χ^2 distribution with the degrees of freedom corresponding to s_e^2, $(a-1)(b-1)$.

As usual, the other parameters are estimated by unbiased linear combinations of the observations. Table 6.4 lists the point estimates for the various quantities that might be estimated, and their variances obtained in the usual way for use in forming confidence intervals. It should be noted that the quantities of primary interest are the treatment means $(\mu+\alpha_i)$ and various comparisons among the treatments ($\sum_{i=1}^{a}h_i\alpha_i$, where $\sum_{i=1}^{a}h_i=0$). Additional parameters may also be of some interest, however, and have been included in the table.

TABLE 6.4. ESTIMATES OF PARAMETERS AND CONFIDENCE INTERVALS FOR RANDOMIZED COMPLETE BLOCK DESIGN, MODEL I*

Parameter (1)	Point Estimate (2)	Variance of Point Estimate (3)
μ	$\bar{Y}_{..}$	σ_e^2/ab
α_i	$\bar{Y}_{i.} - \bar{Y}_{..}$	$\sigma_e^2(a - 1)/ab$
$\mu + \alpha_i$	$\bar{Y}_{i.}$	σ_e^2/b
$\mu + \alpha_i + \beta_j$	$\bar{Y}_{i.} + \bar{Y}_{.j} - \bar{Y}_{..}$	$\sigma_e^2(a + b - 1)/ab$
$\alpha_i - \alpha_i'$	$\bar{Y}_{i.} - \bar{Y}_{i'.}$	$2\ \sigma_e^2/b$
$\sum_{i=1}^{a} h_i\alpha_i$	$\sum_{i=1}^{a} h_i\bar{Y}_{i.}$	$\sum_{i=1}^{a} h_i^2\ \sigma_e^2/b$
σ_e^2	s_e^2	. . . .

*Confidence intervals

The interval for σ_e^2 is

$$\frac{(a-1)(b-1)s_e^2}{\chi^2[1-\alpha/2;(a-1)(b-1)]} < \sigma_e^2 < \frac{(a-1)(b-1)s_e^2}{\chi^2[\alpha/2;(a-1)(b-1)]} .$$

Intervals for other parameters are formed from the corresponding items in columns (2) and (3), as for example

$$\bar{Y}_{..} - t[1-\alpha/2m;(a-1)(b-1)]\sqrt{s_e^2/ab} < \mu <$$

$$\bar{Y}_{..} + t[1-\alpha/2m;(a-1)(b-1)]\sqrt{s_e^2/ab} ,$$

where m is the number of intervals made with an overall level of $1-\alpha$.

For example, a single 95% confidence interval for $\alpha_1 - \alpha_2$ is

$$(4.432 - 5.696) \pm t[.975; 8]\sqrt{2(0.190)/5} \quad \text{or} \quad -1.900 < \alpha_1 - \alpha_2 < -0.628.$$

From this interval we conclude that the population mean for the control treatment is lower than for treatment 1.

In Table 6.4 we find that the estimate of the population mean for the ith treatment and jth block is $\bar{Y}_{i.} + \bar{Y}_{.j} - \bar{Y}_{..}$. At first glance it may be surprising that Y_{ij} is not used as the estimate. Under the present model, the expected value of $\bar{Y}_{i.} + \bar{Y}_{.j} - \bar{Y}_{..}$ is $\mu + \alpha_i + \beta_j$; therefore, it is an unbiased estimate of the population mean for the ith treatment and jth block. The Y_{ij} is also an unbiased estimate of the mean, but its variance (σ_e^2) is larger than the variance of $\bar{Y}_{i.} + \bar{Y}_{.j} - \bar{Y}_{..}$ ($\sigma_e^2(a+b-1)/ab$); thus $\bar{Y}_{i.} + \bar{Y}_{.j} - \bar{Y}_{..}$ is the preferred statistic.

EXTENSIONS OF THE RANDOMIZED COMPLETE BLOCK DESIGN

The block design can be extended to problems involving more than one factor. It can also be extended to problems involving more treatments than can easily be used in a single block. The balanced incomplete block design is the simplest design of this type (see Davies [6.5] and Bennett [6.2]).

HIERARCHICAL OR NESTED DESIGN

It is often necessary to measure the response to a treatment on each individual of a subsample of a unit rather than on the entire unit to which the treatment is applied. In a fertilizer experiment, for example, the treatment may have been applied to plots of ground; from each plot several plants can be picked at random and the response (perhaps weight) of each plant can be measured. Such a procedure is called a hierarchical or nested design.

It is convenient to introduce the term *experimental unit*, which designates the unit to which the treatment is applied. In the fertilizer experiment, the experimental unit is the plot of ground. When drugs are given to patients, the experimental unit is the individual patient. When a single factor such as fertilizer level is studied, if the response is measured on the entire experimental unit, a simple one-way classification is appropriate. For example, if the yield of the entire plot is measured, the model of Chapter 5 is appropriate. If, however, we draw a random sample (called a subsample) from each of the experimental units and measure the response from the observations of the subsamples, a heirarchical design is used. In a study on

TABLE 6.5. DATA AND PRELIMINARY CALCULATIONS ON NITROGEN CONCENTRATION WITH THREE TYPES OF SPRAY

	Tree	1		2		3		4		
	Leaf									
1	1	4.50	$\sum_{K=1}^{6} Y_{11K} = 35.74$	5.78	$\sum_{k=1}^{6} Y_{12K} = 40.19$	13.32	$\sum_{k=1}^{6} Y_{13K} = 76.99$	11.59	$\sum_{k=1}^{6} Y_{14K} = 64.64$	$\overline{Y}_{1..} = 9.0650$
	2	7.04		7.69		15.05		8.96		
	3	4.98	$\overline{Y}_{11.} = 5.96$	12.68	$\overline{Y}_{12.} = 6.70$	12.67	$\overline{Y}_{13.} = 12.83$	10.95	$\overline{Y}_{14.} = 10.77$	
	4	5.48		5.89		12.42		9.87		
	5	6.54	$\sum_{K=1}^{6} Y^2_{11K} = 219.254$	4.07	$\sum_{k=1}^{6} Y^2_{12K} = 321.230$	10.03	$\sum_{k=1}^{6} Y^2_{13K} = 1001.561$	10.48	$\sum_{k=1}^{6} Y^2_{14K} = 705.344$	
	6	7.20		4.08		13.50		12.79		
2	1	15.32	$\sum_{K=1}^{6} Y_{21K} = 91.25$	14.53	$\sum_{k=1}^{6} Y_{22K} = 86.21$	10.89	$\sum_{k=1}^{6} Y_{23K} = 68.03$	15.12	$\sum_{k=1}^{6} Y_{24K} = 84.05$	$\overline{Y}_{2..} = 13.7308$
	2	14.97		14.51		10.27		13.79		
	3	14.81	$\overline{Y}_{21.} = 15.21$	12.61	$\overline{Y}_{22.} = 14.37$	12.21	$\overline{Y}_{23.} = 11.34$	15.32	$\overline{Y}_{24.} = 14.01$	
	4	14.26		16.13		12.77		11.95		
	5	15.88	$\sum_{k-1}^{6} Y^2_{21K} = 1389.982$	13.65	$\sum_{k=1}^{6} Y^2_{22K} = 1245.621$	10.45	$\sum_{k=1}^{6} Y^2_{23K} = 776.298$	12.56	$\sum_{k=1}^{6} Y^2_{24K} = 1188.433$	
	6	16.01		14.78		11.44		15.31		
3	1	7.18	$\sum_{K=1}^{6} Y_{31K} = 41.34$	6.70	$\sum_{k=1}^{6} Y_{32K} = 40.36$	5.94	$\sum_{k=1}^{6} Y_{33K} = 39.77$	4.08	$\sum_{k=1}^{6} Y_{34K} = 36.34$	$\overline{Y}_{3..} = 6.575$
	2	7.98		8.28		5.78		5.46		
	3	5.51	$\overline{Y}_{31.} = 6.89$	6.99	$\overline{Y}_{32.} = 6.73$	7.59	$\overline{Y}_{33.} = 6.63$	5.40	$\overline{Y}_{34.} = 6.06$	
	4	7.48		6.40		7.21		6.85		
	5	7.55	$\sum_{k=1}^{6} Y^2_{31K} = 290.355$	4.96	$\sum_{k=1}^{6} Y^2_{32K} = 277.291$	6.12	$\sum_{k=1}^{6} Y^2_{33K} = 266.576$	7.74	$\sum_{k=1}^{6} Y^2_{34K} = 228.824$	
	6	5.64		7.03		7.13		6.81		

$\overline{Y}_{...} = 9.79$

dyes, the experimental unit might be the batch of wool; from each large batch we might choose several small batches at random and measure some response such as resistance to fading. When response to several drugs is measured on animals, the experimental unit is the animal. However, if the investigator wished to have several chemical determinations for each animal, these would constitute the subsample.

As an example of subsampling, we consider a study of three different sprays used on trees. Each of the three sprays was applied to four trees. After one week the concentration of nitrogen was measured in each of six leaves picked in a random way from each tree. Here the experimental units are the 12 trees; we consider them to have been chosen at random from a large (∞) population of trees. The leaves thus form four subsamples, each consisting of six leaves from a large (∞) population of leaves on the particular tree. Table 6.5 presents the data and the preliminary calculations. Note that in this experiment 12 trees were sprayed and 72 leaves were picked. To clarify the situation, we have the following diagram:

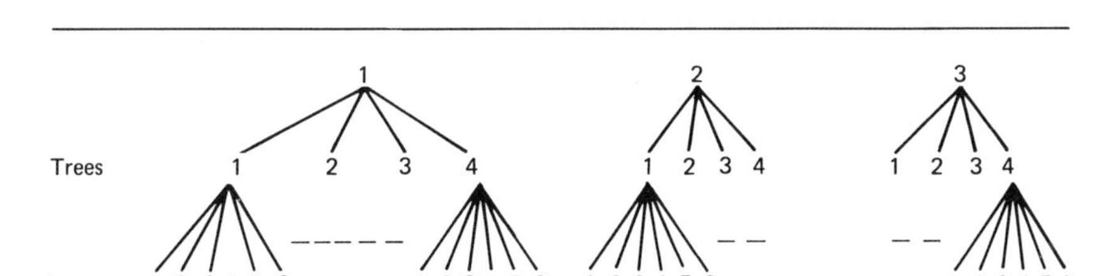

THE MODEL

Each measurement of nitrogen concentration is denoted by Y_{ijk}, where i, the number of the treatment, runs from 1 to a; j, the number of the experimental unit for the ith treatment, runs from 1 to b; and k, the observation number from the jth experimental unit on the ith treatment, runs from 1 to n. In the example, Y_{ijk} is the nitrogen concentration of the kth leaf from the jth tree that received spray i.

In our model, each response can be written

$$Y_{ijk} = \mu + \alpha_i + \beta_{(i)j} + \varepsilon_{(ij)k}, \qquad i=1,\ldots,a; \qquad j=1,\ldots,b; \qquad k=1,\ldots,n, \tag{6.6}$$

where

$$\sum_{i=1}^{a} \alpha_i = 0, \qquad \beta_{(i)j} \text{ IND}(0,\sigma_b^2), \qquad \varepsilon_{(ij)k}\text{IND}(0,\sigma_e^2),$$

and the $\beta_{(i)j}$'s and $\varepsilon_{(ij)k}$'s are mutually independent of one another. In other words, the ab β's and abn ε's are all independent of one another. We define σ_a^2 to be $\sum_{i=1}^{a} \alpha_i^2/(a-1)$.

We have thus divided a nitrogen concentration measurement into four parts. The first is the overall mean μ, which is the mean nitrogen concentration for all three sprays. The second is α_i, the part "due to the particular spray"; if α_i is positive, the ith spray produces a higher nitrogen concentration than the average. The third part, $\beta_{(i)j}$, is due to the particular tree that we happened to use as the jth tree for the ith spray. The fourth part, $\varepsilon_{(ij)k}$, represents the contribution of the particular leaf that we happened to pick.

The following assumptions are implied by the model

1. The ab $\beta_{(i)j}$'s are independently normally distributed; the abn $\varepsilon_{(ij)k}$'s are also independently normally distributed.

2. All the $\beta_{(i)j}$'s come from distributions having the same variance; that is, the distribution of the $\beta_{(1)j}$'s has the same variance as the distribution of the $\beta_{(2)j}$'s. Similarly, all the ε's come from distributions having the same variance.

3. The β's and the ε's are independent of one another.

The statement that the mean of the distribution of the β's is zero is not an assumption inasmuch as we can make the mean zero by a proper choice of μ and the α_i's. Nor is it an assumption to say that the mean of the distribution of the ε's is zero. Assumptions 1 and 2, plus the fact that the mean of the β's is zero, are equivalent to the statement that the β's form a random sample of size ab from a single normally distributed population. Similarly, the ε's form a random sample of size abn from a single normally distributed population. The two random samples are independent of each other, from assumption 3.

With respect to our particular example, the assumptions imply the following statements. We assume that the variation in nitrogen concentration from one tree to another is the same no matter which spray we use. This clearly is an assumption, for it is quite possible that the effects of one spray might vary considerably from tree to tree, whereas there might be little variation among trees when another spray is used. We also assume that the variation in nitrogen from leaf to leaf is the same, no matter which spray and which tree we consider. These are the implications of assumption 2.

Assumption 3 tells us that any knowledge we have concerning the β's gives us no information concerning the ε's. Knowing that the first tree for which the second spray is used has a particularly high nitrogen concentration tells us nothing about whether the third leaf we examine will have a concentration that is high or low for that particular tree. All the assumptions implied by the model seem to be quite reasonable for the spray example.

ANALYSIS OF VARIANCE TABLE

We now divide the basic sum of squares into meaningful parts and make an analysis of variance table. The difference $Y_{ijk}-\bar{Y}_{...}$ can be expressed as the sum of three parts:

$$Y_{ijk}-\bar{Y}_{...}=\left(\bar{Y}_{i..}-\bar{Y}_{...}\right)+\left(\bar{Y}_{ij.}-\bar{Y}_{i..}\right)+\left(Y_{ijk}-\bar{Y}_{ij.}\right). \qquad (6.7)$$

Note that (6.7) is an algebraic identity. In addition, the three quantities in parentheses on the right-hand side of (6.7) provide estimates of the population parameters α_i, $\beta_{(i)j}$, and $\varepsilon_{(ij)k}$, respectively; $\bar{Y}_{...}$ furnishes an estimate of μ. Thus we have in (6.7) a sample estimate of the terms in

$$Y_{ijk}-\mu=\alpha_i+\beta_{(i)j}+\varepsilon_{(ij)k}, \qquad (6.8)$$

which is obtained by subtracting μ from both sides of (6.6).

As before, we now square both sides of (6.7). When we sum over all the observations, the cross-product terms sum to zero, leaving

$$\sum_{i=1}^{a}\sum_{j=1}^{b}\sum_{k=1}^{n}\left(Y_{ijk}-\bar{Y}_{...}\right)^2=bn\sum_{i=1}^{a}\left(\bar{Y}_{i..}-\bar{Y}_{...}\right)^2+n\sum_{i=1}^{a}\sum_{j=1}^{b}\left(\bar{Y}_{ij.}-\bar{Y}_{i..}\right)^2$$

$$+\sum_{i=1}^{a}\sum_{j=1}^{b}\sum_{k=1}^{n}\left(Y_{ijk}-\bar{Y}_{ij.}\right)^2. \qquad (6.9)$$

This, then, is the basic sum of squares. The first sum of squares on the right-hand side of the equation represents the sum "due treatment" or, in the example, due spray. The second sum of squares is said to be within treatment but among samples or, in our example, "within spray among trees." The third sum of squares is called the residual sum of squares; it is within treatment and within sample but among subsample, or "within spray and within tree among leaves."

From Table 6.5, the three sums of squares are calculated to be

$$\text{SS}_a = 24\sum_{i=1}^{3}(\bar{Y}_{i..}-\bar{Y}_{...})^2 = 24[(9.0650-9.7904)^2+\cdots+(6.5754-9.7904)^2]$$

$$=633.344$$

$$\text{SS}_b = 6\sum_{i=1}^{3}\sum_{j=1}^{4}(\bar{Y}_{ij.}-\bar{Y}_{i..})^2 = 6[(5.9567-9.0650)^2+\cdots+(6.0567-6.5754)^2]$$

$$=292.903$$

$$\text{SS}_r = \sum_{i=1}^{3}\sum_{j=1}^{4}\sum_{k=1}^{6}(Y_{ijk}-\bar{Y}_{ij.})^2 = (4.50-5.9567)^2+\cdots+(6.81-6.0583)^2$$

$$=129.143.$$

In Table 6.6, we have the analysis of variance table for the nested design using formulae; Table 6.7 gives the table as it would be used in practice for the spray experiment.

The sum of squares column in Table 6.6 is taken from (6.9) and does not need discussion. The degrees of freedom for the total sum of squares are 1 less than the total number of observations—in this case, $abn-1$. The residual or "among subsample" sum of squares is simply the sum that would be used in a pooled estimate of the variance in which ab samples of size n are pooled; there are $n-1$ degrees of freedom from each of the ab samples, and therefore the degrees of freedom for SS_r are $ab(n-1)$. The among sample sum of squares SS_b uses a sums of squares, each with $b-1$ degrees of freedom; thus the degrees of freedom for this sum of squares are $a(b-1)$. With a treatments, the due to treatment sum of squares SS_a has $a-1$ degrees of freedom. Again, the various degrees of freedom, when added, equal the total degrees of freedom.

In the EMS column we find that the "residual" mean square s_e^2 is used as an estimate of σ_e^2; this mean square measures variability in nitrogen from leaf to leaf and nothing else. The among samples mean square MS_b reflects variation of two types: first, differences between one leaf and another, and second, differences among trees. These two sources of variation appear in the EMS column, where σ_e^2 has coefficient one (every Y_{ijk} is based on one observation) and σ_b^2 is multiplied by n (every $\bar{Y}_{ij.}$ is based on n observations). Similarly, the size of the due to treatment mean square

TABLE 6.6. ANALYSIS OF VARIANCE TABLE FOR ONE-WAY CLASSIFICATION, NESTED DESIGN, EQUAL NUMBERS OF OBSERVATIONS, MODEL I

Source of Variation (1)	Sum of Squares (2)	d.f. (3)	Mean Square (4)	EMS (5)	F Computed (6)	F Tabled (7)
Due to Treatment	$SS_a = bn \sum_{i=1}^{a} (\overline{Y}_{i..} - \overline{Y}_{...})^2$	a-1	$MS_a = SS_a/(a-1)$	$\sigma_e^2 + n\sigma_b^2 + bn\sigma_a^2$	MS_a/MS_b	$F[1-\alpha;a-1,a(b-1)]$
Within Treatment Among Samples	$SS_b = n \sum_{i=1}^{a} \sum_{j=1}^{b} (\overline{Y}_{ij.} - \overline{Y}_{i..})^2$	a(b-1)	$MS_b = SS_b/a(b-1)$	$\sigma_e^2 + n\sigma_b^2$	MS_b/s_e^2	$F[1-\alpha;a(b-1),ab(n-1)]$
Residual	$SS_r = \sum_{i=1}^{a} \sum_{j=1}^{b} \sum_{k=1}^{n} (Y_{ijk}-\overline{Y}_{ij.})^2$	ab(n-1)	$s_e^2 = SS_r/ab(n-1)$	σ_e^2		
Total	$SS_t = \sum_{i=1}^{a} \sum_{j=1}^{b} \sum_{k=1}^{n} (Y_{ijk}-\overline{Y}_{...})^2$	abn-1				

TABLE 6.7. ANALYSIS OF VARIANCE TABLE FOR NITROGEN CONTENT OF LEAVES

Source of Variation (1)	Sum of Squares (2)	d.f. (3)	Mean Square (4)	EMS (5)	F Computed (6)	F Tabled (7)
Due to Sprays	633.402	2	316.701	$\sigma_e^2 + 6\sigma_b^2 + 24\sigma_a^2$	11.54	4.26
Within Sprays Among Trees	246.903	9	27.434	$\sigma_e^2 + 6\sigma_b^2$	12.75	2.04
Residual	129.143	60	2.152	σ_e^2		
Total	1009.406	71				

depends on three considerations—variation among leaves, variation among trees, and the differences among nitrogen levels—and we find three terms in the expected value of the mean square.

F TESTS

First we test the null hypothesis that there are no differences among the treatment means $H_0: \alpha_1 = \cdots = \alpha_a = 0$. From the EMS column we see that if the null hypothesis is true, MS_a and MS_b are both estimates of $\sigma_e^2 + n\sigma_b^2$. Therefore, the test statistic is $F = MS_a / MS_b$, which we calculate and compare with $F[1-\alpha; a-1, a(b-1)]$ from Table A.4. For the leaf example, we have from Table 6.7

$$\frac{MS_a}{MS_b} = 11.54 \qquad \text{and} \qquad F[.95; 2, 9] = 4.26.$$

At an $\alpha = .05$ level of significance, the hypothesis of equal nitrogen content using the three sprays is rejected; we conclude that nitrogen concentration differs among the sprays.

It is also possible to make a second F test from Table 6.7. We might wonder whether differences really exist from tree to tree; perhaps all the sampling variation seen in the data is actually from leaf to leaf. In this case, the null hypothesis to be tested is $H_0: \sigma_b^2 = 0$, and again from the EMS column we can find an appropriate F test. If H_0 is true and $\sigma_b^2 = 0$, we have two independent estimates of σ_e^2 (i.e., MS_b and s_e^2). The appropriate statistic is therefore MS_b / s_e^2, which will be compared with $F[1-\alpha; a(b-1), ab(n-1)]$. In our example, $MS_b / s_e^2 = 12.75$ and $F[.95; 9, 60] = 2.04$. We reject the null hypothesis and conclude that there is variation among trees in nitrogen concentration.

ESTIMATION OF PARAMETERS

As unbiased estimates of the means, we use the overall sample mean $\overline{Y}_{...}$ to estimate μ and the treatment mean $\overline{Y}_{i..}$ to estimate $\mu + \alpha_i$. The various treatment means or linear combinations of treatment means we may wish to estimate appear in Table 6.8, with the statistics used to estimate them. The variance of each statistic is given in column 3 for use in forming the confidence intervals. The degrees of freedom to be used in entering the t table for these confidence intervals is that of MS_b, the estimate of $\sigma_e^2 + n\sigma_b^2$; from the analysis of variance table, we find it to be $a(b-1)$.

We now need estimates of the two variances σ_e^2 and σ_b^2; these can be found from Table 6.6. For σ_e^2 we have the point estimate s_e^2; for $\sigma_e^2 + n\sigma_b^2$ we

TABLE 6.8. ESTIMATES OF PARAMETERS AND CONFIDENCE INTERVALS, NESTED ONE-WAY DESIGN MODEL I, EQUAL NUMBERS*

Parameter (1)	Point Estimate (2)	Variance of Estimate (3)
μ	$\overline{Y}_{...}$	$(\sigma_e^2 + n\sigma_b^2)/abn$
$\mu + \alpha_i$	$\overline{Y}_{i..}$	$(\sigma_e^2 + n\sigma_b^2)/bn$
α_i	$\overline{Y}_{i..} - \overline{Y}_{...}$	$(\sigma_e^2 + n\sigma_b^2)(a-1)/abn$
$\alpha_i - \alpha_{i'}$	$\overline{Y}_{i..} - \overline{Y}_{i'..}$	$2(\sigma_e^2 + n\sigma_b^2)/bn$
$\sum_{i=1}^{a} h_i(\mu + \alpha_i)$	$\sum_{i=1}^{a} h_i\overline{Y}_{i..}$	$(\sigma_e^2 + n\sigma_b^2)\left(\sum_{i=1}^{a} h_i^2\right)/bn$
σ_e^2	s_e^2	--
$\sigma_e^2 + n\sigma_b^2$	MS_b	--
σ_b^2	$(MS_b - s_e^2)/n$	--

*Confidence intervals:

The 1-α level interval for σ_e^2 is

$$\frac{ab(n-1)s_e^2}{\chi^2[1-\alpha/2;ab(n-1)]} < \sigma_e^2 < \frac{ab(n-1)s_e^2}{\chi^2[\ \alpha/2;ab(n-1)]} \quad .$$

Intervals for the means and linear combination of means are formed from columns (2) and (3), as for example

$$(\overline{Y}_{i..} - \overline{Y}_{...}) - t[1-\alpha/2m;a(b-1)]\sqrt{MS_b(a-1)/abn} < \alpha_i < (\overline{Y}_{i..} - \overline{Y}_{...}) + t[1-\alpha/2m;a(b-1)]\sqrt{MS_b(a-1)/abn} \quad .$$

have the estimate MS_b. Subtracting s_e^2 from MS_b and dividing by n, we estimate σ_b^2 by

$$s_b^2 = \frac{\mathrm{MS}_b - s_e^2}{n}, \tag{6.10}$$

if this quantity is positive, and by zero otherwise.

From the general rule for obtaining χ^2 variables from the mean square, we have as a $1-\alpha$ level confidence interval for σ_e^2:

$$\frac{ab(n-1)s_e^2}{\chi^2[1-\alpha/2; ab(n-1)]} < \sigma_e^2 < \frac{ab(n-1)s_e^2}{\chi^2[\alpha/2; ab(n-1)]}. \tag{6.11}$$

From a combination of two confidence intervals, one for σ_e^2 and one for $\sigma_e^2 + n\sigma_b^2$, we obtain a $1-\alpha$ level confidence interval for σ_b^2:

$$0 < \sigma_b^2 < \frac{a(b-1)\mathrm{MS}_b/\chi^2[\alpha/2; a(b-1)] - ab(n-1)s_e^2/\chi^2[1-\alpha/2; ab(n-1)]}{n}. \tag{6.12}$$

RELATIVE EFFICIENCY

The efficiency of an experimental design is often judged in terms of the variance of a treatment mean. The smaller the variance of the treatment mean, the more efficient is the design.

In the example on sprays, the variance of a treatment mean is

$$\operatorname{Var} \overline{Y}_{i..} = \frac{\sigma_e^2 + n\sigma_b^2}{bn}.$$

After the experiment has been performed, σ_e^2 is estimated by $s_e^2 = 2.152$ and σ_b^2 by $s_b^2 = (\mathrm{MS}_b - s_e^2)/n = (27.434 - 2.152)/6 = 4.214$. We may now wonder if it would have been better to have used 5 trees and 4 leaves from each tree instead of 4 trees and 6 leaves. (Possibly these two designs would be approximately equal in difficulty.)

With $b=5$, $n=4$, we estimate the variance of a treatment mean by

$$\operatorname{Var} \overline{Y}_{i..} \cong \frac{2.152 + 4(4.214)}{4\times 5} = \frac{19.008}{20} = 0.950.$$

With $b=4$, $n=6$, our estimate is

$$\operatorname{Var} \overline{Y}_{i..} \cong \frac{2.152 + 6(4.214)}{6\times 4} = \frac{27.434}{24} = 1.143.$$

Other things being equal, we would prefer 4 leaves from each of 5 trees to 6 leaves from each of 4 trees. We consider 4 leaves from 5 trees to be a more efficient design because the variance of the treatment mean has been estimated to be smaller with 4 leaves from 5 trees.

Data of the type just analyzed could have been handled by using the methods of Chapter 5. The measurements on the 6 leaves from each tree could have been averaged to obtain a single measure of nitrogen concentration for that tree. We then would have a simple one-way analysis of variance problem. For each $Y_{ijk} = \mu + \alpha_i + \beta_{(i)j} + \varepsilon_{(ij)k}$ as in (6.6), the mean for the jth tree on the ith spray is then $\bar{Y}_{ij.} = \mu + \alpha_i + \beta_{(i)j} + \bar{\varepsilon}_{(ij).}$. If we now denote $\bar{Y}_{ij.}$ by Z_{ij} and $\beta_{(i)j} + \bar{\varepsilon}_{(ij).}$ by η_{ij}, we have

$$Z_{ij} = \mu + \alpha_i + \eta_{ij}. \tag{6.13}$$

From $\operatorname{Var}\beta_{(i)j} = \sigma_b^2$ and $\operatorname{Var}\varepsilon_{(ij)k} = \sigma_e^2$, we obtain $\operatorname{Var}\eta_{ij} = \sigma_\eta^2 = (\sigma_e^2 + n\sigma_b^2)/n$, and we have exactly the model of Chapter 5.

The confidence intervals for $\mu + \alpha_i$, $\alpha_i - \alpha_{i'}$, and so on, will be exactly the same when the determinations are averaged this way. The test of $H_0: \alpha_1 = \cdots = \alpha_a = 0$ is precisely the same. The advantage in analyzing the data as a nested design as in the present chapter is that we obtain both an estimate of the variance among leaves and an estimate of the variance among trees.

POOLING

Sometimes (usually when the available degrees of freedom for the denominator of an F statistic are small), an investigator is tempted to obtain a more sensitive F test by a process called *pooling*. As an example, we consider a nested one-way analysis of variance design with three treatments ($a = 3$), samples of size 2 on each treatment ($b = 2$), and subsamples of size 5 ($n = 5$). The analysis of variance table is Table 6.9.

With these data, the test of $H_0: \sigma_b^2 = 0$ at $\alpha = .05$ ($F = 11/10$ compared with $F[.95; 3, 24]$) fails to reject the null hypothesis, and the usual test of $H_0: \alpha_i = 0$ at $\alpha = .05$ ($F = 40/11 = 3.6$ compared with $F[.95; 2, 3] = 9.55$) fails to establish differences among the three treatments. If we really believe that σ_b^2 is zero or that it is at any rate of negligible size, the σ_b^2 term can be canceled in the EMS column from both the "due to treatment" row and the "within treatment among sample" row. Since the "within treatment" mean square in column 4 then becomes an estimate of σ_e^2, it can be pooled with the "residual" mean square to obtain a combined estimate of σ_e^2. The new estimate is

$$s_e^2 = \frac{3(11) + 24(10)}{3 + 24} = \frac{33 + 240}{3 + 24} = 10.11.$$

TABLE 6.9. ILLUSTRATION OF POOLING: ANALYSIS OF VARIANCE TABLE FOR ONE-WAY NESTED PROBLEM

Source of Variation (1)		Sums of Squares (2)		d.f. (3)		Mean Square (4)		EMS (5)
Due to Treatment		80		2		40		$\sigma_e^2 + (n\sigma_b^2) + bn\sigma_a^2$
Within Treatment Among Samples	Res	33	273	3	27	11	10.11	$\sigma_e^2 + (n\sigma_b^2)$
Residual		240		24		10		σ_e^2
Total		353		29				

This estimate of σ_e^2 can be calculated by adding the appropriate entries in the sums of squares column and dividing by the sum of the corresponding degrees of freedom. We have thus "pooled" the within treatment and residual sums of squares. A .05 level F test of $H_0: \alpha_i = 0$ can now be made by comparing $F = 40/10.11 = 3.96$ with $F[.95; 2, 27] = 3.37$, and the null hypothesis is rejected.

Often the statistician has no reason to believe, before examining the data, that σ_b^2 is zero or close to zero. If his test of $H_0: \sigma_b^2 = 0$ fails to reject the null hypothesis, should he then make a pooled test of $H_0: \sigma_a^2 = 0$? There is no clear answer to this question, unfortunately.

The objective in pooling is to make a more powerful test and at the same time to keep the level of significance at the stated value, here $\alpha = .05$. If there is no variation among samples ($\sigma_b^2 = 0$), the objective is attained; $\alpha = .05$, and if among treatment differences exist, they are more likely to be detected with 27 degrees of freedom in the denominator than with just 3. On the other hand, if σ_b^2 is actually sizable, the calculated F statistic does not have an F distribution with 2 and 27 degrees of freedom, and the actual significance level of the test may be considerably larger than .05.

There is considerable variation in opinion among statisticians in regard to policy on pooling. Some statisticians feel that there is no justification for pooling in the absence of knowledge, either theoretical or from experience, that the variance σ_b^2 is zero or very close to zero. These statisticians decide in advance whether to pool certain sums of squares. Statisticians who routinely pool vary in the significance level of their preliminary test. One suggested procedure [6.16] is to pool if the F statistic of the preliminary test is less than $2F[.50; \nu_1, \nu_2]$.

Pooling after examination of the data is usually done when the degrees of freedom for the denominator of the F statistic are very small, so that a very large calculated value of F is needed to reject the null hypothesis. Sometimes the experiment can be redesigned in such a way that the denominator degrees of freedom are larger for the tests considered most important. If $a=3$, $b=5$, and $n=2$, are used in the example, we have the same number of observations as with $a=3$, $b=2$, and $n=5$, but there are 12 degrees of freedom for the within treatment among sample mean square. When practical considerations necessitate small denominator degrees of freedom, pooling could be termed a desperation move, and the stated F levels must be viewed with skepticism.

**UNEQUAL SAMPLE SIZES

In experiments involving subsampling, it is important to try to keep the sizes of the subsamples equal. When the sizes of the subsamples are unequal, the expressions for the variances of the treatment means and for the expected values of the mean squares in the analysis of variance table become quite complicated. Table 6.10 is the analysis of variance table for the one-way nested design with unequal sample and subsample sizes, Model I. The notation in Table 6.10 is

$a=$ number of treatments

$n_{ij}=$ size of the subsample from the jth member of the sample on the ith treatment

$b_i=$ size of the sample from the ith treatment, $i=1,\ldots,a$

$n_i=\sum_{j=1}^{b_i} n_{ij}=$ the number of observations on the ith treatment

$N=\sum_{i=1}^{a} n_i=$ the total number of observations in the entire set of data.

From Table 6.10 it is clear that there is no simple F test of $H_0:\alpha_i=0$, $i=1,\ldots,a$. Under the null hypothesis we do not have two mean squares in the analysis of variance table which estimate the same quantity; all three mean squares estimate different things. An approximate test can be made using Satterthwaite's procedure of Chapter 9.

We are also in a complicated situation when we wish to construct confidence intervals. The variance of a treatment mean

$$\bar{Y}_{i..}=\frac{\sum_{j=1}^{b_i}\sum_{k=1}^{n_{ij}} Y_{ijk}}{n_i},$$

TABLE 6.10. ANALYSIS OF VARIANCE TABLE, NESTED DESIGN, ONE-WAY CLASSIFICATION, MODEL I UNEQUAL SAMPLE SIZES

Source of Variation (1)	Sums of Squares (2)	Degrees of Freedom (3)	Mean Square (4)	EMS (5)
Due Treatment	$SS_a = \sum_{i=1}^{a} \sum_{j=1}^{b_i} n_{ij}(\overline{Y}_{i..} - \overline{Y}_{...})^2$	$a-1$	$MS_a = SS_a/(a-1)$	$\sigma_e^2 + \left\{\frac{\sum_{i=1}^{a}\left(\sum_{j=1}^{b_i} n_{ij}^2/n_i\right) - \sum_{i=1}^{a}\sum_{j=1}^{b_i} n_{ij}^2/N}{a-1}\right\}\sigma_b^2 + \left\{\frac{\sum_{i=1}^{a} n_i^2\ \alpha_i^2}{a-1}\right\}$
Within Treatments Among Samples	$SS_b = \sum_{i=1}^{a} \sum_{j=1}^{b_i} n_{ij}(\overline{Y}_{ij.} - \overline{Y}_{i..})^2$	$\sum_{i=1}^{a} b_i - a$	$MS_b = SS_b/\left(\sum_{i=1}^{a} b_i - a\right)$	$\sigma_e^2 + \left\{\frac{N - \sum_{i=1}^{a}\left(\sum_{j=1}^{b_i} n_{ij}^2/n_i\right)}{\sum_{i=1}^{a}(b_i - 1)}\right\}\sigma_b^2$
Residual	$SS_e = \sum_{i=1}^{a} \sum_{j=1}^{b_i} \sum_{k=1}^{n_{ij}} (Y_{ijk} - \overline{Y}_{ij.})^2$	$N - \sum_{i=1}^{a} b_i$	$MS_e = SS_e/\left(N - \sum_{i=1}^{a} b_i\right)$	σ_e^2
Total	$SS_t = \sum_{i=1}^{a} \sum_{j=1}^{b_i} \sum_{k=1}^{n_{ij}} (Y_{ijk} - \overline{Y}_{...})^2$	$N-1$		

is

$$\operatorname{Var} \overline{Y}_{i..} = \frac{n_i\sigma_e^2 + \left(\sum_{j=i}^{b_i} n_{ij}^2\right)\sigma_b^2}{n_i^2}. \tag{6.14}$$

The variance of the overall mean $\overline{Y}_{...}$ is

$$\operatorname{Var} \overline{Y}_{...} = \frac{N\sigma_e^2 + \left(\sum_{i=1}^{a}\sum_{j=1}^{b_i} n_{ij}^2\right)\sigma_b^2}{N^2}. \tag{6.15}$$

The variance of $\overline{Y}_{i..} - \overline{Y}_{i'..}$ is simply twice the variance of $\overline{Y}_{i..}$. The variance of $\overline{Y}_{i..} - \overline{Y}_{...}$ is

$$\frac{N-n_i}{n_iN}\sigma_e^2 + \left(\frac{N-2n_i}{n_i^2N}\sum_{j=1}^{b_i} n_{ij}^2 + \frac{1}{N^2}\sum_{i=1}^{a}\sum_{j=1}^{b_i} n_{ij}^2\right)\sigma_b^2. \tag{6.16}$$

Comparing these variances with the expressions in Table 6.10 for the expected mean square, we see that we have no simple χ^2 estimates of the variances. Satterthwaite's approximate χ^2 procedure (see Chapter 9) can be used to obtain confidence intervals.

For unequal subsample sizes, the method of averaging over the subsamples and analyzing the sample means is also theoretically unsatisfactory. For if we do this as before and let $Z_{ij} = \overline{Y}_{ij.} = \mu + \alpha_i + \beta_{(i)j} + \bar{\varepsilon}_{(ij).} = \mu + \alpha_i + \eta_{ij}$, we have $\operatorname{Var}\bar{\varepsilon}_{(ij).} = \sigma_e^2/n_{ij}$; therefore, $\operatorname{Var}\eta_{ij} = \sigma_e^2/n_{ij} + \sigma_b^2$. We now have violated the assumption that our observations have equal variances.

In planning an experiment involving subsamples, we suggest always using subsamples of equal size. This is often as easy as taking an equal number of leaves from each tree. Equal numbers in each sample are also advisable, although not necessary, in the one-way analysis of variance. If during the course of the experiment certain observations are lost, and the numbers in the various subsamples are no longer quite equal (e.g., if the measurements on one or two leaves are lost in some way), then after some consideration of why the observations were lost, the data can be analyzed simply by using means of the subsamples. The assumption of equal variances has been violated, but the variances are not far from being equal. A disadvantage in analyzing the means is that we lose the estimates of σ_e^2 and σ_b^2. Since F tests are relatively robust against inequalities in variances, the inequalities introduced by slightly different n_{ij}'s are unimportant. An advantage of this procedure is that the use of cell means helps ensure normality.

If, however, we are reasonably sure of normality in the underlying populations and we wish to estimate σ_e^2 and σ_b^2, an alternate procedure can be followed. Missing observations are replaced by the corresponding cell means and the analysis is carried out as for equal n_{ij}. The degrees of freedom for the residual mean square is decreased by 1 for each missing observation.

The consideration of why the data were lost must always be made. We are safe in analyzing the remaining measurements only if the loss of the missing measurements can be considered to have occurred at random.

The nested design as given here can be extended to problems involving subsubsampling and also to more complicated models.

THE LATIN SQUARE

In the randomized complete block design, the effect of a *single* factor was removed. It is occasionally possible to remove the effects of two factors simultaneously in the same experiment by using the *Latin square* design. In order to use the Latin square design, however, it is necessary to assume that no interaction exists between the treatment effect and either block effect. In addition, the number of treatments must be equal to the number of categories for each of the two factors. We might, for instance, wish to test four detergents, using four methods of application, at four hospitals. A 4×4 Latin square design could then be employed, using each detergent exactly once with each method and exactly once in each hospital. The assignment of detergent could be made as in Table 6.11; the roman numeral in the ith row and jth column indicates the detergent that will be used by the ith application method in the jth hospital. As assigned in Table 6.11, the first detergent is used in hospital 1 by method 1, in hospital 2 by

TABLE 6.11. LATIN SQUARE DESIGN

		Hospital Number			
		1	2	3	4
Method Number	1	I	II	III	IV
	2	II	III	IV	I
	3	III	IV	I	II
	4	IV	I	II	III

TABLE 6.12. DATA AND PRELIMINARY CALCULATIONS ON FOUR DETERGENTS

	1	2	3	4	$\bar{Y}_{i..}$
1	8.7 (I)	9.2 (II)	11.6 (III)	9.1 (IV)	9.650
2	7.5 (II)	12.7 (III)	4.6 (IV)	7.3 (I)	8.025
3	14.0 (III)	9.2 (IV)	5.1 (I)	6.7 (II)	8.750
4	11.3 (IV)	8.7 (I)	4.0 (II)	12.9 (III)	9.225
$\bar{Y}_{.j.}$	10.375	9.950	6.325	9.000	$\bar{Y}_{...}$ = 8.9125

$$\bar{Y}_{..1} = \frac{(8.7 + 8.7 + 5.1 + 7.3)}{4} = 7.45$$

$$\bar{Y}_{..2} = \frac{(7.5 + 9.2 + 4.0 + 6.7)}{4} = 6.85$$

$$\bar{Y}_{..3} = \frac{(14.0 + 12.7 + 11.6 + 12.9)}{4} = 12.80$$

$$\bar{Y}_{..4} = \frac{(11.3 + 9.2 + 4.6 + 9.1)}{4} = 8.55$$

method 4, in hospital 3 by method 3, and hospital 4 by method 2. Each detergent is tested exactly four times. Only 16 observations are needed because of the balanced arrangement used *and* because of the assumption of no interaction.

Table 6.12 gives the measurements obtained from using the four detergents.

THE MODEL

It is assumed that each observation Y_{ijk} can be expressed as follows:

$$Y_{ijk} = \mu + \alpha_i + \beta_j + \gamma_k + \varepsilon_{ijk}, \qquad i = 1,\ldots,p; \qquad j = 1,\ldots,p; \qquad k = 1,\ldots,p, \tag{6.17}$$

where

$$\sum_{i=1}^{p} \alpha_i = \sum_{j=1}^{p} \beta_j = \sum_{k=1}^{p} \gamma_k = 0 \qquad \text{and} \qquad \varepsilon_{ijk}\ \text{IND}(0, \sigma_e^2).$$

Here μ denotes the overall mean response for all p treatments using all p^2 combinations of the two factors; there are thus p^3 population means altogether. In the detergent example, μ is the average of $p^3=64$ population means. Each α_i is the part of the mean that is due to the ith row (method); β_j is the part of the mean due to the jth column (hospital); γ_k is the part due to the kth treatment (detergent).

There are p^3 populations, but we have economized by making an observation on only p^2 populations. In the detergent example, instead of 64 observations, we have only 16.

The assumptions implied by the model are as follows:

1. The p^3 populations are normally distributed.
2. They have equal variances.
3. There is no interaction.
4. The ε's are independent of one another.

The assumption of no interaction is implicit in the statement of the model. We have stated that the mean of Y_{ijk} is $\mu+\alpha_i+\beta_j+\gamma_k$; this indicates that the first detergent has the same effect no matter which hospital is involved and no matter which method is used. Each hospital performs equally well with each method.

The assumption of independent ε's would be violated if, for example, half the experiment were conducted at one time and half were conducted six months later.

ANALYSIS OF VARIANCE TABLE AND F TESTS

We can estimate

$$Y_{ijk}-\mu=\alpha_i+\beta_j+\gamma_k+\varepsilon_{ijk}$$

by

$$Y_{ijk}-\bar{Y}_{...}=\left(\bar{Y}_{i..}-\bar{Y}_{...}\right)+\left(\bar{Y}_{.j.}-\bar{Y}_{...}\right)+\left(\bar{Y}_{..k}-\bar{Y}_{...}\right)$$

$$+\left(Y_{ijk}-\bar{Y}_{i..}-\bar{Y}_{.j.}-\bar{Y}_{..k}+2\bar{Y}_{...}\right),$$

an algebraic identity. Squaring each side and summing over the entire set of data, we have (the cross-product terms add to zero):

$$\sum_{i=1}^{p}\sum_{j=1}^{p}\sum_{k=1}^{p}\left(Y_{ijk}-\overline{Y}_{...}\right)^2=p\sum_{i=1}^{p}\left(\overline{Y}_{i..}-\overline{Y}_{...}\right)^2$$
$$+p\sum_{j=1}^{p}\left(\overline{Y}_{.j.}-\overline{Y}_{...}\right)^2$$
$$+p\sum_{k=1}^{p}\left(\overline{Y}_{..k}-\overline{Y}_{...}\right)^2$$
$$+\sum_{i=1}^{p}\sum_{j=1}^{p}\sum_{k=1}^{p}\left(Y_{ijk}-\overline{Y}_{i..}-\overline{Y}_{.j.}-\overline{Y}_{..k}+2\overline{Y}_{...}\right)^2. \tag{6.18}$$

Equation 6.18 looks exactly as if there were p^3 observations rather than only p^2. This is because for each pair (i, j) only one treatment was used, rather than all p treatments. In making the summations, we define any Y_{ijk} that was not included in the design to be zero. With this convention, the row mean $\overline{Y}_{i..}$ (which should equal the mean of the p observations in the ith row) is defined by $\overline{Y}_{i..}=\sum_{j=1}^{p}\sum_{k=1}^{p}Y_{ijk}/p$. Similar definitions hold for the jth column mean and for the kth treatment mean.

In the Latin square being used for the detergent, the treatment mean for the first detergent is given by Table 6.12 as

$$\overline{Y}_{..1}=\frac{\sum_{i=1}^{p}\sum_{j=1}^{p}Y_{ij1}}{4}=\frac{\sum_{i=1}^{p}\left(Y_{i11}+Y_{i21}+Y_{i31}+Y_{i41}\right)}{4}$$
$$=\frac{(Y_{111}+0+0+0)+(0+0+0+Y_{241})+(0+0+Y_{331}+0)+(0+Y_{421}+0+0)}{4}$$
$$=\frac{Y_{111}+Y_{241}+Y_{331}+Y_{421}}{4}$$
$$=\frac{8.7+7.3+5.1+8.7}{4}$$
$$=7.45.$$

TABLE 6.13. ANALYSIS OF VARIANCE TABLE FOR LATIN SQUARE DESIGN

Source of Variation (1)	Sums of Squares (2)	df (3)	Mean Square (4)	EMS (5)	F Computed (6)	F Tabled (7)
Due Rows	$SS_a = p \sum_{i=1}^{p} (\overline{Y}_{i..} - \overline{Y}_{...})^2$	$p - 1$	$MS_a = SS_a/(p-1)$	$\sigma_e^2 + p\sigma_a^2$	MS_a/s_e^2	$F[1-\alpha; p-1, (p-1)(p-2)]$
Due Columns	$SS_b = p \sum_{j=1}^{p} (\overline{Y}_{.j.} - \overline{Y}_{...})^2$	$p - 1$	$MS_b = SS_b/(p-1)$	$\sigma_e^2 + p\sigma_b^2$	MS_b/s_e^2	$F[1-\alpha; p-1, (p-1)(p-2)]$
Due Treatments	$SS_c = p \sum_{k=1}^{p} (\overline{Y}_{..k} - \overline{Y}_{...})^2$	$p - 1$	$MS_c = SS_c/(p-1)$	$\sigma_e^2 + p\sigma_c^2$	MS_c/s_e^2	$F[1-\alpha; p-1, (p-1)(p-2)]$
Residual	$SS_r = \sum_{i=1}^{p} \sum_{j=1}^{p} \sum_{k=1}^{p} (Y_{ijk} - \overline{Y}_{i..} - \overline{Y}_{.j.} - \overline{Y}_{..k} + 2\overline{Y}_{...})^2$	$(p-1)(p-2)$	$s_e^2 = SS_r/(p-1)(p-2)$	σ_e^2	--	--
Total	$\sum_{i=1}^{p} \sum_{j=1}^{p} \sum_{k=1}^{p} (Y_{ijk} - \overline{Y}_{...})^2$	$p^2 - 1$				

The overall mean $\overline{Y}_{...}$, the mean of the p^2 observations, is defined, using the same convention, by

$$\overline{Y}_{...} = \frac{\sum_{i=1}^{p}\sum_{j=1}^{p}\sum_{k=1}^{p} Y_{ijk}}{p^2}.$$

Thus $\overline{Y}_{...} = (8.7+7.5+14.0+\cdots+7.3+6.7+12.9)/16 = 8.9125$.

The analysis of variance table is given in Table 6.13. We define

$$\sigma_a^2 = \frac{\sum_{i=1}^{p}\alpha_i^2}{p-1}; \qquad \sigma_b^2 = \frac{\sum_{j=1}^{p}\beta_j^2}{p-1}; \qquad \sigma_c^2 = \frac{\sum_{k=1}^{p}\gamma_k^2}{p-1}.$$

The degrees of freedom add to the total degrees of freedom (p^2-1), and because the first three sums of squares must have $p-1$ degrees of freedom, the degrees of freedom for the residual sum of squares must be $(p^2-1)-3(p-1)$ or $(p-1)(p-2)$.

The sums of squares for rows, columns, and treatments are calculated as in the previous chapter; for the detergent data, they are:

$$p\sum_{i=1}^{p}\left(\overline{Y}_{i..}-\overline{Y}_{...}\right)^2 = 4\left[(9.65-8.9125)^2+\cdots+(9.225-8.9125)^2\right]$$

$$= 5.822$$

$$p\sum_{j=1}^{p}\left(\overline{Y}_{.j.}-\overline{Y}_{...}\right)^2 = 4\left[(10.375-8.9125)^2+\cdots+(9.000-8.9125)^2\right]$$

$$= 39.672$$

$$p\sum_{k=1}^{p}\left(\overline{Y}_{..k}-\overline{Y}_{...}\right)^2 = 4\left[(7.450-8.9125)^2+\cdots+(8.55-8.9125)^2\right]$$

$$= 86.548.$$

Usually only the F test for the existence of treatment effects is performed, just as in the randomized block. To test $H_0: \gamma_k = 0$, $k = 1,\ldots,p$, we use the statistic

$$F = \frac{\text{MS}_c}{s_e^2},$$

which is compared with

$$F[1-\alpha; p-1, (p-1)(p-2)].$$

However, we can test $H_0: \alpha_i = 0$ and $H_0: \beta_j = 0$, using $F = \text{MS}_a / s_e^2$ and $F = \text{MS}_b / s_e^2$, respectively, which are compared with $F[1-\alpha; p-1, (p-1)(p-2)]$.

Table 6.14 shows the analysis of variance for the detergent data, with the three possible F tests. We conclude, comparing $F = 42.68$, $F = 9.781$, and $F = 1.436$ with $F[.95; 3, 6] = 4.76$ that (*a*) there are differences in performance among the four detergents, (*b*) there are differences among the hospitals, and (*c*) there may be no differences among the four application methods.

If $p = 4$, there are only $3 \times 2 = 6$ degrees of freedom in the residual term. For a 3×3 Latin square ($p = 3$), there are $2 \times 1 = 2$ degrees of freedom. For small values of p, a very large treatment effect relative to the residual is necessary in order to reject a null hypothesis.

TABLE 6.14. ANALYSIS OF VARIANCE TABLE FOR DETERGENT DATA

Source of Variation (1)	SS (2)	df (3)	MS (4)	EMS (5)	Computed F (6)	Tabled F (7)
Due Method	5.822	3	1.941	$\sigma_e^2 + 4\sigma_a^2$	2.87	4.76
Due Hospital	39.672	3	13.224	$\sigma_e^2 + 4\sigma_b^2$	19.56	4.76
Due Detergent	86.548	3	28.849	$\sigma_e^2 + 4\sigma_c^2$	42.68	4.76
Residual	4.055	6	0.676	σ_e^2		
Total	136.097	15				

ESTIMATES OF PARAMETERS

Table 6.15 sets forth the point estimates for various parameters of the problem. Because estimates of the α_i's and β_j's are of the same form as the estimates of the γ_k's, only the latter have been included in the table.

TABLE 6.15. ESTIMATES OF PARAMETERS AND CONFIDENCE INTERVALS, LATIN SQUARE, MODEL I*

Model I*

Parameter (1)	Estimate (2)	Variance of Estimate (3)
μ	$\bar{Y}_{...}$	σ_e^2/p^2
$\mu + \gamma_k$	$\bar{Y}_{..k}$	σ_e^2/p
γ_k	$\bar{Y}_{..k} - \bar{Y}_{...}$	$\left(\frac{p-1}{p^2}\right)\sigma_e^2$
$\gamma_k - \gamma_{k'}$	$\bar{Y}_{..k} - \bar{Y}_{..k'}$	$2\sigma_e^2/p$
$\sum_{k=1}^{p} h_k\gamma_k,\ \sum_{k=1}^{p} h_k = 0$	$\sum_{k=1}^{p} h_k\bar{Y}_{..k}$	$\left(\sum_{k=1}^{p} h_k^2\right)\sigma_e^2/p$
σ_e^2	s_e^2	---

*Confidence intervals:

The $1-\alpha$ level confidence interval for σ_e^2 is

$$\frac{(p-1)(p-2)s_e^2}{\chi^2[1-\alpha/2;(p-1)(p-2)]} < \sigma_e^2 < \frac{(p-1)(p-2)s_e^2}{\chi^2[\alpha/2;(p-1)(p-2)]}$$

Intervals for other parameters are formed from columns (2) and (3), as for example

$$(\bar{Y}_{..k} - \bar{Y}_{..k'}) - t[1-\alpha/2m;(p-1)(p-2)]\sqrt{2s_e^2/p} < \gamma_k - \gamma_{k'} <$$

$$(\bar{Y}_{..k} - \bar{Y}_{..k'}) + t[1-\alpha/2m;(p-1)(p-2)]\sqrt{2s_e^2/p}\ ,$$

where m is the number of intervals made with an overall level of $1-\alpha$.

The estimates of the means and their linear combinations are obtained by beginning with

$$Y_{ijk} = \mu + \alpha_i + \beta_j + \gamma_k + \varepsilon_{ijk}.$$

Then we put

$$\bar{Y}_{..k} = \frac{\sum_{i=1}^{p}\sum_{j=1}^{p}(\mu + \alpha_i + \beta_j + \gamma_k + \varepsilon_{ijk})}{p},$$

where zeros replace all terms where Y_{ijk} was not measured. Because of the balanced design, we have

$$\bar{Y}_{..k} = \frac{p\mu + \sum_{i=1}^{p} \alpha_i + \sum_{j=1}^{p} \beta_j + p\gamma_k + \sum_{i=1}^{p}\sum_{j=1}^{p} \varepsilon_{ijk}}{p}$$

$$= \mu + \gamma_k + \bar{\varepsilon}_{..k}$$

and

$$\bar{Y}_{...} = \mu + \bar{\varepsilon}_{...}.$$

The estimate σ_e^2 is the usual residual mean square s_e^2, and the confidence interval follows as in the table.

From the detergent data, the best point estimate of $\gamma_1 - \gamma_2$ is

$$\bar{Y}_{..1} - \bar{Y}_{..2} = 7.45 - 6.85 = 0.60.$$

A 95% confidence interval for $\gamma_1 - \gamma_2$ is thus

$$0.60 \pm t[.975;6]\sqrt{2(0.676)/4} \qquad \text{or} \qquad -1.142 < \gamma_1 - \gamma_2 < 2.342.$$

DISCUSSION

The advantage of a Latin square design over the randomized complete block design is that the effect of a second factor is eliminated without increasing the size of the experiment, provided always that no interactions exist. However, if such interactions exist, the model is inappropriate, the residual mean square error is too large, and treatment effects are confused. When interactions are suspected, the methods of Chapter 7 must be employed.

Many different $p \times p$ Latin squares are possible. A randomization procedure consists in choosing at random one of the properly sized Latin squares from among those available; see Fisher [6.7].

The idea of a Latin square can be extended in various ways. For example, it is possible to remove the effect of three factors. Incomplete Latin squares also exist, and it is possible in a single experiment to use more than one Latin square. For variations on the basic Latin square design, see Davies [6.5] and Bliss [6.3].

ILLUSTRATIVE EXAMPLE

In planning an experiment, we must first decide our objectives. Next we consider the various types of outcome that might be helpful in attaining the objectives. For example, if the experiment is for the purpose of studying four treatments for diabetes, possible outcome variables to be measured during the course of the experiment would include length of life, length of disability-free life, and blood glucose level. Whatever outcome variables are chosen, the experimenter attempts to plan his study in a way that will enable him to separate the differences due to different treatments from differences due to extraneous factors.

In the diabetes study, it may become obvious to the investigator that he cannot obtain enough diabetic patients of the proper type from a single hospital. It is possible for him to use any of three designs presented in the present chapter (he might also consider designs of later chapters, in particular those of Chapters 7 and 13).

If the investigator has enough hospitals, he can use a nested design, with hospitals nested within the treatments as follows:

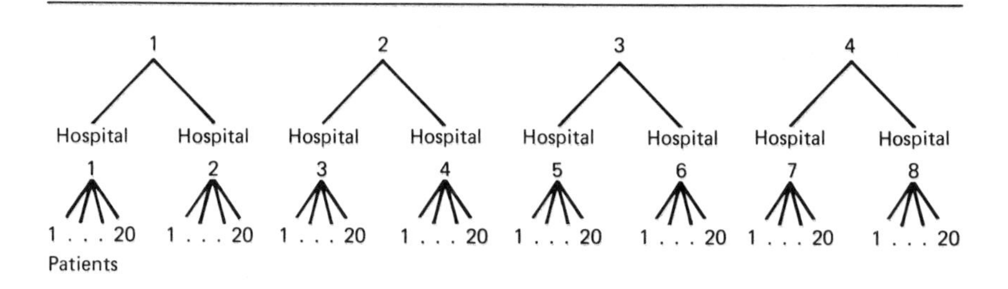

Here the experimental unit is the hospital, but response is measured on the patients or subsamples from the hospitals. The model used is that of (6.6). The problems of administration for the hospital in this design would be very simple because each hospital uses just one treatment. Nevertheless, the design seems to be a rather poor one for this particular example. With four treatments to be studied and eight hospitals, the size of each sample of hospitals is only two. Treatment 1 may easily be used at two hospitals whose patients tend to be especially poor risks. Any interaction that might exist between treatment and hospital could not be measured. Another disadvantage of the design is that the results could not be easily analyzed unless the same number of patients was used in each hospital and the number of hospitals used in the study was a multiple of 4; fulfilling these conditions might be exceedingly inconvenient.

In this example, the randomized block design may be better. To use the design exactly as given in (6.1), blocks consisting of four patients each are formed within each hospital. Now any number of hospitals can be used and the only restriction on the number of patients used in each hospital is that it be a multiple of 4. The patients in a given hospital could be placed at random in blocks of four; perhaps, however, the blocks can be made more homogeneous without too much effort by grouping patients according to severity of illness.

If the investigator wishes to remove the effect of both hospitals and of some other factor—say, method of administering treatment—he can consider the Latin square design of (6.17), although it is unlikely that he will find it to be the best approach. To use this model precisely as given here, he needs exactly four methods of treatment administration and four hospitals; he uses the same number of patients in each hospital and divides them at random into four equal groups, which are assigned at random to the four administrative methods. The patients then receive their treatment as given by the particular Latin square chosen in advance, and the cell means are analyzed.

The foregoing example illustrates the type of thought that must go into the choice of model for an experiment. The decision on whether to consider block effects must be made on the basis of careful thought, helped perhaps by a series of small experiments and a thorough literature search. The experimenter who understands the conditions that may effect the measured response and is familiar with a number of analysis of variance designs can usually find a good match between his problem and available designs. Frequently an experiment is designed and redesigned many times before a good solution is reached. Designing experiments is not simply a skill that can be learned in a class and applied automatically later in real life. If measurements are expensive or hazardous to the individuals or animals involved, time can be well spent in planning.

SUMMARY

This chapter introduced three additional designs that are useful in single factor analysis of variance. The first is the completely randomized block design, which can be used to remove an unwanted effect. The second, the nested design, involves subsampling. The third design introduced is the Latin square design, which can sometimes be used to remove the effects of two unwanted factors when effects are additive. Pooling in the analysis of variance table was discussed.

REFERENCES

APPLIED STATISTICS

6.1. Anderson, R. L., and T. A. Bancroft, *Statistical Theory in Research*, McGraw-Hill, New York, 1952.

6.2. Bennett, C. A., and N. L. Franklin, *Statistical Analysis in Chemistry and the Chemical Industry*, Wiley, New York, 1954.

6.3. Bliss, C. I., *Statistics in Biology*, McGraw-Hill, New York, 1967.

6.4. Brownlee, K. A., *Statistical Theory and Methodology in Science and Engineering*, 2nd ed., Wiley, New York, 1965.

6.5. Davies, O. L. (Ed.), *The Design and Analysis of Industrial Experiments*, 2nd ed., Oliver & Boyd, Edinburgh, 1956.

6.6. Dixon, W. J., and F. J. Massey, Jr., *Introduction to Statistical Analysis*, 3rd ed., McGraw-Hill, New York, 1969.

6.7. Fisher, R. A., and F. Yates, *Statistical Tables for Biological, Agricultural, and Medical Research*, 6th ed., Oliver & Boyd, London, 1963.

6.8. Fisher, R. A., *Statistical Methods for Research Workers*, Oliver & Boyd, Edinburgh, 1925.

6.9. Kempthorne, O., *The Design and Analysis of Experiments*, Wiley, New York, 1952.

6.10. Ostle, B., *Statistics in Research*, Iowa State University Press, Ames, Iowa, 1963.

6.11. Snedecor, G. W., and W. G. Cochran, *Statistical Methods*, 6th ed., Iowa State University Press, Ames, Iowa, 1967.

6.12. Winer, B. J., *Statistical Principles in Experimental Design*, McGraw-Hill, New York, 1962.

6.13. Peng, K. E., *The Design and Analysis of Scientific Experiments*, Addison-Wesley, Reading, Mass., 1967.

MATHEMATICAL STATISTICS

6.14. Scheffé, H., *Analysis of Variance*, Wiley, New York, 1967.

6.15. Mann, H. B., *Analysis and Design of Experiments*, Dover, New York, 1949.

POOLING

6.16. Paull, A. E., "On a Preliminary Test for Pooling Mean Squares in the Analysis of Variance," *Annals of Mathematical Statistics*, Vol. 21 (1950), pp. 539–556.

PROBLEMS

6.1. Six 4-month-old rats (USC strain) were used to compare the plasma cholesterol for diets involving three levels of butter fat. Two technicians were used. The six rats were assigned at random to the three butterfat levels with two rats at each level. For each butterfat level, the two rats were assigned at random to the two technicians. The data were as follows:

	Technicians	
Percent	I	II
0	69.3	70.3
10	84.8	80.5
20	85.0	72.6

(*a*) State an appropriate model.
(*b*) What assumptions must be made in order to use this model?
(*c*) Make an analysis of variance table.
(*d*) Test with $\alpha = .05$ whether there are differences among the three levels of butterfat.
(*e*) Test with $\alpha = .05$ whether there is a difference between the two technicians.
(*f*) Make confidence intervals for $\alpha_1 - \alpha_2$, $\alpha_1 - \alpha_3$, and $\alpha_1 - (\alpha_2 + \alpha_3)/2$ using an overall level of .95. Interpret the intervals.

6.2. Three diets were given to a group of adult volunteers to assess their effects on serum cholesterol levels (mg%). The volunteers were randomly assigned to the three diets, and there were four volunteers for each diet. Duplicate determinations of cholesterol level were made at the laboratory on each individual. The data were as follows:

	Individual							
Diet	1		2		3		4	
1	349	360	400	369	262	326	401	384
2	269	340	390	359	354	333	306	441
3	133	146	183	180	171	169	220	147

Preliminary calculations gave the following:

Diet	1	2	3	4	$\bar{Y}_{i..}$
1	$\bar{Y}_{11.} = 354.5$	$\bar{Y}_{12.} = 384.5$	$\bar{Y}_{13.} = 294.0$	$\bar{Y}_{14.} = 392.5$	356.375
2	$\bar{Y}_{21.} = 304.5$	$\bar{Y}_{22.} = 374.5$	$\bar{Y}_{23.} = 343.5$	$\bar{Y}_{24.} = 373.5$	349.0
3	$\bar{Y}_{31.} = 139.5$	$\bar{Y}_{32.} = 181.5$	$\bar{Y}_{33.} = 170.0$	$\bar{Y}_{34.} = 183.5$	168.625

$$\bar{Y}_{...} = 291.333$$

(*a*) Fill out an analysis of variance table and make a test at $\alpha = .05$ level to decide whether there are differences in effect among the three diets.

(*b*) Obtain confidence intervals with overall 95% level for the means of the three diets and for $\alpha_3-(\alpha_1+\alpha_2)/2$.

(*c*) Obtain a 95% confidence interval for the variance σ_e^2.

6.3. Three cleansing agents for the skin were used on three persons. For each person, three patches of skin were exposed to a contaminant and subsequently cleansed using each of three cleansing agents, one on each patch. After 8 hr the residual contaminant was measured. The measurements of residual contamination were:

Cleansing Agent	Individual 1	2	3
A	1.25	2.30	1.45
B	0.87	1.45	0.65
C	1.11	1.38	2.99

(*a*) Write down an appropriate model and state the necessary assumptions in terms of the problem.

(*b*) Make the analysis of variance table.

(*c*) Test for treatment effects with $\alpha=.01$.

(*d*) Give a 99% confidence interval for σ_e^2.

(*e*) Estimate the following by confidence intervals with an overall 95% level:

(i) The overall mean.

(ii) The population mean difference between residual contaminant using cleansing agents A and B.

(iii) The population mean residual contaminant for individual 1 using cleansing agent C.

6.4. Total liver cholesterol was measured on rats fed three types of diet. The diets were I: basal, II: basal + 10% corn oil, III: basal + 10% butter fat. Three technicians were used; rats were fed the diets for three different time periods, 2 weeks, 4 weeks, and 6 weeks. You wish to decide whether there is a significant difference in liver cholesterol depending on diet. The measurements obtained were as follows:

	Time on Diet		
	1	2	3
Technician	2 weeks	4 weeks	6 weeks
1	1.44 (I)	1.71 (II)	2.07 (III)
2	1.72 (II)	1.83 (III)	1.97 (I)
3	1.84 (III)	1.56 (I)	1.60 (II)

Preliminary computations gave the following:

	1	2	3
$\bar{Y}_{i..}$	1.740	1.840	1.667
$\bar{Y}_{.j.}$	1.667	1.700	1.880
$\bar{Y}_{..k}$	1.657	1.677	1.913

$\bar{Y}_{...} = 1.749$

(*a*) State the model.
(*b*) What assumptions are being made in terms of this problem?
(*c*) Make an analysis of variance table.
(*d*) Test whether there are differences in liver cholesterol among the three diets.
(*e*) Test whether there are differences in liver cholesterol among the different lengths of time.
(*f*) Test whether there are differences among technicians.
(*g*) Find a 95% confidence interval for σ_e^2.
(*h*) Give 95% confidence interval for the difference between liver cholesterol using basal alone and the average of liver cholesterol for the two diets using basal plus a supplement.

6.5. In a study on the effect of four fertilizers (A, B, C, D) on yields of wheat, four types of wheat were used and four plots of ground. Each plot was divided into four equal subplots and a Latin square design was used.

Type of wheat	Plot 1	2	3	4
1	35.5 (A)	24.5 (B)	14.7 (C)	35.5 (D)
2	14.3 (B)	6.2 (C)	13.7 (D)	24.5 (A)
3	14.1 (C)	16.2 (D)	34.3 (A)	19.7 (B)
4	15.0 (D)	64.5 (A)	34.6 (B)	19.0 (C)

Preliminary computations gave the following results:

	1	2	3	4
$\bar{Y}_{i..}$	27.55	14.67	21.07	33.27
$\bar{Y}_{.j.}$	19.72	27.85	24.32	24.67
$\bar{Y}_{..k}$	39.70	23.27	13.50	20.10

$\bar{Y}_{...} = 24.14$

(*a*) Make an analysis of variance table.
(*b*) State in terms of this example the assumptions underlying the use of this model.
(*c*) Test with $\alpha = .01$ whether there are differences in yields due to fertilizers.
(*d*) Test with $\alpha = .01$ whether there are differences in yields among the four types of wheat.
(*e*) Give point estimates for the mean yield for each type of wheat.
(*f*) Estimate the mean yield for each fertilizer.
(*g*) Find a 99% confidence interval for $(\alpha_1 + \alpha_3)/2 - (\alpha_2 + \alpha_4)/2$.
(*h*) Find a 99% confidence interval for σ_e^2.

CHAPTER 7

CROSS CLASSIFICATIONS

Often a researcher can use a single experiment advantageously to study two or more different kinds of treatment. For example, in investigating performance of two types of seed, he may wish to vary the level of fertilizer used during the experiment. If he chose three levels of fertilizer—low, medium, and high—one factor would be "type of seed," the second factor, "level of fertilizer." A factorial design, with two factors, would consist of employing all six treatments formed by using each type of seed with each level of fertilizer. Factorial designs can involve more than two factors; however, we consider first the case of two factors.

A factorial design can also be used in a survey. For example, we might wish to compare three methods of teaching grade school mathematics, and at the same time compare the first four grades. We might have records on standardized tests for two classes in each grade taught by each method. The class mean improvement from initial test to final test could be the measure of success. Our data would then consist of two observations on each of 12 (3×4) different treatment combinations.

The characteristic of the factorial design is that every level of one factor is used in combination with every level of the other factor. The design is effective for studying the two factors in combination.

Some factors can be measured quantitatively, and different levels for them are chosen on an ordered scale; level of fertilizer, dosage level, and temperature are all factors of this type. Other factors involve no obvious underlying continuum and can be said to be qualitative; drug and type of seed are factors of the second type.

CROSS CLASSIFICATIONS

THE TWO-WAY CLASSIFICATION

AN EXAMPLE

As an example of a two-factor design, let us take a study on rye yields involving two types of seed, each used at three fertilizer levels—low, medium, and high. There were available 24 small plots of ground, and the six treatment combinations were assigned at random to 24 plots, 4 plots receiving each treatment. Table 7.1 shows the notation. A response (in this case yield) is denoted by Y_{ijk}, where i indicates the seed type, j indicates the fertilizer level, and k is the observation number. For example, Y_{213} is the yield in the third of the four plots that used seed type 2 and a low fertilizer level. The *cell means*, denoted by $\overline{Y}_{ij.}$ are the means for each treatment combination. The mean of all 12 observations on the ith seed type is $\overline{Y}_{i..}$; the mean of all 8 observations on the jth fertilizer is $\overline{Y}_{.j.}$; the

TABLE 7.1. NOTATION FOR TWO-WAY CROSS CLASSIFICATION

Seed Type	Fertilizer Level Low (1)	Medium (2)	High (3)	
1	Y_{111}	Y_{121}	Y_{131}	
	Y_{112}	Y_{122}	Y_{132}	
	Y_{113}	Y_{123}	Y_{133}	
	Y_{114}	Y_{124}	Y_{134}	
	$\overline{Y}_{11.}$	$\overline{Y}_{12.}$	$\overline{Y}_{13.}$	$\overline{Y}_{1..}$
2	Y_{211}	Y_{221}	Y_{231}	
	Y_{212}	Y_{222}	Y_{232}	
	Y_{213}	Y_{223}	Y_{233}	
	Y_{214}	Y_{224}	Y_{234}	
	$\overline{Y}_{21.}$	$\overline{Y}_{22.}$	$\overline{Y}_{23.}$	$\overline{Y}_{2..}$
	$\overline{Y}_{.1.}$	$\overline{Y}_{.2.}$	$\overline{Y}_{.3.}$	$\overline{Y}_{...}$

TABLE 7.2. YIELDS OF RYE AND THEIR MEANS (BUSHELS/ACRE)

Seed Type	Fertilizer Level Low	Medium	High	
1	14.3	18.1	17.6	
	14.5	17.6	18.2	
	11.5	17.1	18.9	
	13.6	17.6	18.2	
	$\bar{Y}_{11.} = 13.475$	$\bar{Y}_{12.} = 17.600$	$\bar{Y}_{13.} = 18.225$	$\bar{Y}_{1..} = 16.433$
2	12.6	10.5	15.7	
	11.2	12.8	17.5	
	11.0	8.3	16.7	
	12.1	9.1	16.6	
	$\bar{Y}_{21.} = 11.725$	$\bar{Y}_{22.} = 10.175$	$\bar{Y}_{23.} = 16.625$	$\bar{Y}_{2..} = 12.842$
	$\bar{Y}_{.1.} = 12.600$	$\bar{Y}_{.2.} = 13.888$	$\bar{Y}_{.3.} = 17.425$	$\bar{Y}_{...} = 14.638$

overall mean of the 24 observations is $\bar{Y}_{...}$. With the observations themselves replacing their symbols, we have Table 7.2.

THE MODEL

We turn now to an underlying model. We divide each observation Y_{ijk} into four parts, which together form the mean of a population of responses to the ijth treatment combination, and a fifth part, which is the deviation of the particular observation from the population mean. We assume that

$$Y_{ijk} = \mu + \alpha_i + \beta_j + (\alpha\beta)_{ij} + \varepsilon_{ijk}, \quad \text{for} \quad i = 1, \ldots, a; \quad j = 1, \ldots, b; \quad k = 1, \ldots, n, \tag{7.1}$$

where

$$\sum_{i=1}^{a} \alpha_i = \sum_{j=1}^{b} \beta_j = \sum_{i=1}^{a} (\alpha\beta)_{ij} = \sum_{j=1}^{b} (\alpha\beta)_{ij} = 0$$

and ε_{ijk}IND$(0,\sigma_e^2)$. Here a is the number of levels of the first factor (number of types of seed, in the example), b is the number of levels of the second factor, and n is the number of observations on each treatment combination. The parameters are as follows:

μ = overall mean response; that is, the average of the mean responses for the ab populations

α_i = effect of the ith level of the first factor; averaged over the b levels of the second factor, the ith level of the first factor adds α_i to the overall mean μ

β_j = effect of the jth level of the second factor

$(\alpha\beta)_{ij}$ = interaction between the ith level of the first factor and the jth level of the second factor; the population mean for the ijth treatment minus $\mu+\alpha_i+\beta_j$

ε_{ijk} = deviation of Y_{ijk} from the population mean response for the ijth population

The terms α_i and β_j are called *main effects*. They are average effects for each type of seed and for each level of fertilizer. The term $(\alpha\beta)_{ij}$ is an *interaction*. If seed and fertilizer levels behave in a strictly additive way—that is, if a high level of fertilizer adds a certain amount to the average yield, regardless of the seed type—the $(\alpha\beta)_{ij}$'s are all zero. On the other hand, if a high level of fertilizer increases yield more with seed type 1 than with seed type 2, $(\alpha\beta)_{13}$ is positive and $(\alpha\beta)_{23}$ is negative.

Implicit in the statement of the model are three assumptions.

1. The ε_{ijk} are independently distributed. If, for example, several of the plots were planted a month before all the remaining plots, the deviations from the population means could not be said to be independent of one another. The several plots planted earlier might all tend to have particularly high yields.

2. The ε_{ijk} are normally distributed.

3. The ε_{ijk} all come from a population with the same variance. (We might suspect that in the presence of a high fertilizer level, the yields will vary more widely from one observation to another; in this case we cannot claim equal variances for the ε_{ijk}.)

These are the only assumptions we must make for the design. In the randomized complete block design, an additional assumption was necessary; we assumed that the means could be expressed as $\mu+\alpha_i+\beta_j$. Here the addition of a fourth term has removed the necessity of assuming anything at all concerning the means. It can be shown that *any* set of ab population means can be expressed in the form $\mu+\alpha_i+\beta_j+(\alpha\beta)_{ij}$ in such a way that the following conditions are fulfilled: all the α_i's sum to zero; all the β_j's sum to zero; for each particular j, all the $(\alpha\beta)_{ij}$'s sum to zero; and for each fixed i, all the $(\alpha\beta)_{ij}$'s sum to zero.

INTERACTION

To illustrate the meaning of interaction, we write down a possible set of population means:

Seed Type	Fertilizer Level 1	2	3
1	14	18	19
2	12	10	17

We note that the difference in yield between the two seed types is larger for the middle level of fertilizer than for the low and high levels (18 − 10 versus 14 − 12 and 19 − 17). We see also that the differences between any two fertiliizer levels differ for the two seed types ($18-14 \neq 10-12$, etc.) With unequal differences in population means, we say that interactions exist between fertilizer and seed type. These population means, expressed as $\mu+\alpha_i+\beta_j+(\alpha\beta)_{ij}$, are as follows:

Seed Type	Fertilizer Level 1	2	3
1	15+2−2−1	15+2−1+2	15+2+3−1
2	15−2−2+1	15−2−1−2	15−2+3+1

Here $\mu=15$, $\alpha_1=2$, $\alpha_2=-2$, $\beta_1=-2$, $\beta_2=-1$, $\beta_3=+3$, $(\alpha\beta)_{11}=-1$, $(\alpha\beta)_{12}=2$, $(\alpha\beta)_{13}=-1$, $(\alpha\beta)_{21}=+1$, $(\alpha\beta)_{22}=-2$, $(\alpha\beta)_{23}=+1$. Note: $(\alpha\beta)_{11}+(\alpha\beta)_{12}+(\alpha\beta)_{13}=0$, and so forth.

A set of population means with no interaction is the following:

Seed Type	Fertilizer Level 1	2	3
1	14	19	12
2	12	17	10

We observe that the change from one fertilizer level is the same for each seed type ($19-14=17-12$, etc.) and that the difference between yields is 2 at each level of fertilizer. The interactions are therefore all zero, and the means can be expressed with no $(\alpha\beta)_{ij}$ term but with $\mu=14$, $\alpha_1=1$, $\alpha_2=-1$, $\beta_1=-1$, $\beta_2=+4$, $\beta_3=-3$.

ANALYSIS OF VARIANCE TABLE AND F TESTS

In order to analyze the variance, we divide the deviation of each observation from the overall mean $Y_{ijk}-\overline{Y}_{...}$ into meaningful parts. We do so with the following equation in mind:

$$Y_{ijk}-\mu=\alpha_i+\beta_j+(\alpha\beta)_{ij}+\varepsilon_{ijk}. \tag{7.2}$$

The left-hand side of (7.2) is clearly to be estimated by $Y_{ijk}-\overline{Y}_{...}$. On the right-hand side, α_i should be estimated by $\overline{Y}_{i..}-\overline{Y}_{...}$, and β_j by $\overline{Y}_{.j.}-\overline{Y}_{...}$. Because we have made no assumptions at all concerning the population means, ε_{ijk} is estimated by $Y_{ijk}-\overline{Y}_{ij.}$. By subtracting the estimates of α_i, β_j, and ε_{ijk} from the estimate of $Y_{ijk}-\mu$, we obtain $\overline{Y}_{ij.}-\overline{Y}_{i..}-\overline{Y}_{.j.}+\overline{Y}_{...}$ as an estimate of $(\alpha\beta)_{ij}$. Thus we have

$$Y_{ijk}-\overline{Y}_{...}=\left(\overline{Y}_{i..}-\overline{Y}_{...}\right)+\left(\overline{Y}_{.j.}-\overline{Y}_{...}\right)$$

$$+\left(\overline{Y}_{ij.}-\overline{Y}_{i..}-\overline{Y}_{.j.}+\overline{Y}_{...}\right)+\left(Y_{ijk}-\overline{Y}_{ij.}\right). \tag{7.3}$$

If we express each deviation $Y_{ijk}-\overline{Y}_{...}$ in this way, square both sides, and add over all the observations, the cross-product terms add to zero and we have

$$\sum_{i=1}^{a}\sum_{j=1}^{b}\sum_{k=1}^{n}\left(Y_{ijk}-\overline{Y}_{...}\right)^2=bn\sum_{i=1}^{a}\left(\overline{Y}_{i..}-\overline{Y}_{...}\right)^2$$

$$+an\sum_{j=1}^{b}\left(\overline{Y}_{.j.}-\overline{Y}_{...}\right)^2+n\sum_{i=1}^{a}\sum_{j=1}^{b}\left(\overline{Y}_{ij.}-\overline{Y}_{i..}-\overline{Y}_{.j.}+\overline{Y}_{...}\right)^2$$

$$+\sum_{i=1}^{a}\sum_{j=1}^{b}\sum_{k=1}^{n}\left(Y_{ijk}-\overline{Y}_{ij.}\right)^2. \tag{7.4}$$

The first term on the right-hand side of (7.4) is said to be due to the first factor, or the main effect of the first factor. We often call it the sum of squares "due A," or the "A effect." The second term, similarly, is the "B effect." The third term is called the interaction term, the "due AB term" or the "AB effect." The last term is the residual.

Table 7.3 gives the analysis of variance table for the two-way classification Model I. The quantity σ_a^2 is defined by $\sigma_a^2 = \sum_{i=1}^{a}\alpha_i^2/(a-1)$; similarly, we have

$$\sigma_b^2 = \frac{\sum_{j=1}^{b}\beta_j^2}{b-1} \qquad \text{and} \qquad \sigma_{ab}^2 = \frac{\sum_{i=1}^{a}\sum_{j=1}^{b}(\alpha\beta)_{ij}^2}{(a-1)(b-1)}.$$

The degrees of freedom for the A sum of squares, the B sum of squares, and the total sum of squares are easily obtained as $a-1, b-1$, and $abn-1$. The residual sum of squares is simply the pooled estimate of the variance from ab independent samples, and thus the degrees of freedom are $abn - ab = ab(n-1)$. By subtracting $a-1, b-1$, and $ab(n-1)$ from the total degrees of freedom $abn-1$, we obtain $(a-1)(b-1)$ as the degrees of freedom for the interaction term, the only remaining sum of squares.

The EMS column of Table 7.3 follows the same form as in the one-way classification of Chapter 5, Model I. Now MS_a estimates $\sigma_e^2 + bn\sigma_a^2$, where bn is the number of observations on which a single $\overline{Y}_{i..}$ is based; similarly, MS_b estimates $\sigma_e^2 + an\sigma_b^2$. The interaction mean square estimates $\sigma_e^2 + n\sigma_{ab}^2$, and we note that n is the number of observations on which each $\overline{Y}_{ij.}$ is based.

By considering the EMS column, we see that it is possible to make three tests. First we can test to determine whether interaction exists. The null hypothesis for this test is $H_0: (\alpha\beta)_{ij} = 0,\ i = 1,\ldots,a;\ j = 1,\ldots,b$. If H_0 is true, MS_{ab} and s_e^2 are two independent estimates of σ_e^2. Therefore, the test statistic MS_{ab}/s_e^2 is compared with $F[1-\alpha; (a-1)(b-1), ab(n-1)]$.

Similarly, there are two tests for main effects. The null hypothesis $H_0: \alpha_i = 0, i = 1,\ldots,a$ is tested by comparing MS_a/s_e^2 with $F[1-\alpha; a-1, ab(n-1)]$; $H_0: \beta_j = 0,\ \ j = 1,\ldots,b$ is tested by comparing MS_b/s_e^2 with $F[1-\alpha; b-1, ab(n-1)]$.

In Table 7.4, the analysis of variance table for the data of Table 7.2, the tabled F values were obtained by interpolating in Table A.4 on the reciprocal of the degrees of freedom. A simple, conservative technique is to use the larger F value obtained from Table A.4 rather than bothering to interpolate. For example, we might use $F[.95; 1, 15] = 4.54$ instead of $F[.95; 1, 18] = 4.41$.

TABLE 7.3. ANALYSIS OF VARIANCE, TWO-WAY CROSS CLASSIFICATION, MODEL I

Source of Variation	SS	df	MS	EMS	Computed F	Tabled F
Due A	$SS_a = bn \sum_{i-1}^{a} (\overline{Y}_{i..} - \overline{Y}_{...})^2$	a-1	$MS_a = SS_a/(a-1)$	$\sigma_e^2+bn\sigma_a^2$	MS_a/s_e^2	$F[1-\alpha; a-1, ab(n-1)]$
Due B	$SS_b = an \sum_{j=1}^{b} (\overline{Y}_{.j.} - \overline{Y}_{...})^2$	b-1	$MS_b = SS_b/(b-1)$	$\sigma_e^2+an\sigma_b^2$	MS_b/s_e^2	$F[1-\alpha; b-1, ab(n-1)]$
Due AB	$SS_{ab} = n \sum_{i=1}^{a} \sum_{j=1}^{b} (\overline{Y}_{ij.} - \overline{Y}_{i..} - \overline{Y}_{.j.} + \overline{Y}_{...})^2$	(a-1)(b-1)	$MS_{ab} = SS_{ab}/(a-1)(b-1)$	$\sigma_e^2+n\sigma_{ab}^2$	MS_{ab}/s_e^2	$F[1-\alpha; (a-1)(b-1), ab(n-1)]$
Residual	$SS_r = \sum_{i=1}^{a} \sum_{j=1}^{b} \sum_{k=1}^{n} (Y_{ijk} - \overline{Y}_{ij.})^2$	ab(n-1)	$s_e^2 = SS_r/ab(n-1)$	σ_e^2	--	--
Total	$SS_t = \sum_{i=1}^{a} \sum_{j=1}^{b} \sum_{k=1}^{n} (Y_{ijk} - \overline{Y}_{...})^2$	abn-1				

TABLE 7.4. ANALYSIS OF VARIANCE TABLE FOR EXPERIMENT ON TWO TYPES OF SEED WITH THREE FERTILIZER LEVELS

Source of Variation (1)	SS (2)	df (3)	MS (4)	EMS (5)	Computed F (6)	Tabled F (7)
Due Seeds	77.4	1	77.4	$\sigma_e^2+12\sigma_a^2$	63.3	4.41
Due Fertilizer Levels	99.9	2	49.9	$\sigma_e^2+8\sigma_b^2$	40.9	3.55
Seed Fertilizer Interaction	44.1	2	22.1	$\sigma_e^2+4\sigma_{ab}^2$	18.0	3.55
Residual	22.0	18	1.2	σ_e^2		
Total	243.4	23				

From the last two columns of the table, we conclude, (comparing 18.0 with 3.55) that interactions exist between seed type and fertilizer level. Similarly, we conclude that differences exist between yields from the two seed types (63.33 compared with 4.41) and among yields from the three levels of fertilizer (40.86 compared with 3.55).

ESTIMATES OF PARAMETERS AND CONFIDENCE INTERVALS

Table 7.5 shows the point estimates, the variances of the estimates, and the confidence intervals for the parameters of interest in the two-way classification Model I. The estimate for σ_e^2 and its confidence interval follow as usual from the analysis of variance table. The other rows in the table can be checked by beginning with

$$Y_{ijk}=\mu+\alpha_i+\beta_j+(\alpha\beta)_{ij}+\varepsilon_{ijk}. \tag{7.5}$$

To check the third line of the table, for instance, we average over k to obtain

$$\overline{Y}_{ij.}=\mu+\alpha_i+\beta_j+(\alpha\beta)_{ij}+\bar{\varepsilon}_{ij}. \tag{7.6}$$

From (7.6) we see that the expected value of the cell mean $\overline{Y}_{ij.}$ is $\mu+\alpha_i+\beta_j+(\alpha\beta)_{ij}$, the population mean for the ijth treatment combination.

TABLE 7.5. TWO-WAY CLASSIFICATION, MODEL I; ESTIMATES OF PARAMETERS AND CONFIDENCE INTERVALS*

Parameter (1)	Point Estimate (2)	Variance of Estimate (3)
μ	$\overline{Y}_{...}$	σ_e^2/abn
$\mu + \alpha_i$	$\overline{Y}_{i..}$	σ_e^2/bn
$\mu + \alpha_i + \beta_j + (\alpha\beta)_{ij}$	$\overline{Y}_{ij.}$	σ_e^2/n
α_i	$\overline{Y}_{i..} - \overline{Y}_{...}$	$\sigma_e^2(a-1)/abn$
$(\alpha\beta)_{ij}$	$\overline{Y}_{ij.} - \overline{Y}_{i..} - \overline{Y}_{.j.} + \overline{Y}_{...}$	$\sigma_e^2(a-1)(b-1)/abn$
$\alpha_i - \alpha_{i'}$	$\overline{Y}_{i..} - \overline{Y}_{i'..}$	$2\sigma_e^2/bn$
$\beta_j - \beta_{j'}$	$\overline{Y}_{.j.} - \overline{Y}_{.j'.}$	$2\sigma_e^2/an$
$\sum_{i=1}^{a} h_i\alpha_i,\ \sum_{i=1}^{a} h_i = 0$	$\sum_{i=1}^{a} h_i\overline{Y}_{i..}$	$\sum_{i=1}^{a} h_i^2\ \sigma_e^2/bn$
σ_e^2	s_e^2	-

*Confidence intervals:

The $1-\alpha$ level confidence interval for σ_e^2 is

$$\frac{ab(n-1)s_e^2}{\chi^2[1-\alpha/2;ab(n-1)]} < \sigma_e^2 < \frac{ab(n-1)s_e^2}{\chi^2[\alpha/2;ab(n-1)]} .$$

Intervals for other parameters are formed from the corresponding items in columns (2) and (3), as for example

$$(\overline{Y}_{i..} - \overline{Y}_{...}) - t[1-\alpha/2m;ab(n-1)]\sqrt{s_e^2(a-1)/abn} < \alpha_i <$$

$$(\overline{Y}_{i..} - \overline{Y}_{...}) + t[1-\alpha/2m;ab(n-1)]\sqrt{s_e^2(a-1)/abn} \quad ,$$

where m is the number of intervals made with an overall level of $1-\alpha$.

The variance of $\bar{Y}_{ij.}$ is simply σ_e^2/n, because $\bar{\varepsilon}_{ij.}$ is the mean of n independent ε_{ijk}'s. The confidence interval for $\mu+\alpha_i+\beta_j+(\alpha\beta)_{ij}$ follows immediately. The other lines in the table are checked in a similar way; we also use the conditions

$$\sum_{i=1}^{a}\alpha_i=\sum_{j=1}^{b}\beta_j=\sum_{i=1}^{a}(\alpha\beta)_{ij}=\sum_{j=1}^{b}(\alpha\beta)_{ij}=0.$$

From the data on rye yields, for example, the researcher may wish to estimate the following parameters: μ, the overall mean; $\mu+\alpha_1$ and $\mu+\alpha_2$, the means for the two varieties; $\mu+\beta_j$, $j=1,\ldots,3$, the means for the three levels of fertilizer; all six mean yields $\mu+\alpha_i+\beta_j+(\alpha\beta)_{ij}$; the three interactions $(\alpha\beta)_{11}, (\alpha\beta)_{12}$, and $(\alpha\beta)_{13}$ [a confidence interval for $(\alpha\beta)_{1j}$ automatically yields one for $(\alpha\beta)_{2j}$]; $\alpha_1-\alpha_2$; $\beta_1-\beta_2$, and $\beta_2-\beta_3$. There are thus $m=18$ confidence intervals he wishes to make using Table 7.5. With $1-\alpha=.95$, the interval for $(\alpha\beta)_{12}$ is

$$(17.600-16.433-13.888+14.638)$$

$$\pm t[1-.05/(2\times 18);18]\sqrt{1.222(2-1)(3-1)/4}$$

or

$$-0.819<(\alpha\beta)_{12}<4.653.$$

THE INTERPRETATION OF INTERACTIONS

When the data obtained indicate that large interactions exist, it is important to consider whether large interactions actually are present in the population means or whether there may be some other explanation for the occurrence of the interactions in the data. Perhaps the two categories of treatments have a catalytic effect on each other with the result that the treatment mean response cannot be expressed in terms of the overall mean plus the main effects. The question of whether interactions exist can often be decided by considering the treatments themselves.

Sometimes, however, when the investigator believes that no interactions exist, the data obtained point to sizable interactions. This could occur, of course, by chance alone. On the other hand, such unexpected interactions may be caused by a problem in the data—there may be an outlier (an observation far removed from the rest of the data) or an erroneous response. If the observations were not made in a random order, an appreciable time effect may be included in the response. In such a case, the ε_{ij} can no longer be said to be independent random observations. Some other uncontrolled variable may be affecting the observations. In an experiment involving animals, for example, the position of the cage may have an effect on the outcome variable, and if we neither formed blocks according to cage position nor randomly placed the animals in the room, we might obtain an apparent interaction when none exists.

Good test procedure consists in randomizing on every factor that is not controlled in some other way (see Fisher [7.5]). Random assignment of the experimental units to the treatment combinations is especially important.

Thus an unexpected interaction may be a clue to failure in meeting the assumptions of the model being used in analysis of variance. Further discussion on the effects of various failures in assumptions can be found in Scheffé [7.10] or Bennett and Franklin [7.1]. Chapter 14, which includes methods of screening data to test whether assumptions are met, also discusses possible transformations.

It has been noticed, too, that interactions often occur when the main effects are *very* large. Interactions frequently disappear if the investigator lessens the differences among the levels of a treatment, making the main effects less pronounced.

If the factors are quantitative, graphing the cell means can be an aid in interpreting the interaction. For example, consider an agricultural experiment that involves three levels of water and three fertilizer levels. If lines joining cell means appear roughly parallel, as in Figure 7.1, we do not expect a large interaction term. In Figure 7.2 two sets of cell means are graphed which would lead us to expect a large interaction. In the first graph of Figure 7.2, a high fertilizer level interacts positively with a high water level. In the second graph, high levels of water and fertilizer together result in an unexpectedly low response in comparison with the response to the low and medium water levels. In this experiment, it is reasonable to expect the effects of water level and fertilizer level to be additive if the levels are restricted in range. If the water level is very low, however, very high levels of fertilizer may result in particularly bad yields.

THE TWO-WAY CLASSIFICATION WITH $n=1$

When interaction exists, it is necessary in order to estimate the variance to have more than one observation on each treatment combination. The residual sum of squares is zero when $n=1$ and we have no estimate of σ_e^2.

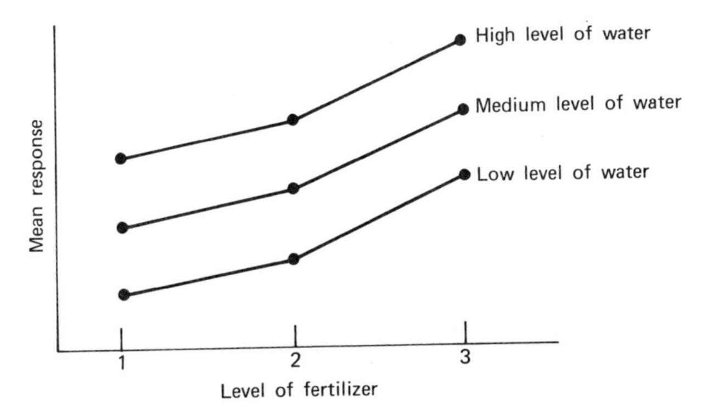

Figure 7.1. Mean response without interaction.

We omit the subscript k because it is always 1, and the model becomes

$$Y_{ij}=\mu+\alpha_i+\beta_j+(\alpha\beta)_{ij}+\varepsilon_{ij}, \qquad i=1,\ldots,a; \qquad j=1,\ldots,b \tag{7.7}$$

$$\sum_{i=1}^{a}\alpha_i=\sum_{j=1}^{b}\beta_j=\sum_{i=1}^{a}(\alpha\beta)_{ij}=\sum_{j=1}^{b}(\alpha\beta)_{ij}=0; \qquad \varepsilon_{ij}\ \mathrm{IND}(0,\sigma_e^2).$$

The analysis of variance table is then given in Table 7.6. Under this model, we cannot test either $H_0:\alpha_i=0,\ i=1,\ldots,a$ or $H_0:\beta_j=0,\ j=1,\ldots,b$, because there is no way of separating MS_{ab} into two parts—one to estimate σ_e^2 and the other to estimate σ_{ab}^2. Because of the impossibility of obtaining an estimate of σ_e^2, we cannot make any of the confidence intervals given in Table 7.5. (A testing procedure given by Tukey [7.13] allows nonadditivity to be tested with 1 degree of freedom.)

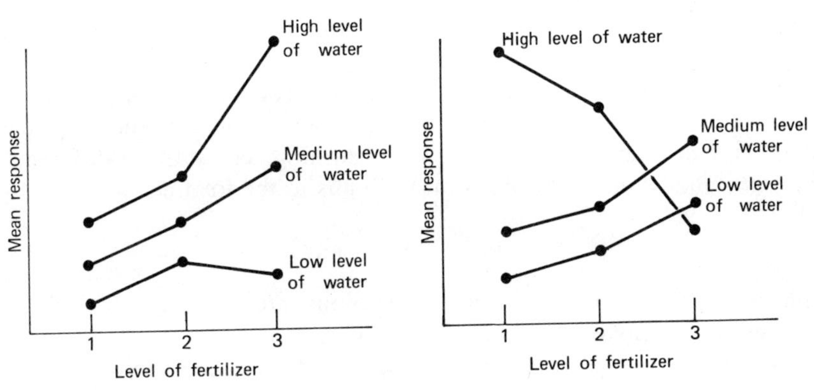

Figure 7.2. Mean response with interaction.

TABLE 7.6. TWO-WAY CROSS CLASSIFICATION WHEN $n=1$

Source of Variation (1)	Sums of Squares (2)	df (3)	MS (4)	EMS (5)
Due A	$SS_a = b\sum_{i=1}^{a}(\bar{Y}_{i.} - \bar{Y}_{..})^2$	a-1	$MS_a = SS_a/(a-1)$	$\sigma_e^2 + b\sigma_a^2$
Due B	$SS_b = a\sum_{j=1}^{b}(\bar{Y}_{.j} - \bar{Y}_{..})^2$	b-1	$MS_b = SS_b/(b-1)$	$\sigma_e^2 + a\sigma_b^2$
Interaction	$SS_{ab} = \sum_{i=1}^{a}\sum_{j=1}^{b}(Y_{ij} - \bar{Y}_{i.} - \bar{Y}_{.j} + \bar{Y}_{..})^2$	(a-1)(b-1)	$MS_{ab} = \frac{SS_{ab}}{(a-1)(b-1)}$	$\sigma_e^2 + \sigma_{ab}^2$
Total	$SS_t = \sum_{i=1}^{a}\sum_{j=1}^{b}(Y_{ij} - \bar{Y}_{..})^2$	ab-1		

When $n=1$, the assumption of no interaction is frequently made, and Model I becomes

$$Y_{ij} = \mu + \alpha_i + \beta_j + \varepsilon_{ij}, \qquad i=1,\ldots,a; \qquad j=1,\ldots,b \tag{7.8}$$

$$\sum_{i=1}^{a}\alpha_i = \sum_{j=1}^{b}\beta_j = 0, \qquad \varepsilon_{ij}\ \text{IND}(0,\sigma_e^2).$$

We note that this is precisely the model given in Chapter 6 for a single factor, randomized complete block design. In Table 7.6, if we set $\sigma_{ab}^2=0$ and change the word "interaction" to "residual," we have the same analysis of variance table as Table 6.2. Estimation and tests are the same whether we have a completely randomized design with two factors or a single factor and a randomized complete block design, even though in the first instance we picked the levels of the two factors and assigned the experimental units to them at random, whereas in the second instance we blocked the experimental units and we then made a random assignment within each block to the treatments. A difference between these two designs is that in the two-way completely randomized design we wish to study both factors; in the one-way randomized complete block design, on the other hand, we usually wish to study one factor and eliminate the other. This distinction is not hard and fast, however; the true difference lies in the allocation of the experimental units to the treatments.

MISSING VALUES

With $n=1$, when one or more observations are lost, the analysis is considerably complicated. The missing values are estimated from the available observations and from the model, and then the analysis is made using all the data, actual and estimated. An unbiased estimate of σ_e^2 is obtained (we are assuming that $\sigma_{ab}^2=0$), but the estimates of the main effects σ_a^2 and σ_b^2 are biased. For every missing value estimated, the degrees of freedom for the residual mean square are reduced by 1.

When a single observation is missing, it is estimated by

$m_{ij}=\dfrac{aT_{i.}+bT_{.j}-T_{..}}{(a-1)(b-1)}$, where $T_{i.}$ denotes the total of $b-1$ observations in the ith row, $T_{.j}$ denotes the total of $a-1$ observations in the jth column, and $T_{..}$ denotes the sum of all $ab-1$ observations.

When more than one observation is missing, see Bennett and Franklin [7.1] or Töcher [7.11].

For n larger than 1, and when only a few values are missing, any missing values in the ijth treatment combination are replaced by the mean for that treatment combination $\bar{Y}_{ij.}$. For each missing value replaced, the residual degrees of freedom are reduced by 1. If the number of observations varies widely from one treatment combination to another, we can no longer use the analysis of variance techniques given in this chapter and must resort to a more general procedure (see Chew [7.9]) which is available in the BMD programs.

If at all possible, it is advantageous to use n larger than 1. Replication decreases the variance of the cell mean as a function of $1/n$. It also frees us from assuming that there is no interaction. It makes an occasional missing value less troublesome. Finally, it gives us a check on the assumptions underlying the analysis of variance (see Chapter 14).

THE THREE-WAY CLASSIFICATION

Experiments and surveys frequently involve more than two factors. The techniques just given for analyzing the two-way classification extend directly to three or more classifications. We present the three-way classification in some detail because it is simpler and because it enables us to see how more than three factors are analyzed.

In the previous section we discussed an example in which both type of seed and level of fertilizer were varied. If in addition we now vary the level of water, we have a three-way classification. One factor is qualitative, two are quantitative.

Table 7.7 contains data representing yields of rye for an experiment involving three types of seed, two fertilizer levels, and two water levels. There are 12 treatment combinations, and we assume that 48 plots were assigned at random to the 12 treatment combinations.

TABLE 7.7. RYE YIELDS FROM THREE SEED TYPES

	Low Water Level				High Water Level			
Seed Type \ Fertilizer Level	Low		High		Low		High	
1	18.6 18.8 15.8 17.9	17.775	10.2 10.5 8.5 10.6	9.95	12.3 10.9 10.7 11.8	11.425	12.3 10.7 12.0 11.8	11.700
2	19.4 18.9 18.4 18.9	18.900	14.7 17.8 15.6 16.5	16.150	13.2 15.5 11.0 11.8	12.875	10.2 8.5 7.1 6.5	8.075
3	15.9 16.5 17.2 16.5	16.525	20.9 21.0 21.3 20.4	20.900	19.4 21.2 20.4 20.3	20.325	13.2 13.5 12.8 15.2	13.675

Seed Type \ Fertilizer Level	Low	High	
1	14.600	10.825	12.713
2	15.888	12.113	14.000
3	18.425	17.287	17.856
	16.304	13.408	14.856

Seed Type \ Water Level	Low	High	
1	13.863	11.563	12.713
2	17.525	10.475	14.000
3	18.713	17.000	17.856
	16.700	13.013	14.856

Fertilizer Level \ Water Level	Low	High	
Low	17.733	14.875	16.304
High	15.667	11.150	13.408
	16.700	13.013	14.856

The symbol Y_{ijkl} is now used to denote lth observation of the ijkth treatment combination (i.e., the yield of the lth plot that used the ith seed type, the jth fertilizer level, and the kth water level). Table 7.8 presents the observations and their means in symbols. We can imagine the observations in a three-way classification problem as lying in a box built of square blocks. In the seed example there are three rows of blocks, two columns of blocks, and two layers of blocks. Because paper is two-dimensional, it is more convenient to write out the data as in Tables 7.7 and 7.8, where the two layers of blocks have been separated and set side by side. The upper part of the tabulation contains the observations and the means for each cell. Means formed from the cell means by averaging over a single factor are entered into three two-way tables in the lower part of Tables 7.7 and 7.8.

TABLE 7.8. THREE-WAY CLASSIFICATION, THREE LEVELS OF FACTOR A, TWO LEVELS OF B, TWO LEVELS OF C

	First Level C				Second Level C			
	B Level				B Level			
A Level	1		2		1		2	
1	Y_{1111}		Y_{1211}		Y_{1121}		Y_{1221}	
	Y_{1112}	$\bar{Y}_{111.}$	Y_{1212}	$\bar{Y}_{121.}$	Y_{1122}	$\bar{Y}_{112.}$	Y_{1222}	$\bar{Y}_{122.}$
	Y_{1113}		Y_{1213}		Y_{1123}		Y_{1223}	
	Y_{1114}		Y_{1214}		Y_{1124}		Y_{1224}	
2	Y_{2111}		Y_{2211}		Y_{2121}		Y_{2221}	
	Y_{2112}	$\bar{Y}_{211.}$	Y_{2212}	$\bar{Y}_{221.}$	Y_{2122}	$\bar{Y}_{212.}$	Y_{2222}	$\bar{Y}_{222.}$
	Y_{2113}		Y_{2213}		Y_{2123}		Y_{2223}	
	Y_{2114}		Y_{2214}		Y_{2124}		Y_{2224}	
3	Y_{3111}		Y_{3211}		Y_{3121}		Y_{3221}	
	Y_{3112}	$\bar{Y}_{311.}$	Y_{3212}	$\bar{Y}_{321.}$	Y_{3122}	$\bar{Y}_{312.}$	Y_{3222}	$\bar{Y}_{322.}$
	Y_{3113}		Y_{3213}		Y_{3123}		Y_{3223}	
	Y_{3114}		Y_{3214}		Y_{3124}		Y_{3224}	

A \ B	1	2	
1	$\bar{Y}_{11..}$	$\bar{Y}_{12..}$	$\bar{Y}_{1...}$
2	$\bar{Y}_{21..}$	$\bar{Y}_{22..}$	$\bar{Y}_{2...}$
3	$\bar{Y}_{31..}$	$\bar{Y}_{32..}$	$\bar{Y}_{3...}$
	$\bar{Y}_{.1..}$	$\bar{Y}_{.2..}$	$\bar{Y}_{....}$

A \ C	1	2	
1	$\bar{Y}_{1.1.}$	$\bar{Y}_{1.2.}$	$\bar{Y}_{1...}$
2	$\bar{Y}_{2.1.}$	$\bar{Y}_{2.2.}$	$\bar{Y}_{2...}$
3	$\bar{Y}_{3.1.}$	$\bar{Y}_{3.2.}$	$\bar{Y}_{3...}$
	$\bar{Y}_{..1.}$	$\bar{Y}_{..2.}$	$\bar{Y}_{....}$

B \ C	1	2	
1	$\bar{Y}_{.11.}$	$\bar{Y}_{.12.}$	$\bar{Y}_{.1..}$
2	$\bar{Y}_{.21.}$	$\bar{Y}_{.22.}$	$\bar{Y}_{.2..}$
	$\bar{Y}_{..1.}$	$\bar{Y}_{..2.}$	$\bar{Y}_{....}$

THE MODEL

In Model I of the three-way classification, we express each observation Y_{ijkl} as follows:

$$Y_{ijkl}=\mu+\alpha_i+\beta_j+\gamma_k+(\alpha\beta)_{ij}+(\alpha\gamma)_{ik}+(\beta\gamma)_{jk}+(\alpha\beta\gamma)_{ijk}+\varepsilon_{ijkl},$$

$$i=1,\ldots,a; \qquad j=1,\ldots,b; \qquad k=1,\ldots,c; \qquad l=1,\ldots,n, \qquad (7.9)$$

where

$$\sum_{i=1}^{a}\alpha_i=\sum_{j=1}^{b}\beta_j=\sum_{k=1}^{c}\gamma_k=\sum_{i=1}^{a}(\alpha\beta)_{ij}=\sum_{j=1}^{b}(\alpha\beta)_{ij}=\sum_{i=1}^{a}(\alpha\gamma)_{ik}$$

$$=\sum_{k=1}^{c}(\alpha\gamma)_{ik}=\sum_{j=1}^{b}(\beta\gamma)_{jk}=\sum_{k=1}^{c}(\beta\gamma)_{jk}=\sum_{i=1}^{a}(\alpha\beta\gamma)_{ijk}=\sum_{j=1}^{b}(\alpha\beta\gamma)_{ijk}$$

$$=\sum_{k=1}^{c}(\alpha\beta\gamma)_{ijk}=0 \qquad \text{and} \qquad \varepsilon_{ijkl}\,\text{IND}(0,\sigma_e^2).$$

As given in (7.9), there are n observations on each of abc treatment combinations, a levels of factor A, b levels of factor B, and c levels of factor C. The various parts comprising each population mean are:

- μ — overall mean: average of the population mean response for the abc treatments
- α_i — average effect of the ith level of factor A; average of the population means for the bc treatments involving level i of factor A minus μ
- β_j — average effect of the jth level of factor B
- γ_k — average effect of the kth level of factor C
- $(\alpha\beta)_{ij}$ — two-factor interaction of the ith level of factor A with the jth level of factor B; average of the population means for the c treatments that involve the ith level of A and the jth level of B minus $\mu+\alpha_i+\beta_j$.
- $(\alpha\gamma)_{ik}$ — two-factor interaction of the ith level of factor A with the kth level of factor C; the average of the population means for the b treatments involving the ith level of A and the kth level of C minus $\mu+\alpha_i+\gamma_k$.
- $(\beta\gamma)_{jk}$ — two-factor interaction of the jth level of factor B with the kth level of factor C; average of the population means for the a treatments which involve jth level of B and the kth level of C minus $\mu+\beta_j+\gamma_k$.

TABLE 7.9. 12 POPULATION MEANS FOR THREE-WAY CLASSIFICATION

Water Level 1

Seed Type \ Fertilizer Level	1	2
1	23: $\mu = 30$, $(\alpha\beta)_{11} = 3$ $\alpha_1 = 2$, $(\alpha\gamma)_{11} = ?$ $\beta_1 = -5$, $(\beta\gamma)_{11} = -3$ $\gamma_1 = -4$, $(\alpha\beta\gamma)_{111} = -2$	37: $\mu = 30$, $(\alpha\beta)_{12} = -3$ $\alpha_1 = 2$, $(\alpha\gamma)_{11} = 2$ $\beta_2 = 5$, $(\beta\gamma)_{21} = 3$ $\gamma_1 = -4$, $(\alpha\beta\gamma)_{121} = 2$
2	12: $\mu = 30$, $(\alpha\beta)_{21} = 1$ $\alpha_2 = -5$, $(\alpha\gamma)_{21} = -1$ $\beta_1 = -5$, $(\beta\gamma)_{11} = -3$ $\gamma_1 = -4$, $(\alpha\beta\gamma)_{211} = -1$	28: $\mu = 30$, $(\alpha\beta)_{22} = -1$ $\alpha_2 = -5$, $(\alpha\gamma)_{21} = -1$ $\beta_2 = 5$, $(\beta\gamma)_{21} = 3$ $\gamma_1 = -4$, $(\alpha\beta\gamma)_{221} = 1$
3	19: $\mu = 30$, $(\alpha\beta)_{31} = -4$ $\alpha_3 = 3$, $(\alpha\gamma)_{31} = -1$ $\beta_1 = -5$, $(\beta\gamma)_{11} = -3$ $\gamma_1 = -4$, $(\alpha\beta\gamma)_{311} = 3$	37: $\mu = 30$, $(\alpha\beta)_{32} = 4$ $\alpha_3 = 3$, $(\alpha\gamma)_{31} = -1$ $\beta_2 = 5$, $(\beta\gamma)_{21} = 3$ $\gamma_1 = -4$, $(\alpha\beta\gamma)_{321} = -3$

Water Level 2

Seed Type \ Fertilizer Level	1	2
1	37: $\mu = 30$, $(\alpha\beta)_{11} = 3$ $\alpha_1 = 2$, $(\alpha\gamma)_{12} = -2$ $\beta_1 = -5$, $(\beta\gamma)_{12} = 3$ $\gamma_2 = 4$, $(\alpha\beta\gamma)_{112} = 2$	31: $\mu = 30$, $(\alpha\beta)_{12} = -3$ $\alpha_1 = 2$, $(\alpha\gamma)_{12} = -2$ $\beta_2 = 5$, $(\beta\gamma)_{22} = -3$ $\gamma_2 = 4$, $(\alpha\beta\gamma)_{122} = -2$
2	30: $\mu = 30$, $(\alpha\beta)_{21} = 1$ $\alpha_2 = -5$, $(\alpha\gamma)_{22} = 1$ $\beta_1 = -5$, $(\beta\gamma)_{12} = 3$ $\gamma_2 = 4$, $(\alpha\beta\gamma)_{212} = 1$	30: $\mu = 30$, $(\alpha\beta)_{22} = -1$ $\alpha_2 = -5$, $(\alpha\gamma)_{22} = 1$ $\beta_2 = 5$, $(\beta\gamma)_{22} = -3$ $\gamma_2 = 4$, $(\alpha\beta\gamma)_{222} = -1$
3	29: $\mu = 30$, $(\alpha\beta)_{31} = -4$ $\alpha_3 = 3$, $(\alpha\beta)_{32} = 1$ $\beta_1 = -5$, $(\beta\gamma)_{12} = 3$ $\gamma_2 = 4$, $(\alpha\beta\gamma)_{312} = -3$	47: $\mu = 30$, $(\alpha\beta)_{32} = 4$ $\alpha_3 = 3$, $(\alpha\gamma)_{32} = 1$ $\beta_2 = 5$, $(\beta\gamma)_{22} = -3$ $\gamma_2 = 4$, $(\alpha\beta\gamma)_{322} = 3$

$(\alpha\beta\gamma)_{ijk}$ three-factor interaction between the ith level of factor A, the jth level of factor B, and the kth level of factor C; population mean response for the ijkth treatment combination minus $\mu+\alpha_i+\beta_j+\gamma_k+(\alpha\beta)_{ij}+(\alpha\gamma)_{ik}+(\beta\gamma)_{jk}$.

Table 7.9 gives as an example a set of population means for three seed types, two fertilizer levels, and two water levels; by each mean are the eight terms whose sum is the mean. Notice that all the summations listed in (7.9) are zero; for instance, we can write

$$(\alpha\gamma)_{11}+(\alpha\gamma)_{12}=2-2=0;$$

$$(\alpha\gamma)_{11}+(\alpha\gamma)_{21}+(\alpha\gamma)_{31}=2-1-1=0.$$

THE ANALYSIS OF VARIANCE TABLE AND F TESTS

In order to obtain a meaningful division of the basic sum of squares, we first separate the deviation of each observation from the overall mean into eight parts—the estimates of α_i, β_j, γ_k, $(\alpha\beta)_{ij}$, $(\alpha\gamma)_{ik}$, $(\beta\gamma)_{jk}$, $(\alpha\beta\gamma)_{ijk}$, and ε_{ijkl}. We have

$$\begin{aligned}
Y_{ijkl}-\bar{Y}_{....}=&\left(\bar{Y}_{i...}-\bar{Y}_{....}\right)+\left(\bar{Y}_{.j..}-\bar{Y}_{....}\right)+\left(\bar{Y}_{..k.}-\bar{Y}_{....}\right)\\
&+\left(\bar{Y}_{ij..}-\bar{Y}_{i...}-\bar{Y}_{.j..}+\bar{Y}_{....}\right)\\
&+\left(\bar{Y}_{i.k.}-\bar{Y}_{i...}-\bar{Y}_{..k.}+\bar{Y}_{....}\right)\\
&+\left(\bar{Y}_{.jk.}-\bar{Y}_{.j..}-\bar{Y}_{..k.}+\bar{Y}_{....}\right)\\
&+\left(\bar{Y}_{ijk.}-\bar{Y}_{ij..}-\bar{Y}_{i.k.}-\bar{Y}_{.jk.}+\bar{Y}_{i...}+\bar{Y}_{.j..}+\bar{Y}_{..k.}-\bar{Y}_{....}\right)\\
&+\left(Y_{ijkl}-\bar{Y}_{ijk.}\right).
\end{aligned}\tag{7.10}$$

The first six terms on the right-hand side of (7.10) require little explanation; they are the estimates of main effects and two-factor interactions used in the two-way classification, now with an added dot representing the extra subscript over which each mean was taken. The eighth term also presents no problem. The seventh term, the estimate of the three-factor interaction $(\alpha\beta\gamma)_{ijk}$, can be obtained by subtracting all the other terms on the right-hand side of (7.10) from $Y_{ijkl}-\bar{Y}_{....}$.

When both sides of (7.10) are squared and summed over the entire set of data, we obtain the various sums of squares in the SS column of the analysis of variance table (Table 7.10). The degrees of freedom are easily written down except for that of the ABC interaction; this is obtained by subtracting all the other degrees of freedom from the total, $abcn-1$.

From columns 6 and 7 of the table, we can now make F tests of seven null hypotheses. For main effects, we test $H_0:\alpha_i=0$, $i=1,\ldots,$a; $H_0:\beta_j=0$, $j=1,\ldots,b$; and $H_0:\gamma_k=0$, $k=1,\ldots,c$. For two-factor interactions, we have H_0: $(\alpha\beta)_{ij}=0$, $i=1,\ldots,a$, $j=1,\ldots,b$; $H_0:(\alpha\gamma)_{ik}=0$, $i=1,\ldots,a$, $k=1,\ldots,c$; and $H_0:(\beta\gamma)_{jk}=0$, $j=1,\ldots,b$, $k=1,\ldots,c$. For three-factor interactions, we have $H_0:(\alpha\beta\gamma)_{ijk}=0$, $i=1,\ldots,a$, $j=1,\ldots,b$, $k=1,\ldots,c$. If we wish to make m of these tests with an overall significance level of $1-\alpha$, we use α/m as the significance level for each F test.

Table 7.11 is the analysis of variance for the data on rye yields. The calculation of the nine sums of squares in the table can be performed in many ways (see Bennett and Franklin [7.1]). Here they have been calculated using the basic formulae of Table 7.10. Typical calculations are as follows:

Due A:

$$bcn\sum_{i=1}^{a}\left(\bar{Y}_{i...}-\bar{Y}_{....}\right)^2$$

$$=(2)(2)(4)\left[(12.713-14.856)^2+(14.000-14.856)^2+(17.856-14.856)^2\right]$$

$$=16(4.592+0.733+9.000)$$

$$=229.20.$$

Due AB:

$$cn\sum_{i=1}^{a}\sum_{j=1}^{b}\left(\bar{Y}_{ij..}-\bar{Y}_{i...}-\bar{Y}_{.j..}+\bar{Y}_{....}\right)^2$$

$$=(2)(4)\left[(14.600-12.713-16.304+14.856)^2+\cdots+(17.287-17.856-13.408+14.856)^2\right]$$

$$=8(0.193+\cdots+0.773)$$

$$=18.51.$$

TABLE 7.10. ANALYSIS OF VARIANCE TABLE, THREE-WAY CLASSIFICATION, MODEL I

Source of Variation (1)	SS (2)	df (3)	MS (4)	EMS (5)	Computed F (6)	Tabled F (7)
Main Effects						
A	$SS_a = bcn \sum_{i=1}^{a} (\overline{Y}_{i...} - \overline{Y}_{....})^2$	a-1	$MS_a = SS_a/(a-1)$	$\sigma_e^2 + bcn\sigma_a^2$	MS_a/s_e^2	$F[1-\alpha;a-1,abc(n-1)]$
B	$SS_b = acn \sum_{j=1}^{b} (\overline{Y}_{.j..} - \overline{Y}_{....})^2$	b-1	$MS_b = SS_b/(b-1)$	$\sigma_e^2 + acn\sigma_b^2$	MS_b/s_e^2	$F[1-\alpha;b-1,abc(n-1)]$
C	$SS_c = abn \sum_{k=1}^{c} (\overline{Y}_{..k.} - \overline{Y}_{....})^2$	c-1	$MS_c = SS_c/(c-1)$	$\sigma_e^2 + abn\sigma_c^2$	MS_c/s_e^2	$F[1-\alpha;c-1,abc(n-1)]$
2 Factor Interactions						
AB	$SS_{ab} = cn \sum_{i=1}^{a}\sum_{j=1}^{b} (\overline{Y}_{ij..} - \overline{Y}_{i...} - \overline{Y}_{.j..} + \overline{Y}_{....})^2$	(a-1)(b-1)	$MS_{ab} = SS_{ab}/(a-1)(b-1)$	$\sigma_e^2 + cn\sigma_{ab}^2$	MS_{ab}/s_e^2	$F[1-\alpha;(a-1)(b-1),abc(n-1)]$
AC	$SS_{ac} = bn \sum_{i=1}^{a}\sum_{k=1}^{c} (\overline{Y}_{i.k.} - \overline{Y}_{i...} - \overline{Y}_{..k.} + \overline{Y}_{....})^2$	(a-1)(c-1)	$MS_{ac} = SS_{ac}/(a-1)(c-1)$	$\sigma_e^2 + bn\sigma_{ac}^2$	MS_{ac}/s_e^2	$F[1-\alpha;(a-1)(c-1),abc(n-1)]$
BC	$SS_{bc} = an \sum_{j=1}^{b}\sum_{k=1}^{c} (\overline{Y}_{.jk.} - \overline{Y}_{.j..} - \overline{Y}_{..k.} + Y_{....})^2$	(b-1)(c-1)	$MS_{bc} = SS_{bc}/(b-1)(c-1)$	$\sigma_e^2 + an\sigma_{bc}^2$	MS_{bc}/s_e^2	$F[1-\alpha;(b-1)(c-1),abc(n-1)]$
3 Factor Interaction						
ABC	$SS_{abc} = n \sum_{i=1}^{a}\sum_{j=1}^{b}\sum_{k=1}^{c} (\overline{Y}_{ijk.} - \overline{Y}_{ij..} - \overline{Y}_{i.k.} - \overline{Y}_{.jk.} + \overline{Y}_{i...} + \overline{Y}_{.j..} + \overline{Y}_{..k.} - \overline{Y}_{....})^2$	(a-1)(b-1)(c-1)	$MS_{abc} = SS_{abc}/(a-1)(b-1)(c-1)$	$\sigma_e^2 + n\sigma_{abc}^2$	MS_{abc}/s_e^2	$F[1-\alpha;(a-1)(b-1)(c-1),abc(n-1)]$
Residual	$SS_r = \sum_{i=1}^{a}\sum_{j=1}^{b}\sum_{k=1}^{c}\sum_{\ell=1}^{n} (Y_{ijk\ell} - \overline{Y}_{ijk.})^2$	abc(n-1)	$s_e^2 = SS_r/abc(n-1)$	σ_e^2		
Total	$SS_t = \sum_{i=1}^{a}\sum_{j=1}^{b}\sum_{k=1}^{c}\sum_{\ell=1}^{n} (Y_{ijk\ell} - \overline{Y}_{....})^2$	abcn-1				

TABLE 7.11. ANALYSIS OF VARIANCE TABLE FOR RYE YIELDS, THREE-WAY CLASSIFICATION, MODEL I

Source of Variation (1)	Sum of Squares (2)	df (3)	MS (4)	EMS (5)	Computed F (6)	Tabled F (7)
Due A	229.200	2	114.600	$\sigma_\ell^2 + 16\sigma_a^2$	94.85	3.27
Due B	100.642	1	100.642	$\sigma_\ell^2 + 24\sigma_b^2$	83.30	4.12
Due C	163.128	1	163.128	$\sigma_\ell^2 + 24\sigma_c^2$	135.02	4.12
Due AB	18.543	2	9.272	$\sigma_\ell^2 + 8\sigma_{ab}^2$	7.67	3.27
Due AC	68.521	2	34.260	$\sigma_\ell^2 + 8\sigma_{ac}^2$	28.36	3.27
Due BC	8.257	1	8.257	$\sigma_\ell^2 + 12\sigma_{bc}^2$	6.83	4.12
Due ABC	183.211	2	91.606	$\sigma_\ell^2 + 4\sigma_{abc}^2$	75.35	3.27
Residual	43.496	36	1.208	σ_ℓ^2		
Total	814.998	47				

Due ABC:

$$n\sum_{i=1}^{a}\sum_{j=1}^{b}\sum_{k=1}^{c}\left(\bar{Y}_{ijk.}-\bar{Y}_{ij..}-\bar{Y}_{i.k.}-\bar{Y}_{.jk.}+\bar{Y}_{i...}+\bar{Y}_{.j..}+\bar{Y}_{..k.}-\bar{Y}_{....}\right)^2$$

$$=4\Big[(17.775-14.600-13.863-17.733+12.713+16.304+16.700-14.856)^2$$

$$+\cdots$$

$$+(13.675-17.287-17.000-11.150+17.856+13.408+13.013-14.857)^2\Big]$$

$$=4(5.954+\cdots+5.485)$$

$$=183.25.$$

Residual: $$\sum_{i=1}^{a}\sum_{j=1}^{b}\sum_{k=1}^{c}\sum_{l=1}^{n}\left(Y_{ijkl}-\bar{Y}_{ijk.}\right)^2$$

$$=(18.6-17.775)^2+\cdots+(15.2-13.675)^2$$

$$=0.681+\cdots+2.326$$

$$=43.496.$$

Total: $$\sum_{i=1}^{a}\sum_{j=1}^{b}\sum_{k=1}^{c}\sum_{l=1}^{n}\left(Y_{ijkl}-\bar{Y}_{....}\right)^2$$

$$=(18.6-14.856)^2+\cdots+(15.2-14.856)^2$$

$$=14.018+\cdots+0.118$$

$$=814.998.$$

As a check on calculations, the sum of squares were added to obtain the total sum of squares.

The F values in column 6 were obtained by dividing each MS in column 4 by the residual mean square 1.208. For single .05-level tests, these are compared with the tabled F values of column 7, which were obtained by interpolating on the reciprocal of the degrees of freedom. The F tests result in rejecting all null hypotheses; examination of the treatment means in Table 7.7 clearly reveals the presence of interactions.

ESTIMATES OF PARAMETERS

Table 7.12 gives point estimates and confidence intervals for various means and linear combinations of means in the three-way classification in which all effects are fixed. Estimates of β_j, γ_k, $(\alpha\gamma)_{ik}$, and other quantities that have been omitted from the table can be obtained by exchanging subscripts in the estimates in the table.

Some of the point estimates for parameters have already been set down in Table 7.7 for the rye yield data. Others can be calculated readily using Table 7.12. We find, for example, that α_1 is estimated by $\bar{Y}_{1...}-\bar{Y}_{....}=-2.143$ and that $(\alpha\beta)_{12}$ is estimated by $\bar{Y}_{12..}-\bar{Y}_{1...}-\bar{Y}_{.2..}+\bar{Y}_{....}=10.825-12.713-13.408+14.856=-0.44$.

TABLE 7.12. ESTIMATES OF PARAMETERS AND CONFIDENCE INTERVALS FOR PARAMETERS, THREE-WAY CLASSIFICATION, MODEL I*

Parameter (1)	Point Estimate of Parameter (2)	Variance of Estimate (3)
$\mu + \alpha_i + \beta_j + \gamma_k + (\alpha\beta)_{ij} + (\alpha\gamma)_{ik} + (\beta\gamma)_{jk} + (\alpha\beta\gamma)_{ijk}$	$\overline{Y}_{ijk.}$	σ_e^2/n
$\mu + \alpha_i + \beta_j + (\alpha\beta)_{ij}$	$\overline{Y}_{ij..}$	σ_e^2/cn
$\mu + \alpha_i$	$\overline{Y}_{i...}$	σ_e^2/bcn
μ	$\overline{Y}_{....}$	$\sigma_e^2/abcn$
α_i	$\overline{Y}_{i...} - \overline{Y}_{....}$	$\sigma_e^2(a-1)/abcn$
$(\alpha\beta)_{ij}$	$(\overline{Y}_{ij..} - \overline{Y}_{i...} - \overline{Y}_{.j..} + \overline{Y}_{....})$	$\sigma_e^2(a-1)(b-1)/abcn$
$(\alpha\beta\gamma)_{ijk}$	$(\overline{Y}_{ijk.} - \overline{Y}_{ij..} - \overline{Y}_{i.k.} - \overline{Y}_{.jk.} + \overline{Y}_{i...} + \overline{Y}_{.j..} + \overline{Y}_{..k.} - \overline{Y}_{....})$	$\sigma_e^2(a-1)(b-1)(c-1)/abcn$
$\alpha_i - \alpha_{i'}$	$\overline{Y}_{i...} - \overline{Y}_{i'...}$	$2\sigma_e^2/bcn$
$\sum_{i=1}^{a} h_i\alpha_i,\ \sum_{i=1}^{a} h_i = 0$	$\sum_{i=1}^{a} h_i\overline{Y}_{i...}$	$\left(\sum_{i=1}^{a} h_i^2\right)\sigma_e^2/bcn$
σ_e^2	s_e^2	--

*Confidence Intervals:

The 1-α level confidence interval for σ_e^2 is $\frac{abc(n-1)s_e^2}{\chi^2[1-\alpha/2;abc(n-1)]} < \sigma_e^2 < \frac{abc(n-1)s_e^2}{\chi^2[\alpha/2;abc(n-1)]}$. Intervals for other parameters are formed from the corresponding items in columns (2) and (3), as for example

$\overline{Y}_{i...} - t[1-\alpha/2m;abc(n-1)]\sqrt{s_e^2/bcn} < \mu + \alpha_i < \overline{Y}_{i...} + t[1-\alpha/2m;abc(n-1)]\sqrt{s_e^2/bcn}$ where m is the number of intervals made with an overall level of 1-α.

A 95% confidence interval for $(\alpha\beta\gamma)_{111}$ (using $m=1$) is

$$(17.775-14.600-13.863-17.733+12.713+16.304+16.700-14.856)\pm$$

$$t[.975;36]\sqrt{1.208(3-1)(2-1)(2-1)/4}$$

$$\text{or} \qquad .861<(\alpha\beta\gamma)_{111}<4.017.$$

ILLUSTRATIVE EXAMPLE

In this section we discuss an example of a two-way cross-classification problem and point out the advantages of using this design instead of an older, more traditional design.

An investigator wishes to study the distribution of labeled cholesterol in tissues of rats following various diets. To assess cholesterol levels in the plasma, liver, and other locations, rats are fed one of three diets, and additional dosages of cholesterol are administered within 24 hours of sacrifice. The investigator assigns the treatments to the following numbers of rats:

	Additional Dosage of Cholesterol		
Basic Diet	None	One Large Dose	Four Equal Small Doses
1. Standard + no fat	n rats	n rats	n rats
2. Standard + saturated fat	n rats	n rats	n rats
3. Standard + unsaturated fat	n rats	n rats	n rats

In a single experiment, the investigator is able to assess the main effects of three diets, the main effects of the three additional dosages of cholesterol prior to sacrifice, and the interactions between diet and cholesterol dosage. His analysis of variance table for any one measure—say, cholesterol level in the liver—is as follows:

	df	EMS
Due diet	2	$\sigma_e^2+3n\sigma_{\text{diet}}^2$
Due cholesterol	2	$\sigma_e^2+3n\sigma_{\text{chol}}^2$
Due interaction	4	$\sigma_e^2+n\sigma_{\text{diet}\times\text{chol}}^2$
Residual	$9(n-1)$	σ_e^2
Total	$9n-1$	

Often, in concentrating on doing the various F tests and obtaining contrasts, investigators ignore the numerical value of s_e^2. The estimate s_e^2 of σ_e^2 can be a very useful measure in assessing the results of the present experiment and in planning future experiments. The basic variation of cholesterol level in the liver of rats (excluding treatment and interaction effects) is measured by s_e^2 or its square root s_e. If in the present experiment s_e was much higher than in the past experiments, the discrepancy could indicate a loss in accuracy in the present study that should be investigated before further analyses are done on the data. The value of s_e^2 is also used in making confidence intervals for differences between pairs of diets and in the denominator of the F test statistics for tests of no diet effect, no cholesterol effect, and no interaction.

The estimate of σ_e^2 also serves in planning future experiments to assist in determining what sample size to use in each treatment combination. If σ_e^2 is large, a larger sample size n is needed in order to predict accurately the mean level in each treatment group. The length of the confidence intervals is inversely related to the square root of the sample size and directly related to s_e. In an approximate fashion, we could find the value of n that would result in $t[1-\alpha/2; ab(n-1)]\sqrt{\sigma_e^2/n}$ (see Table 7.5 for confidence intervals) of the desired size. More precise techniques for estimating sample size for analysis of variance are given in Dixon and Massey [7.4]. The estimate of σ_e^2 (s_e^2) tends to be quite variable for small degrees of freedom (the variance of s_e^2 is $2\sigma_e^4/\text{d.f.}$). Since it is highly desirable to have a relatively stable estimate of σ_e^2, the investigator wishes to use a limited number of rats to achieve as many degrees of freedom as possible for the mean square residual. If he uses five rats in each treatment group, he uses 45 rats altogether and has $9(5-1)=36$ degrees of freedom for his estimate of σ_e^2.

An investigator who is unfamiliar with the two-way analysis of variance may plan to experiment with diets using just two diets at a time. He may follow a procedure that is still widely used and consists of several experiments, each comparing one treatment with a standard treatment. He may begin using the two diets (1) standard + no fat and (2) standard + saturated fat; from the first experiment he can assess the effect of the addition of saturated fat to the standard diet. His one-way analysis of variance table is as follows:

	df	EMS
Due diet	1	$\sigma_e^2 + n\sigma_{\text{diet}}^2$
Residual	$2(n-1)$	σ_e^2
Total	$2n-1$	

With five rats on each diet, he has only $2(5-1)=8$ degrees of freedom for his estimate of σ_e^2.

We also notice that the "due diet" mean square is an estimate $\sigma_e^2 + n\sigma_{\text{diet}}^2$ in the one-way analysis of variance, as compared with $\sigma_e^2 + 3n\sigma_{\text{diet}}^2$ in the two-way analysis of variance. This mean square appears in the numerator of the F statistic in testing for differences due to diet. The mean square estimates a smaller quantity in the one-way analysis of variance than in the two-way, provided n and σ_{diet}^2 are the same in both experiments; therefore, it seems reasonable to suggest that the F test is less sensitive in the one-way analysis to differences among diets. The experimenter using the one-way plan will probably wish to increase the number of rats on each treatment in order to have a more sensitive F test, more degrees of freedom, more accurate estimates of σ_e^2, and confidence intervals that are less variable in length. With 10 rats on each treatment, he has 18 degrees of freedom and has used 20 rats.

After the first one-way analysis of variance, the investigator can perform a similar experiment to compare (1) standard diet with (2) standard diet + unsaturated fat. He can then make statements comparing these two diets relative to the control diet. But with 10 rats on each treatment in each experiment, he has already used 40 rats (only 5 less than in the two-way analysis of variance, which had 5 rats on each treatment combination), and he still has no information on the effect of additional dosage of cholesterol *or* on interaction of diet and cholesterol.

The two and three-way analysis of variance designs are among the most widely used approaches, particularly in laboratory work. If the investigator wishes to examine several factors, he can obtain more information from a given number of observations than from successive one-way analysis of variance experiments. Furthermore, since these designs are relatively simple, analysis and interpretation are not too difficult. Computer programs are widely available for performing the computations (see BMD02V, BMD05V, or BMDX64, or IBMSSP). After an investigator has learned to interpret data in his field from these designs, he finds them a tremendous help in his research. If at first he has difficulty in interpretation, he will find it relatively easy to obtain assistance from statisticians and from statistically trained persons in his field; the necessary techniques are widely taught.

In defense of the one-way analysis of variance approach comparing two treatments at a time, it should be noted that there are situations in which performing one small experiment after another has advantages. For example, if the second factor is dosage of a chemical that may or may not be lethal (instead of cholesterol dosage), it may be better to begin with several small one-way experiments, each comparing just two dosage levels. Sometimes a series of small one-way experiments is advantageous when little is known about the response being measured and the investigator does not at first know how to pick levels or sample sizes.

In the past, one-way analysis of variance had some advantage over the two and three-way designs in that missing values seemed to be a less vexing problem. Observations lost because of errors or accidents could be simply omitted and the analysis done with unequal numbers. With the two- and three-way designs, the situation presented difficulties when more than a few observations are missing. Now, however, when many observations are lost, the analysis can still be carried out using programs such as BMD05V or BMDX64 with professional help. If just a few observations are missing, the mean for that treatment combination can be substituted for the missing observation and a degree of freedom subtracted from the residual degrees of freedom. As mentioned earlier in the chapter, with a moderate number of observations missing, it is possible to analyze the means of the cells, assuming that there is no interaction. Finally, if all else fails, we can analyze the two- or three-way design as if it were a one-way design and make comparisons among different treatment combinations. In the example, if a number of rats died and we were convinced that their deaths were not caused by diet or cholesterol dosage, we would omit the missing observations rather than replace them. We could then compare any two treatments and make an overall F test of whether there are differences in liver cholesterol among the nine treatments.

Thus the two and three-way analysis of variance designs are among the most widely used designs. They possess a happy degree of complexity, combining ease in interpretation with the ability to study several factors simultaneously.

REVIEW OF DESIGNS

It is comparatively easy to understand one model at a time in the analysis of variance. It is more difficult to decide on an appropriate model in practice, however, and the student may initially experience difficulty in distinguishing between nested and crossed classifications. In this section, therefore, we summarize the models already given and add slightly more complicated models. We also show how a computer program for a cross classification can be used for other designs.

CROSSED AND NESTED CLASSIFICATIONS

It is often impossible to decide between crossed and nested classifications by looking at the data. Let us suppose, for example, that four treatments are studied at each of two hospitals, with 10 patients given each treatment at each hospital. From this information, the appropriate model may be either a two-way cross classification or a one-way nested classification, depending on the number of hospitals involved. If the data were gathered from just two hospitals (i.e., of 40 patients at each hospital, 10 were assigned randomly to each treatment), a two-way classification is appropriate. Hospitals are crossed with treatment, and the model is as given in (7.1).

Another possible way of evaluating four treatments might be to choose eight hospitals rather than two, allowing the use of each treatment on 10 patients at each of two hospitals. In this case, hospitals are nested within treatment, and the model is as given in (6.6). Here it is no longer important to talk about the average effect of, say, the first hospital, for there are four first hospitals, and being a "first" hospital is of no significance. Instead, we speak of the effect of the first hospital that used a particular treatment. In the foregoing examples we have $a=4, b=2, n=10$. If the number of hospitals used in the study is 2 (b), then hospitals have been crossed with treatment; the same hospitals were used for each treatment. If the number of hospitals used is 8 (ab), hospitals are nested within treatment; different hospitals were used for each treatment.

We can consider the one-way nested design as a two-way design in which we estimate $\beta_j+(\alpha\beta)_{ij}=\beta_{(i)j}$ instead of β_j and $(\alpha\beta)_{ij}$ separately. For example, $\beta_{(3)1}$ is the effect of the first hospital assigned to treatment 3. We estimate it rather than breaking it into two parts—β_3, the average effect of the two hospitals on treatment 3, and $(\alpha\beta)_{31}$, the difference between $\beta_{(3)1}$ and β_3. We have

$$
\begin{aligned}
Y_{ijk}-\mu &= \alpha_i + \underbrace{\beta_j + (\alpha\beta)_{ij}} + \varepsilon_{(ij)k} \\
&= \alpha_i + \beta_{(i)j} + \varepsilon_{(ij)k}. \qquad (7.11)
\end{aligned}
$$

In terms of sample estimates, these are

$$
\begin{aligned}
Y_{ijk}-\bar{Y}_{\ldots} &= (\bar{Y}_{i..}-\bar{Y}_{\ldots}) + \underbrace{(\bar{Y}_{.j.}-\bar{Y}_{\ldots}) + (\bar{Y}_{ij.}-\bar{Y}_{i..}-\bar{Y}_{.j.}+\bar{Y}_{\ldots})} + (Y_{ijk}-\bar{Y}_{ij.}) \\
&= (\bar{Y}_{i..}-\bar{Y}_{\ldots}) \quad + \quad (\bar{Y}_{ij.}-\bar{Y}_{i..}) \quad + \quad (Y_{ijk}-\bar{Y}_{ij.}). \qquad (7.12)
\end{aligned}
$$

It follows that we can obtain the sums of squares and degrees of freedom for the nested design by adding the two corresponding terms in the two-way cross classification.

Suppose, on the other hand, that the two factors are crossed but we decide that there is no interaction; in this case, we combine the AB with

TABLE 7.13. SUMS OF SQUARES AND DEGREES OF FREEDOM FOR TWO-WAY CROSSED AND ONE-WAY NESTED DESIGNS

	Two-Way Cross Classification				One-Way Nested Design	
	No interaction		Interaction			
	SS	d.f.	SS	d.f.	SS	d.f.
A	$bn\sum_i(\bar{Y}_{i..}-\bar{Y}_{...})^2$	$a-1$	$bn\sum_i(\bar{Y}_{i..}-\bar{Y}_{...})^2$	$a-1$	$bn\sum_i(\bar{Y}_{i..}-\bar{Y}_{...})^2$	$a-1$
B	$an\sum_j(\bar{Y}_{.j.}-\bar{Y}_{...})^2$	$b-1$	$an\sum_j(\bar{Y}_{.j.}-\bar{Y}_{...})^2$	$b-1$	$n\sum\sum_{ij}(\bar{Y}_{ij.}-\bar{Y}_{i..})^2$ (B + AB)	$a(b-1)$
AB			$n\sum\sum_{ij}(\bar{Y}_{ij.}-\bar{Y}_{i..}-\bar{Y}_{.j.}+\bar{Y}_{...})^2$	$(a-1)(b-1)$		
Residual	$\sum\sum\sum_{ijk}(Y_{ijk}-\bar{Y}_{i..}-\bar{Y}_{.j.}+\bar{Y}_{...})^2$	$abn-a-b+1$	$\sum\sum\sum_{ijk}(Y_{ijk}-\bar{Y}_{ij.})^2$	$ab(n-1)$	$\sum\sum\sum_{ijk}(Y_{ijk}-\bar{Y}_{ij.})^2$	$ab(n-1)$
Total	$\sum\sum\sum_{ijk}(Y_{ijk}-\bar{Y}_{...})^2$	$abn-1$	$\sum\sum\sum_{ijk}(Y_{ijk}-\bar{Y}_{...})^2$	$abn-1$	$\sum\sum\sum_{ijk}(Y_{ijk}-\bar{Y}_{...})^2$	$abn-1$

the residual terms. Table 7.13 gives the sums of squares and degrees of freedom columns for these three designs.

Table 7.13 illustrates how the user of a computer program can proceed after he has decided on the appropriate model. Since he has two factors (treatment and hospital), he uses the program for a two-way cross classification. If hospitals *are* crossed with treatments (b hospitals involved in the study), he may have reason to expect no interaction; he then combines the AB with residual terms. If there are ab hospitals involved, he combines the B and AB terms as printed out for him by the program.

The calculations for several designs can be performed using a three-way cross-classification program. For example, in an experiment using a sprays, b fertilizers, c trees, and n measurements of the response, we would have a three-way cross classification if the same c trees could somehow be used with all ab spray–fertilizer combinations. If, as seems more likely, abc trees are used, with c different trees for each spray–fertilizer combination, we have a two-way cross classification with trees nested within treatment. The expected value of each response must be

$$EY_{ijkl} = \mu + \alpha_i + \beta_j + (\alpha\beta)_{ij} + \gamma_{(ij)k} + \varepsilon_{(ijk)l}, \tag{7.13}$$

and by comparing this with (7.9) we see that $\gamma_{(ij)k}$ replaces $\gamma_k + (\alpha\gamma)_{ik} + (\beta\gamma)_{jk} + (\alpha\beta\gamma)_{ijk}$. The estimate of $\gamma_{(ij)k}$ must be the sum of corresponding estimates; thus we have

$$\begin{aligned}\gamma_{(ij)k} \sim & \left(\bar{Y}_{..k.} - \bar{Y}_{....}\right) + \left(\bar{Y}_{i.k.} - \bar{Y}_{i...} - \bar{Y}_{..k.} + \bar{Y}_{....}\right) \\ & + \left(\bar{Y}_{.jk.} - \bar{Y}_{.j..} - \bar{Y}_{..k.} + \bar{Y}_{....}\right) \\ & + \left(\bar{Y}_{ijk.} - \bar{Y}_{ij..} - \bar{Y}_{i.k.} - \bar{Y}_{.jk.} + \bar{Y}_{i...} + \bar{Y}_{.j..} + \bar{Y}_{..k.} - \bar{Y}_{....}\right) = \bar{Y}_{ijk.} - \bar{Y}_{ij..}.\end{aligned}$$

We check this estimate of $\gamma_{(ij)k}$ with the estimate of $\varepsilon_{(ij)k}$ in the two-way classification and find them to be the same except that now we have an extra dot from averaging over the n observations in the subsample. To obtain the among tree sum of squares, we use the three-way program and add the four sums of squares corresponding to C, AC, BC, and ABC. The degrees of freedom are also obtained by adding the four corresponding degrees of freedom.

The same program could be used for a one-way block design twice nested. Here we might have a sprays used on b trees (ab trees altogether). If c leaves are picked from each tree (a total of abc leaves) and n determinations are made on each leaf ($abcn$ determinations), trees are

TABLE 7.14. PARAMETERS IN $EY-\mu$ AND NUMBERS OF UNITS INVOLVED IN SEVERAL DESIGNS

Design								
1-way	α_i a sprays							$\varepsilon_{(i)j}$ an trees
1-way plus block, n=1	α_i a sprays	β_j b plots						ε_{ij} ab subplots
2-way, crossed	α_i a sprays	β_j b fert	$(\alpha\beta)_{ij}$					$\varepsilon_{(ij)k}$ abn trees
1-way, nested	α_i a sprays	$\beta_{(i)j}$ ab trees						$\varepsilon_{(ij)k}$ abn leaves
2-way, block n=1	α_i a sprays	β_j b fert	$(\alpha\beta)_{ij}$	γ_k c plots				ε_{ijk} abc subplots
3-way, crossed	α_i a sprays	β_j b fert	$(\alpha\beta)_{ij}$	γ_k c plots	$(\alpha\gamma)_{ik}$	$(\beta\gamma)_{jk}$	$(\alpha\beta\gamma)_{ijk}$	$\varepsilon_{(ijk)\ell}$ abcn subplots
2-way, nested	α_i a sprays	β_j b fert	$(\alpha\beta)_{ij}$	$\gamma_{(ij)k}$ abc trees				$\varepsilon_{(ijk)\ell}$ abcn leaves
1-way, twice nested	α_i a sprays	$\beta_{(i)j}$ ab trees		$\gamma_{(ij)k}$ abc leaves				$\varepsilon_{(ijk)\ell}$ abcn determinations

nested within sprays and leaves are nested within trees. The model is

$$EY_{ijkl} = \mu + \alpha_i + \beta_{(i)j} + \gamma_{(ij)k} + \varepsilon_{(ijk)l}. \tag{7.14}$$

We use this model to illustrate a rule for obtaining nested sums of squares.

RULE FOR NESTED SUMS OF SQUARES

To obtain the sum of squares corresponding to any nested effect (say, $\gamma_{(ij)k}$) we form all fully crossed effects from the entire set of subscripts (i,j,k) containing all the subscripts outside the parentheses—in this case, k. Thus $\gamma_{(ij)k} = \gamma_k + (\alpha\gamma)_{ik} + (\beta\gamma)_{jk} + (\alpha\beta\gamma)_{ijk}$. Similarly, $\beta_{(i)j} = \beta_j + (\alpha\beta)_{ij}$.

Table 7.14, which presents the parameters (other than μ) in EY for the designs mentioned thus far, may be helpful. Problems 7.6 to 7.9 are designed as practice in discriminating among models.

SUMMARY

In this chapter, we have presented cross-classification designs for two and three factors. The designs are commonly employed and have been given in detail. Their use allows the investigator to study several factors simultaneously and can result in an appreciable saving of observations as compared with several separate one-way analyses of variance.

REFERENCES

APPLIED STATISTICS

7.1. Bennett, C. A., and N. L. Franklin, *Statistical Analysis in Chemistry and the Chemical Industry*, Wiley, New York, 1954.

7.2. Brownlee, K. A., *Statistical Theory and Methodology in Science and Engineering*, 2nd ed., Wiley, New York, 1965.

7.3. Davies, O. L. (Ed.), *The Design and Analysis of Industrial Experiments*, 2nd ed., Oliver & Boyd, Edinburgh, 1956.

7.4. Dixon, W. J., and F. J. Massey, Jr., *Introduction to Statistical Analysis*, 3rd ed., McGraw-Hill, New York, 1969.

7.5. Fisher, R. A., *The Design of Experiments*, 8th ed., Oliver & Boyd, Edinburgh, 1966.

7.6. Kempthorne, O., *The Design and Analysis of Experiments*, Wiley, New York, 1952.

7.7. Ostle, B., *Statistics in Research*, Iowa State University Press, Ames, Iowa, 1963.

7.8. Snedecor, G. W., and W. G. Cochran, *Statistical Methods*, 6th ed., Iowa State College Press, Ames, Iowa, 1963.

MATHEMATICAL STATISTICS

7.9. Chew, V. (Ed.), *Experimental Design in Industry*, Wiley, New York, 1958.

7.10. Scheffé, H., *The Analysis of Variance*, Wiley, New York, 1967.

7.11. Töcher, K. D., "The Design and Analysis of Block Experiments," *Journal of Royal Statistical Society, Series B*, Vol. 14 (1952), pp. 45–100.

7.12. Gaylor, D. W., and F. N. Happer, "Estimating Degrees of Freedom for Linear Combinations of Mean Squares by Satterthwaite's Formula," *Technometrics*, Vol. 11 (1969), pp. 691–706.

7.13. Tukey, John W., "One Degree of Freedom for Nonadditivity," *Biometrics*, Vol. 5 (1949), pp. 232–242.

PROBLEMS

7.1. Four levels of fertilizer were used in a field experiment, with and without irrigation. The eight treatment combinations were assigned at random to eight plots. Barley yields were in bushels per acre.

	Level of Fertilizer			
Irrigation	None	Low	Medium	High
No	317	341	354	329
Yes	275	304	334	380

(*a*) State the appropriate model.
(*b*) Fill in an analysis of variance table.
(*c*) Test whether fertilizers make any difference.
(*d*) Estimate the overall difference in yield due to irrigation (point estimate and confidence interval estimate).
(*e*) Estimate the mean difference in yield between no fertilizer and fertilizer at a low level (point estimate and confidence interval estimate).

7.2. Eighteen adult males were used in a study to compare the sphygmomanometers from three manufacturers. The subjects were assigned at random into six groups of three each. Three groups had systolic blood pressure measurements (mm Hg) made at entry to the experiment; the other three groups were measured after resting 10 minutes. The data were as follows:

	Sphygmomanometer Manufacturer		
Resting	A	B	C
No	147	156	127
	124	127	122
	113	155	153
Yes	150	110	124
	140	150	149
	122	115	136

Preliminary calculations gave the following results:

$\sum_k Y_{11k}=$ 384.	$\sum_k Y_{12k}=$ 438.	$\sum_k Y_{13k}=$ 402.	$\sum_{jk} Y_{1jk}=$ 1224.
$\sum_k Y_{11k}^2=$ 49754.	$\sum_k Y_{12k}^2=$ 64490.	$\sum_k Y_{13k}^2=$ 54422.	$\sum_{jk} Y_{1jk}^2=$ 168666.
$\sum_k Y_{21k}=$ 412.	$\sum_k Y_{22k}=$ 375.	$\sum_k Y_{23k}=$ 409.	$\sum_{jk} Y_{2jk}=$ 1196.
$\sum_k Y_{21k}^2=$ 56984.	$\sum_k Y_{22k}^2=$ 47825.	$\sum_k Y_{23k}^2=$ 56073.	$\sum_{jk} Y_{2jk}^2=$ 160882.
$\sum_{ik} Y_{i1k}=$ 796.	$\sum_{ik} Y_{i2k}=$ 813.	$\sum_{ik} Y_{i3k}=$ 811.	$\sum_{ijk} Y_{ijk}=$ 2420.
$\sum_{ik} Y_{i1k}^2=$ 106738.	$\sum_{ik} Y_{i2k}^2=$ 112315.	$\sum_{ik} Y_{i3k}^2=$ 110495.	$\sum_{ijk} Y_{ijk}^2=$ 329548.

$i=1,2$ (resting)
$j=1,2,3$ (manufacturer)
$k=1,2,3$ (replicate)

(*a*) State the model.
(*b*) Fill in an analysis of variance table.
(*c*) Test whether there are differences in blood pressure measurements among the three manufacturers.
(*d*) Test whether there is a difference in blood pressure between resting time periods.
(*e*) Find a 95% confidence interval for σ_e^2.
(*f*) Give confidence intervals for $\alpha_1-\alpha_2$, $\beta_1-\beta_3$, using an overall level of .99.

7.3. Thirty-six adults (18 males and 18 females) were used in a study to compare the sphygmomanometers from three manufacturers. Subjects from each sex were assigned at random into six groups of three each. Three groups from each sex had systolic blood pressure measurements made at entry to the experiment; the other three groups were measured after resting 10 minutes. The data were as follows:

Sphygmomanometer

	A		*B*		*C*	
Resting	M	F	M	F	M	F
No	147	122	156	131	127	110
	124	142	127	133	122	115
	113	136	155	146	153	105
Yes	140	108	100	141	114	103
	130	151	140	125	139	135
	112	138	105	139	126	114

Preliminary calculations were as follows:

	$(111l)$	$(112l)$	$(121l)$	$(122l)$	$(131l)$	$(132l)$
$\sum Y$	384.	400.	438.	410.	402.	330.
$\sum Y^2$	49754.	53544.	64490.	56166.	54422.	36350.
	$(211l)$	$(212l)$	$(221l)$	$(222l)$	$(231l)$	$(232l)$
$\sum Y$	382.	397.	345.	405.	379.	352.
$\sum Y^2$	49044.	53509.	40625.	54827.	48193.	41830.

$i=1,2$ (resting)
$j=1,2,3$ (manufacturer)
$k=1,2$ (sex)
$l=1,2,3$ (replicate)

(*a*) State the model and appropriate assumptions.
(*b*) Make an analysis of variance table.
(*c*) Test whether there are differences in blood pressure measurement among the three manufacturers.
(*d*) Test whether there is a difference in blood pressure between resting and no resting.
(*e*) Test whether there is a difference in blood pressure between males and females.
(*f*) Test H_0: No Interaction in blood pressure between resting and sex.
(*g*) Make a 95% confidence interval for σ_e^2.

7.4. Brain weights measured on 28 individuals were as follows:

	Weight (g)	
Age (years)	Men	Women
50–80	1312	1211
	1323	1196
	1325	1207
	1318	1198
	1319	1204
	1342	1191
	1309	1205
20–49	1331	1234
	1330	1204
	1335	1222
	1327	1211
	1338	1228
	1338	1217
	1335	1223

Preliminary calculations were as follows:

$\sum_k Y_{11k}=$	9,248	$\sum_k Y_{12k}=$	8,412	$\sum_{jk} Y_{1jk}=$	17,660
$\sum_k Y_{11k}^2=$	12,218,628	$\sum_k Y_{12k}^2=$	10,109,112	$\sum_{jk} Y_{1jk}^2=$	22,327,740
$\sum_k Y_{21k}=$	9,334	$\sum_k Y_{22k}=$	8,539	$\sum_{jk} Y_{2jk}=$	17,873
$\sum_k Y_{21k}^2=$	12,446,328	$\sum_k Y_{22k}^2=$	10,416,979	$\sum_{jk} Y_{2jk}^2=$	22,863,307
$\sum_{ik} Y_{i1k}=$	18,582	$\sum_{ik} Y_{i2k}=$	16,951	$\sum_{ijk} Y_{ijk}=$	35,533
$\sum_{ik} Y_{i1k}^2=$	24,644,956	$\sum_{ik} Y_{i2k}^2=$	20,526,091	$\sum_{ijk} Y_{ijk}^2=$	45,191,047

$i=1,2$ (age group)
$j=1,2$ (sex)
$k=1,\ldots,7$

(*a*) The appropriate model is $Y_{ijk}=\mu+\alpha_i+\beta_j+(\alpha\beta)_{ij}+\varepsilon_{ijk}$, where

$$\sum_1^2 \alpha_i=0, \qquad \sum_1^2 \beta_j=0, \qquad \sum_{j=1}^2 (\alpha\beta)_{ij}=\sum_{i=1}^2 (\alpha\beta)_{ij}=0, \qquad \varepsilon_{ijk}\ \text{IND}(0,\sigma_e^2).$$

What assumptions are implicit in the model, in terms of the problem?

(*b*) Fill in the analysis of variance table.

(*c*) Make F tests ($\alpha=.05$) of $H_0:\sigma_{ab}^2=0$, $H_0:\sigma_a^2=0$, and $H_0:\sigma_b^2=0$.

(*d*) Make t tests ($\alpha=.05$) of $H_0:(\alpha\beta)_{11}=0$, $H_0:\alpha_1=0$, and $H_0:\beta_1=0$. Note the similarity between the tests in (*c*) and (*d*).

(*e*) Estimate by point estimates and by 95% single level confidence intervals the following quantities: α_1, β_1, $(\alpha\beta)_{11}$, and $\mu+\alpha_1+\beta_2+(\alpha\beta)_{12}$.

7.5. In a study of intelligence of children with heart disease of both acyanotic and cyanotic types, changes in IQ from the first test available to the last available test were classified as follows:

Surgery	Acyanotic	Cyanotic
No	9	2
	−1	1
	−10	−4
	3	−5
	−2	0
Yes	−7	5
	−7	10
	−12	9
	−13	2
	−12	15

(*a*) What model is appropriate here? Give the assumptions underlying the model.
(*b*) Fill in the analysis of variance table.
(*c*) List possible parameters or linear combinations of parameters which the experimenter might wish to estimate. Give their point estimates.
(*d*) Give expressions for the variances of the estimates in (*c*).
(*e*) Give a 95% confidence interval for σ_e^2.

For each of Problems 7.6 to 7.9, answer the following questions:
(*a*) What is an appropriate model?
(*b*) What type of computer program can you use (one-way, two-way, etc.)?
(*c*) Give the source of variation, degrees of freedom, and expected mean square columns of the analysis of variance table.

7.6. Six litters, each consisting of three rats, were available to study three treatments. The treatments were assigned at random within each litter; thus one rat in each litter received each treatment.

7.7. Six litters, each consisting of four rats, were available to study two treatments for two time periods. In each litter, two rats were assigned to each treatment; of the two rats on the same treatment, one was sacrificed after one week and the response measured, the other rat was sacrificed after one month.

7.8. Twelve animals were assigned at random to each of two treatments, with six on each treatment. The animals on each treatment were sacrificed as follows: two after one week, two after two weeks, and two after one month.

7.9. Twelve animals were assigned to each of two treatments, with six on each treatment. Three animals on each treatment were sacrificed after one month, and three after two months. Two determinations were made on each animal.

**CHAPTER 8

FACTORIAL DESIGNS WITH ALL FACTORS AT TWO LEVELS

In this chapter we study factorial designs involving k factors with each factor restricted to two levels. This, the simplest type of factorial design, has been thoroughly developed. Concepts that are somewhat difficult in the more general model are readily understood when each factor has only two levels. In the latter part of the chapter we use this design to illustrate confounding of effects. For additional material on two-level factorial designs and also on factorial designs at more levels, see Davies [8.3], Cochran and Cox [8.2], Bennett and Franklin [8.1], or Kempthorne [8.4].

The factorial design in which there are two levels for each factor is called the 2^k design. It is useful in practice when the investigator wishes to study many factors simultaneously, keeping the number of treatment combinations as small as possible. Frequently a series of small two-level factorial experiments performed sequentially is more satisfactory than a single huge experiment. The later experiments in the series can be designed using information gained from the earlier analyses. Small laboratory experiments can be performed quickly while conditions such as personnel and equipment remain constant. In addition, interest on the part of personnel can be kept high when fast feedback is available. Levels for the various factors can be selected realistically in the later experiments—it is disappointing to find, at the close of a large experiment, that many observations were taken at improperly chosen levels.

EXAMPLE

As our basic example, we use a factorial design having just three factors, each at two levels, and one observation at each treatment combination (i.e., eight observations altogether). This example is large enough to illustrate the techniques.

For this example, Model I as given in Chapter 7 is ($a=b=c=2; n=1$):

$$Y_{ijk}=\mu+\alpha_i+\beta_j+\gamma_k+(\alpha\beta)_{ij}+(\alpha\gamma)_{ik}+(\beta\gamma)_{jk}+(\alpha\beta\gamma)_{ijk}+\varepsilon_{ijk}, \quad (8.1)$$

where

$$\sum_{i=1}^{2}\alpha_i=\sum_{j=1}^{2}\beta_j=\sum_{k=1}^{2}\gamma_k=\sum_{i=1}^{2}(\alpha\beta)_{ij}=\sum_{j=1}^{2}(\alpha\beta)_{ij}=\sum_{j=1}^{2}(\beta\gamma)_{jk}=\sum_{k=1}^{2}(\beta\gamma)_{jk}$$

$$=\sum_{i=1}^{2}(\alpha\gamma)_{ik}=\sum_{k=1}^{2}(\alpha\gamma)_{ik}=\sum_{i=1}^{2}(\alpha\beta\gamma)_{ijk}=\sum_{j=1}^{2}(\alpha\beta\gamma)_{ijk}=\sum_{k=1}^{2}(\alpha\beta\gamma)_{ijk}$$

$$=0$$

and ε_{ijk} IND$(0,\sigma_e^2)$. The subscript l in (7.9) is now unnecessary because n equals one.

The calculations for the estimates of the various parameters and for the various sums of squares in the analysis of variance table can be considerably simplified because $a=b=c=2$. In the first place, we now need to estimate only one main effect for each factor, only one interaction for each pair of factors, and only one interaction for the three factors. If α_2 has been estimated, then from $\sum_{i=1}^{2}\alpha_i=0$ we have $\alpha_1=-\alpha_2$; thus the estimate for α_1 is simply the negative of the estimate for α_2. Similarly, $(\alpha\beta)_{11}=+(\alpha\beta)_{22}$, $(\alpha\beta)_{21}=-(\alpha\beta)_{22}$, $(\alpha\beta)_{12}=-(\alpha\beta)_{22}$, $(\alpha\beta\gamma)_{122}=-(\alpha\beta\gamma)_{222}$, and so on.

Before proceeding to the simpler calculations, we should mention that in the model as presented in (8.1), with only one observation on each treatment combination, we have no estimate of σ_e^2; we cannot, therefore, make tests or confidence intervals for parameters. We can, however, write down point estimates for all parameters except σ_e^2 from the eight observations. To make tests and confidence intervals with just these eight observations, we must modify the model by assuming that certain interactions are zero.

We return now to Table 7.12 for point estimates of the various parameters. Rather than make the calculations using the formula as given, we write out each mean in terms of the original observations. We have

$$\mu\sim\bar{Y}_{...}=\tfrac{1}{8}(Y_{111}+Y_{211}+Y_{121}+Y_{221}+Y_{112}+Y_{212}+Y_{122}+Y_{222}) \quad (8.2)$$

$$\alpha_2 \sim \bar{Y}_{2..} - \bar{Y}_{...} = \frac{Y_{211}+Y_{221}+Y_{212}+Y_{222}}{4}$$

$$-\frac{Y_{111}+Y_{211}+Y_{121}+Y_{221}+Y_{112}+Y_{212}+Y_{122}+Y_{222}}{8}$$

$$=\tfrac{1}{8}(-Y_{111}+Y_{211}-Y_{121}+Y_{221}-Y_{112}+Y_{212}-Y_{122}+Y_{222})$$

$$(\alpha\beta)_{22} \sim \bar{Y}_{22.} - \bar{Y}_{2..} - \bar{Y}_{.2.} + \bar{Y}_{...} = \frac{Y_{221}+Y_{222}}{2}$$

$$-\frac{Y_{211}+Y_{221}+Y_{212}+Y_{222}}{4} - \frac{Y_{121}+Y_{221}+Y_{122}+Y_{222}}{4}$$

$$+\frac{Y_{111}+Y_{211}+Y_{121}+Y_{221}+Y_{112}+Y_{212}+Y_{122}+Y_{222}}{8}$$

$$=\tfrac{1}{8}(Y_{111}-Y_{211}-Y_{121}+Y_{221}+Y_{112}-Y_{212}-Y_{122}+Y_{222}).$$

Similarly, we have

$$(\alpha\beta\gamma)_{222} \sim \tfrac{1}{8}(-Y_{111}+Y_{211}+Y_{121}-Y_{221}+Y_{112}-Y_{212}-Y_{122}+Y_{222}).$$

Expressions for β_2, γ_2, $(\alpha\gamma)_{22}$, and $(\beta\gamma)_{22}$ can be written by simply changing subscripts on the observations. Estimates of $\alpha_1, \ldots, (\alpha\beta\gamma)_{111}$ follow immediately.

We have now considerably reduced the calculations necessary for estimating the parameters. They can be performed by hand. We can also set down the various sums of squares for an analysis of variance table from these expressions. We need only square each estimate and multiply it by 8, the number of observations. For example, the due A sum of squares is

$$\mathrm{SS}_A = 8\left[\tfrac{1}{8}(-Y_{111}+Y_{211}-Y_{121}+Y_{221}-Y_{112}+Y_{212}-Y_{122}+Y_{222})\right]^2. \tag{8.3}$$

In each sum of squares, half the numbers to be squared have minus signs.

SYMBOLIC NOTATION

Thus far we have used the same notation as in Chapter 7. In this section we introduce some new notation particularly well adapted for the 2^k

design. First let us make one additional change. Instead of estimating, for example, α_1 and α_2, we estimate the difference $\alpha_2-\alpha_1$. When there are more than two levels of the A factor, it is reasonable to speak of the main effect of treatment 1, the main effect of treatment 2, and so on. With just two levels, however, we define the main effect of A by the difference between the high and the low level of A, averaged over the levels of B and C. The three main effects are called A, B, and C:

$$\mathsf{A}=\alpha_2-\alpha_1=2\alpha_2$$

[note that $\alpha_1+\alpha_2=0$; thus $\alpha_1=-\alpha_2$ and $\alpha_2-(-\alpha_2)=2\alpha_2$]

$$\mathsf{B}=2\beta_2$$

$$\mathsf{C}=2\gamma_2.$$

The interactions are named similarly. In terms of the separate $(\alpha\beta)_{ij}$ effects, we define the interaction of A and B by the difference between the AB interaction when both A and B are at a high level and when one factor is at a low level. That is, the interactions between pairs of factors are defined by

$$(\mathsf{AB})=(\alpha\beta)_{22}-(\alpha\beta)_{21}=2(\alpha\beta)_{22}$$

$$(\mathsf{BC})=2(\beta\gamma)_{22}$$

$$(\mathsf{AC})=2(\alpha\gamma)_{22}.$$

The interaction term (ABC) is defined by

$$(\mathsf{ABC})=(\alpha\beta\gamma)_{222}-(\alpha\beta\gamma)_{221}$$

$$=2(\alpha\beta\gamma)_{222}.$$

The three factors are denoted as usual by A, B, and C. In an agricultural experiment these may be amounts of nitrogen, potash, and phosphate in fertilizer; in a chemical experiment they may be temperature, pressure, and time. It is convenient to name the eight treatment combinations. In each name, the low level of any factor is denoted by (1), and the high level of the factor by a small letter (a, b, or c). With this convention, and omitting unnecessary (1)'s, the names of the treatment combinations are as in Table 8.1.

In Chapter 7 we used a, b, and c for the number of levels of each factor, but since we no longer need them for this purpose, no confusion results from using them in the names of the treatments. We observe that ab is *not*

TABLE 8.1. DESIGNATION OF TREATMENT COMBINATIONS

		C Low		C High	
		B		B	
		Low	High	Low	High
A	Low	(1) or ---	b or -+-	c or --+	bc or -++
	High	a or +--	ab or ++-	ac or +-+	abc or +++

the product of a and b. It is the name of the treatment consisting of the A factor at the high level, B at the high level, and C at the low level.

Another designation for the treatments consists of using plus signs to indicate a high level and minus signs to indicate a low level. These treatment names are also given in Table 8.1. Thus $++-$ is another way of representing treatment ab. As a numerical example of the 2^3 design, we use the data in Table 8.2.

It is convenient to use a symbolic notation in which we insert the treatment names $[(1),\ldots,abc]$ to replace the observations on the treatment. We return now to the parameters (8.2) which we wish to estimate. We use A for the estimate of A, (AB) for the estimate of (AB), and so forth.

$$2\mu \sim 2\bar{Y}_{\ldots} = \tfrac{1}{4}(\bar{Y}_{111} + Y_{211} + Y_{121} + Y_{221} + Y_{112} + Y_{212} + Y_{122} + Y_{222}).$$

Symbolically, we have

$$2\bar{Y}_{\ldots} = \tfrac{1}{4}[(1) + a + b + ab + c + ac + bc + abc]$$

$$= \tfrac{1}{4}(a+1)(b+1)(c+1). \tag{8.4}$$

The "product" $(a+1)(b+1)(c+1)$ is symbolic: algebraic multiplication of the three factors yields a formula for $2\bar{Y}_{\ldots}$.

Similarly, and again symbolically, we can write

$$\mathsf{A} \sim A = \tfrac{1}{4}(-Y_{111} + Y_{211} - Y_{121} + Y_{221} - Y_{112} + Y_{212} - Y_{122} + Y_{222})$$

$$= \tfrac{1}{4}[-(1) + a - b + ab - c + ac - bc + abc]$$

$$= \tfrac{1}{4}(a-1)(b+1)(c+1) \tag{8.5}$$

TABLE 8.2. NUMERICAL EXAMPLE

		C Low		C High	
		B		B	
		Low	High	Low	High
A	Low	(1):5	b:10	c:10	bc:15
	High	a:6	ab:12	ac:12	abc:20

$$\mathsf{B}\sim B=\tfrac{1}{4}(a+1)(b-1)(c+1)$$

$$\mathsf{C}\sim C=\tfrac{1}{4}(a+1)(b+1)(c-1)$$

$$(\mathsf{AB})\sim(AB)=\tfrac{1}{4}(Y_{111}-Y_{211}-Y_{121}+Y_{221}+Y_{112}-Y_{212}-Y_{122}+Y_{222})$$

$$=\tfrac{1}{4}[(1)-a-b+ab+c-ac-bc+abc]$$

$$=\tfrac{1}{4}(a-1)(b-1)(c+1)$$

$$(\mathsf{AC})\sim(AC)=\tfrac{1}{4}(a-1)(b+1)(c-1)$$

$$(\mathsf{BC})\sim(BC)=\tfrac{1}{4}(a+1)(b-1)(c-1)$$

$$(\mathsf{ABC})\sim(ABC)=\tfrac{1}{4}(a-1)(b-1)(c-1).$$

The symbolic notation serves as a convenience in remembering which observations have plus signs and which have minus signs in forming an effect; it cannot be used for calculations. Note that the symbolic expressions for estimating A, AB, AC, and ABC all contain the factor $a-1$, whereas the expressions estimating B, C, and BC contain $a+1$. For the example, we have

$$2\overline{Y}_{\dots}=\tfrac{1}{4}(a+1)(b+1)(c+1)=\tfrac{1}{4}[(1)+a+b+ab+c+ac+bc+abc]$$

$$=\tfrac{1}{4}(5+6+10+12+10+12+15+20)=\tfrac{90}{4}$$

$$A = \tfrac{1}{4}(a-1)(b+1)(c+1) = \tfrac{1}{4}[-(1)+a-b+ab-c+ac-bc+abc]$$

$$= \tfrac{1}{4}(-5+6-10+12-10+12-15+20) = \tfrac{10}{4}$$

$$B = \tfrac{1}{4}[-(1)-a+b+ab-c-ac+bc+abc]$$

$$= \tfrac{1}{4}(-5-6+10+12-10-12+15+20) = \tfrac{24}{4}$$

$$(AB) = \tfrac{1}{4}(a-1)(b-1)(c+1) = \tfrac{1}{4}[(1)-a-b+ab+c-ac-bc+abc]$$

$$= \tfrac{1}{4}(5-6-10+12+10-12-15+20) = \tfrac{4}{4}$$

$$C = \tfrac{1}{4}[-(1)-a-b-ab+c+ac+bc+abc]$$

$$= \tfrac{1}{4}(-5-6-10-12+10+12+15+20) = \tfrac{24}{4}$$

$$(AC) = \tfrac{1}{4}[(1)-a+b-ab-c+ac-bc+abc]$$

$$= \tfrac{1}{4}(5-6+10-12-10+12-15+20) = \tfrac{4}{4}$$

$$(BC) = \tfrac{1}{4}[(1)+a-b-ab-c-ac+bc+abc]$$

$$= \tfrac{1}{4}(5+6-10-12-10-12+15+20) = \tfrac{2}{4}$$

$$(ABC) = \tfrac{1}{4}(a-1)(b-1)(c-1) = \tfrac{1}{4}[-(1)+a+b-ab+c-ac-bc+abc]$$

$$= \tfrac{1}{4}(-5+6+10-12+10-12-15+20) = \tfrac{2}{4}.$$

Table 8.3 summarizes the signs that are used on each observation in forming any effect. We note that a main effect column such as A has plus signs where a occurs in the "treatment" column and minus signs elsewhere. Furthermore, the proper sign for any interaction column can be obtained by "multiplying" the signs in the main effects columns belonging to that interaction, observing the convention that two like signs "multiply" to give a plus and two unlike signs "multiply" to give a minus: For example, to obtain any sign in the BC column, we multiply the signs in the B and C columns.

When there are k factors, the expressions used to form the various effects are like (8.4) and (8.5), with $1/2^{k-1}$ replacing $\tfrac{1}{4}$; minuses appear in each symbolic factor corresponding to a letter found in the name of the

TABLE 8.3. SIGNS FOR CALCULATING EFFECTS

		Effect							
Observation	Treatment	$2\bar{Y}_{...}$	A	B	C	AB	AC	BC	ABC
$Y_{111} = 5$	(1)	+	-	-	-	+	+	+	-
$Y_{211} = 6$	a	+	+	-	-	-	-	+	+
$Y_{121} = 10$	b	+	-	+	-	-	+	-	+
$Y_{221} = 12$	ab	+	+	+	-	+	-	-	-
$Y_{112} = 10$	c	+	-	-	+	+	-	-	+
$Y_{212} = 12$	ac	+	+	-	+	-	+	-	-
$Y_{122} = 15$	bc	+	-	+	+	-	-	+	-
$Y_{222} = 20$	abc	+	+	+	+	+	+	+	+

effect. If $k=5$, for example, we have

$$(ABDE)=\frac{1}{2^4}(a-1)(b-1)(c+1)(d-1)(e-1).$$

ORTHOGONAL LINEAR COMBINATIONS

It is useful to define orthogonal linear combinations. Two linear functions of the observations $Y_1,\ldots,Y_n$, say

$$\sum_{i=1}^{n} a_i Y_i = a_1 Y_1 + \cdots + a_n Y_n$$

and

$$\sum_{i=1}^{n} b_i Y_i = b_1 Y_1 + \cdots + b_n Y_n$$

are orthogonal if

$$\sum_{i=1}^{n} a_i b_i = 0.$$

Note from Table 8.3 that $2\bar{Y}_{...}$ and all the main effects and interaction effects are linear functions of the observations, with coefficients all equal to $\pm\frac{1}{4}$. By comparing the plus and minus signs of any two columns, we see that each pair of effects is orthogonal.

Orthogonal linear combinations of independent observations are statistically independent of one another. Knowledge of one effect gives no information about another effect. In an orthogonal design, estimates of effects are statistically independent. Orthogonality is achieved by a design balanced in the sense that all treatment combinations occur the same number of times.

COMPUTATIONAL METHOD

We now have point estimates of all parameters in the model (8.1) except σ_e^2. We can write down an analysis of variance making use of the property mentioned earlier—namely, that each observation contributes the same amount to any sum of squares. For example, from (8.3) and (8.5) the due A sum of squares is

$$SS_A = 8\left[\tfrac{1}{8}(-Y_{111}+Y_{211}-Y_{121}+Y_{221}-Y_{112}+Y_{212}-Y_{122}+Y_{222})\right]^2$$

$$=8\left[\tfrac{1}{8}(-(1)+a-b+ab-c+ac-bc+abc)\right]^2$$

$$=8\left[(\tfrac{1}{2}A)\right]^2$$

$$=2A^2.$$

Similarly, $SS_{AB}=2(AB)^2$, $SS_{ABC}=2(ABC)^2$, and so on.

For k factors, the sums of squares are $SS_A = 2^{k-2}A^2 = (1/2^k)[(a-1)(b+1)\ldots]^2$, and so on.

Table 8.4 gives the first three columns of the analysis of variance table for the data in Table 8.2.

In making confidence intervals or testing hypotheses, we usually decide at the outset of the experiment that certain population interactions (usually the higher order ones) are either zero or negligible. For example, if we decide that σ_{ab}^2, σ_{ac}^2, σ_{bc}^2, and σ_{abc}^2 are all equal or close to zero, the due AB, due AC, due BC, and due ABC sums of squares can be pooled to obtain an estimate of σ_e^2. This is accomplished in Table 8.5 by adding the four sums of squares to obtain a residual sum of squares. The corresponding degrees of freedom are also added, and their sum is the degrees of freedom for the residual.

TABLE 8.4. ANALYSIS OF VARIANCE TABLE FOR 2^3 EXAMPLE

Source of Variation (1)	SS (2)	d.f. (3)
A	$2(\frac{10}{4})^2 = \frac{100}{8} = 12.5$	1
B	$2(\frac{24}{4})^2 = \frac{576}{8} = 72.0$	1
AB	$2(\frac{4}{4})^2 = \frac{16}{8} = 2.0$	1
C	$2(\frac{24}{4})^2 = \frac{576}{8} = 72.0$	1
AC	$2(\frac{4}{4})^2 = \frac{16}{8} = 2.0$	1
BC	$2(\frac{2}{4})^2 = \frac{4}{8} = 0.5$	1
ABC	$2(\frac{2}{4})^2 = \frac{4}{8} = 0.5$	1
TOTAL	161.5	7

TABLE 8.5. ANALYSIS OF VARIANCE TABLE FOR NUMERICAL EXAMPLE

Source of Variation (1)	SS (2)	d.f. (3)	MS (4)	EMS (5)	F calc (6)	F tabled (7)
A	12.5	1	12.5	$\sigma_e^2 + 2A^2$	10.0	7.71
B	72.0	1	72.0	$\sigma_e^2 + 2B^2$	57.6	7.71
C	72.0	1	72.0	$\sigma_e^2 + 2C^2$	57.6	7.71
Residual	5.0	4	1.25	σ_e^2		
Total	161.5	7				

In Table 8.5 we have written EMS_a as $\sigma_e^2 + 2A^2$ in place of $\sigma_e^2 + nbc\ \sigma_a^2$, as in Table 7.10, because in the notation of this chapter $n=1$, $b=c=2$, and $nbc\ \sigma_a^2 = 4\sum_{i=1}^2 \alpha_i^2/(a-1) = 8\alpha^2 = 2A^2$. The three tests given in the table are three separate F tests, each having level of significance $\alpha = .05$. The three null hypotheses are $H_0: A = 0$; $H_0: B = 0$; $H_0: C = 0$, and each mean square is compared with $F[1-\alpha; 1,4] = 7.71$. All three hypotheses are rejected.

To write down confidence intervals we need only the variances of the estimates. Because every estimate consists of 8 observations, each multiplied by $\pm\frac{1}{4}$, the variance of each linear combination is simply $8(\frac{1}{4})^2\sigma_e^2 = \sigma_e^2/2$. A $1-\alpha$ confidence interval for $A = \alpha_2 - \alpha_1$ is

$$A \pm t[1-\alpha/2; 4]\sqrt{s_e^2/2}$$

In the example with $1-\alpha = .95$, we have $2.5 \pm 2.78\sqrt{1.25/2}$ or $0.3 < A < 4.7$.

For other values of k, we proceed with the analysis of variance table, F tests, and confidence intervals in a similar way. For any k, the EMS is σ_e^2 plus 2^{k-2} times the square of the corresponding population effect. For example, if $k=4$, $\text{EMS}_{AB} = \sigma_e^2 + 4(AB)^2$ and $\text{EMS}_{ABD} = \sigma_e^2 + 4(ABD)^2$. The variance of any treatment effect reduces to $\sigma_e^2/2^{k-2}$, and a $1-\alpha$ level confidence interval for (ABD), if $k=4$, is

$$(ABD) \pm t[1-\alpha/2; \nu]\sqrt{s_e^2/2^{4-2}},$$

where ν is the number of degrees of freedom for the residual sum of squares obtained by pooling.

If we decide which interactions to pool in order to obtain the residual sum of squares *after* looking at the mean squares obtained from the data, we increase considerably our chances of finding that the nonpooled (larger) effects are nonzero; thus the significance level of the test is meaningless.

YATES'S COMPUTATIONAL METHOD

Yates gives a still simpler method for calculating the effects and the sums of squares. We illustrate his method with the same numerical example. Table 8.6 sets forth the calculations. Beginning with a column of observations, systematic additions and subtractions are performed in order to obtain k columns, each column being obtained from the previous one.

TABLE 8.6. EXAMPLE OF YATES'S COMPUTATIONAL METHOD

Treatment	Observations (0)	(1)	k = 3 (2)	(3)	$(3)^2/2^3$ Sums of Squares
(1)	5	11	33	90	
a	6	22	57	10	100/8 = 12.5
b	10	22	3	24	576/8 = 72.0
ab	12	35	7	4	16/8 = 2.0
c	10	1	11	24	576/8 = 72.0
ac	12	2	13	4	16/8 = 2.0
bc	15	2	1	2	4/8 = .5
abc	20	5	3	2	4/8 = .5
	ΣY = 90				1292/8 =161.5

In Table 8.6, the first half of column 1 is obtained by adding succeeding pairs of observations in the 0 column (Y_{ijk}); the second half of column 1 is obtained by subtracting the first from the second observation in each pair. Thus $5+6=11$, $10+12=22$, $10+12=22$; $15+20=35$; $6-5=1$, $12-10=2$, $12-10=2$, and $20-15=5$. The process is repeated once to obtain column 2 from column 1 and again to obtain column 3 from column 2. The numbers in column 3 are precisely those obtained earlier for $8\bar{Y}_{\ldots}$, $4A$, $4B$, $4(AB)$, $4C$, $4(AC)$, and $4(ABC)$. This is true in general, for if we performed the additions and subtraction in terms of $(1), a, b, \ldots, abc$, we would find $(1)+a+b+ab+c+ac+bc+abc=(a+1)(b+1)(c+1)$ in the first row of column 3, and $-(1)+a-b+ab-c+ac-bc+abc=(a-1)(b+1)(c+1)$ in the second row of column 3. In order to use this procedure, the treatments must be listed in the order given. The scheme is general; for $k=4$, we add to the eight treatments listed in Table 8.6 eight more treatments—*d*, *ad*, *bd*, *abd*, *cd*, *acd*, *bcd*, *abcd*, in that order—and add a column 4.

To obtain the effects, we divide the numbers in the kth column by 2^{k-1}. To obtain the sums of squares for the analysis of variance table, we square the corresponding numbers in the kth column and then divide by 2^k.

THE CASE OF n LARGER THAN 1

In order to avoid the necessity of pooling interaction terms to estimate σ_e^2, it is sometimes possible to have several observations at each treatment combination. If, for example, k types of fertilizer are being used, each at two levels, the 2^k treatment combination might be randomly assigned to $2^k n$ small plots, n plots receiving each treatment. When n is larger than 1 and there are three factors, the model is as given in (7.9). In this case, the cell means are

$$\overline{Y}_{ijk.} = \mu + \alpha_i + \beta_j + \gamma_k + (\alpha\beta)_{ij} + (\beta\gamma)_{jk} + (\alpha\gamma)_{ik}$$

$$+ (\alpha\beta\gamma)_{ijk} + \bar{\varepsilon}_{ijk.} \tag{8.6}$$

where $\bar{\varepsilon}_{ijk.}$ IND$(0, \sigma_e^2/n)$.

We notice that the model for $\overline{Y}_{ijk.}$ is like that for Y_{ijk} from (8.1), although σ_e^2/n replaces σ_e^2. Therefore, we calculate the effects exactly as before with the symbolic notation, but with the cell means in place of the observations. Yates's computational method can be used.

In the analysis of variance table (Table 8.7), each sum of squares contains n as a factor.

F tests can be made using the residual mean square in the denominator; occasionally some of the interaction terms are pooled with the residual.

For arbitrary k, the sums of squares are $\mathrm{SS}_A = 2^{k-2} n A^2$, and so on; the expected mean squares are $\mathrm{EMS}_A = \sigma_e^2 + 2^{k-2} n \mathbf{A}^2$, and so on. The variance of any one of the effects is $\sigma_e^2/2^{k-2}n$; thus, for example, a $1-\alpha$ confidence interval for $(\mathbf{AB})$ is

$$(AB) \pm t[1-\alpha/2; 2^k(n-1)]\sqrt{s_e^2/2^{k-2}n} \, .$$

THE CASE OF r BLOCKS

Another way to obtain an estimate of σ_e^2 in a 2^k factorial design is to use several blocks. In the agricultural experiment involving k fertilizers, each at two levels, it may be convenient to use r large plots, dividing each one into 2^k smaller plots and assigning the 2^k treatment combinations at random within each large plot. A block design is also appropriate when the same experiment involving 2^k treatment combinations is performed r times.

TABLE 8.7. ANALYSIS OF VARIANCE TABLE FOR 2^3 FACTORIAL, $n>1$

Source of Variation (1)	Sum of Squares (2)	d.f. (3)	MS (4)	EMS (5)
Due A	$2nA^2$	1	$2nA^2$	$\sigma_e^2 + 2nA^2$
Due B	$2nB^2$	1	$2nB^2$	$\sigma_e^2 + 2nB^2$
Due AB	$2n(AB)^2$	1	$2n(AB)^2$	$\sigma_e^2 + 2n(AB)^2$
Due C	$2nC^2$	1	$2nC^2$	$\sigma_e^2 + 2nC^2$
Due AC	$2n(AC)^2$	1	$2n(AC)^2$	$\sigma_e^2 + 2n(AC)^2$
Due BC	$2n(BC)^2$	1	$2n(BC)^2$	$\sigma_e^2 + 2n(BC)^2$
Due ABC	$2n(ABC)^2$	1	$2n(ABC)^2$	$\sigma_e^2 + 2n(ABC)^2$
Residual	$\sum_{i=1}^{2}\sum_{j=1}^{2}\sum_{k=1}^{2}\sum_{\ell=1}^{n}(Y_{ijk\ell}-\overline{Y}_{ijk.})^2$	$2^3(n-1)$	s_e^2	σ_e^2
Total	$\sum_{i=1}^{2}\sum_{j=1}^{2}\sum_{k=1}^{2}\sum_{\ell=1}^{n}(Y_{ijk\ell}-\overline{Y}_{....})^2$	2^3n-1		

For $k=3$, if we assume that the block effect does not interact with any of the factors, an appropriate model is

$$Y_{ijkl}=\mu+\alpha_i+\beta_j+\gamma_k+(\alpha\beta)_{ij}+(\alpha\gamma)_{ik}+(\beta\gamma)_{jk}+(\alpha\beta\gamma)_{ijk}$$

$$+\rho_l+\varepsilon_{ijkl}, \qquad i=1,2; \qquad j=1,2, \qquad k=1,2, \qquad l=1,\ldots,r, \quad (8.7)$$

where

$$\varepsilon_{ijkl}\,\text{IND}(0,\sigma_e^2) \qquad \text{and} \qquad \sum_{i=1}^{2}\alpha_i=\cdots=\sum_{l=1}^{r}\rho_l=0.$$

In (8.7) an additional term ρ_l has been added to the model because of a possible block effect, and an analysis of variance table contains a "due R" sum of squares. The estimate from the data for ρ_l is $\overline{Y}_{...l}-\overline{Y}_{....}$; thus the block sum of squares is $2^k\sum(\overline{Y}_{...l}-\overline{Y}_{....})^2$. With no blocks, the residual sum of squares would be

$$\sum_{i=1}^{2}\sum_{j=1}^{2}\sum_{k=1}^{2}\sum_{l=1}^{r}\left(Y_{ijkl}-\overline{Y}_{ijk.}\right)^2,$$

The residual sum of squares must be reduced by the due R sum of squares. The degrees of freedom for the new residual sum of squares must also be reduced; it is

$$2^k(r-1)-(r-1)=(2^k-1)(r-1).$$

Averaging over blocks, we have

$$Y_{ijk.}=\mu+\alpha_i+\beta_j+\gamma_k+(\alpha\beta)_{ij}+(\alpha\gamma)_{ik}+(\beta\gamma)_{jk}+(\alpha\beta\gamma)_{ijk}+\bar{\varepsilon}_{ijk.}, \quad (8.8)$$

with $\bar{\varepsilon}_{ijk.}$ IND$(0,\sigma_e^2/r)$. Since this is like the mean $\overline{Y}_{ijk.}$ from the previous model, with r replacing n, we can now calculate all the effects and the sums of squares for the analysis of variance table just as before (except for dividing the old residual sum of squares into the new residual plus the block sums of squares). Variances of the effects are the same as before (with n replaced by r), and confidence intervals follow as before.

CONFOUNDING BLOCK EFFECTS IN A 2^k FACTORIAL DESIGN

Frequently it is impractical to use all 2^k treatment combinations in a single block. In the first example presented in this chapter ($k=3$ and just eight observations), it may be impossible to obtain all eight observations under the same conditions. If the maximum block size is four and there are eight treatments, it is necessary to *confound* the block effect with another effect. Confounding two or more effects means that the method of obtaining the data prevents us from estimating them separately; they are confounded or confused.

If an investigator wishes to use two batches of material as blocks, he might assign his treatments as follows:

Batch 1	Batch 2
$(1), a, b, ab$	c, ac, bc, abc

Here block 1 contains all the observations taken at a low level of factor C; block 2 contains all observations at a high level of C. If the treatments at a high level of C are more effective than those at a low C level, the investigator has no way of knowing whether the difference is caused by the levels of factor C or by the difference between the two batches. Thus the C effect is confounded with the block effect.

Another way of looking at this is to examine the expression for C. It is

$$C = \tfrac{1}{4}(a+1)(b+1)(c-1)$$
$$= \tfrac{1}{4}[-(1) - a - b - ab + c + ac + bc + abc].$$

When any number X is added to each of the observations in the second block (to represent a block effect), it changes the value of C, which becomes

$$\tfrac{1}{4}[-(1) - a - b - ab + (c+X) + (ac+X) + (bc+X) + (abc+X)]$$
$$= \tfrac{1}{4}[-(1) - a - b - ab + c + ac + bc + abc] + X.$$

In the expressions for the other effects, X cancels; thus if X is added in the expression for ABC, we have

$$(ABC) = \tfrac{1}{4}[-(1) + a + b - ab + (c+X) - (ac+X) - (bc+X) + (abc+X)]$$
$$= \tfrac{1}{4}[-(1) + a + b - ab + c - ac - bc + abc].$$

Therefore, all other effects can be estimated.

It is usually undesirable to confound a main effect. We try to choose an uninteresting or negligible effect, leaving the other effects unconfounded. The highest-order interaction effect is frequently confounded. To confound (ABC), we write

$$(ABC) = \tfrac{1}{4}(a-1)(b-1)(c-1)$$
$$= \tfrac{1}{4}\{(a+b+c+abc) - [ab+bc+ac+(1)]\}.$$

The blocks are then as follows:

Block 1	Block 2
$(1), ab, ac, bc$	a, b, c, abc

To illustrate the procedure, let us assume that the data in Table 8.2 were obtained in two blocks as follows:

Block 1		Block 2	
(1): 5	ab: 12	a: 6	c: 10
ac: 12	bc: 15	b: 10	abc: 20

We calculate the various estimates just as before; $2\bar{Y}_{...}=90/4$, $A=10/4,\ldots,BC=2/4$, $ABC=2/4$. We know that ABC estimates $\mathsf{ABC}+\mathsf{R}$, where R is the difference in response between the two blocks. Our other point estimates are satisfactory. However, we have no estimate of the variance, and we cannot make confidence intervals or tests except by assuming that some of the first-order population interactions are zero and pooling these terms to obtain an estimate of σ_e^2.

To make tests and confidence intervals using the smaller blocks, we shall want each treatment combination used more than once. This might be done if $2r$ blocks are available and r of them (picked at random) are assigned treatments (1), ab, ac, and bc; the other blocks are assigned treatments a, b, c, and abc. On the other hand, perhaps the entire experiment is replicated on r different days, but within each replicate two batches of material are available and four treatment combinations are used in each batch. In either case, for comparison with (8.7), our model can be

$$Y_{ijklm}=\mu+\alpha_i+\beta_j+\gamma_k+(\alpha\beta)_{ij}+(\alpha\gamma)_{ik}+(\beta\gamma)_{jk}+\rho_{ijklm}+\varepsilon_{ijklm}, \quad (8.9)$$

where $\varepsilon_{ijklm}\,\text{IND}(0,\sigma_e^2)$ and

$$\sum_{i=1}^{2}\alpha_i=\cdots=\sum_{k}^{2}(\beta\gamma)_{jk}=\sum_{ijk}\sum_{l=1}^{r}\rho_{ijklm}=0.$$

Here the subscript l refers to the number of the replicate, hence runs from 1 to r. The subscript m is 1 for treatment combinations in blocks $11,21,\ldots,r1$; it is 2 for those in blocks $12,\ldots,r2$. (If we simply have $2r$ small blocks, they can be paired at random; the pairing of the blocks does not enter into our analysis.)

As a numerical example, let us use the following data, where for simplicity we have taken $r=2$:

Replicate 1				Replicate 2			
Block 11		Block 12		Block 21		Block 22	
(1)	14	*a*	12	(1)	5	*a*	31
ab	21	*b*	20	*ab*	7	*b*	36
ac	3	*c*	6	*ac*	0	*c*	21
bc	12	*abc*	11	*bc*	2	*abc*	33

$\bar{Y}_{...11}=50/4$ $\bar{Y}_{...12}=49/4$ $\bar{Y}_{...21}=14/4$ $\bar{Y}_{...22}=121/4$ $\bar{Y}_{.....}=234/16$

$=12.50$ $=12.25$ $=3.50$ $=30.25$ $=14.62$

Our first step is to obtain estimates of the parameters μ, $A=2\alpha_2$, and so on. This can be done exactly as if there were no blocks to be considered at all. Either the treatment totals or the means can be used with the Yates method; in Table 8.8 we use the totals.

Each effect is obtained from column k by dividing by $2^{k-1}r$, where $2r$ is the number of blocks; when means are used in the calculations, the

TABLE 8.8. COMPUTATION OF SUMS OF SQUARES BY THE YATES METHOD

Treatment	Total Y_i	(1)	(2)	(3)	Effect (3)/8	Sums of Squares $(3)^2/16$
(1)	19	62	146	234		
a	43	84	88	2	.25	.25
b	56	30	- 4	50	6.25	156.25
ab	28	58	6	2	.25	.25
c	27	24	22	-58	-7.25	210.25
ac	3	-28	28	10	1.25	6.25
bc	14	-24	-52	6	.75	2.25
abc	44	30	54	106	13.25	702.25
						1077.75

numbers in column k are divided by 2^{k-1}. The sum of squares column is obtained (from totals) by squaring the numbers in column k and dividing by $2^k r$; from means, the numbers in column k are squared and multiplied by $r/2^k$.

We next obtain the total sum of squares from

$$\sum_{i=1}^{2}\sum_{j=1}^{2}\sum_{k=1}^{2}\sum_{l=1}^{r}\left(Y_{ijklm}-\overline{Y}_{.....}\right)^2.$$

We have $\overline{Y}_{.....}=14.62$, and the total sum of squares is

$$(12-14.62)^2+\cdots+(31-14.62)^2=1933.75.$$

If there were no blocks at all, and if an interaction term ABC existed, we could obtain the residual sum of squares either by subtraction $(1933.75-(0.25+\ldots+702.25)=1933.75-1077.75=856.00)$, or from

$$\sum_{i=1}^{2}\sum_{j=1}^{2}\sum_{k=1}^{2}\sum_{l=1}^{r}\left(Y_{ijklm}-\overline{Y}_{ijk..}\right)^2=\left(14-\frac{14+5}{2}\right)^2+\cdots+\left(33-\frac{11+33}{2}\right)^2$$

$$=856.00.$$

The first three columns of the analysis of variance table would be those given in the first three columns of Table 8.9.

TABLE 8.9. COMPUTATION OF SUMS OF SQUARES

No Blocks, An ABC Effect		No Blocks No ABC Effect			2r Blocks No ABC Effect		
Source of Variation	SS	d.f.		SS	SS	d.f.	Source of Variation
A	.25	1			.25	1	A
B	156.25	1			156.25	1	B
AB	.25	1			.25	1	AB
C	210.25	1			210.25	1	C
AC	6.25	1			6.25	1	AC
BC	2.25	1			2.25	1	BC
ABC	702.25	1	$2^k(r-1)+1=9$	1558.25	1512.25	$2r-1=3$	Block
Residual	856.00	$2^k(r-1)=8$			$SS_r=46.00$	$(2^k-2)(r-1)=6$	Residual
Total	1933.75	$2^k r-1=15$			1933.75	$2^k r-1=15$	Total

If we now assume that there is no interaction term **ABC**, we pool the ABC sum of squares with the residual sum of squares to obtain the analysis of variance columns (see the middle columns of Table 8.9). We now have 1558.25 as the residual sum of squares, and from this we must remove the effects of the $2r$ blocks.

The block sum of squares, in the type of notation used in the earlier chapters, is

$$2^{k-1}\sum_{l=1}^{r}\sum_{m=1}^{2}\left(\bar{Y}_{...lm}-\bar{Y}_{.....}\right)^2;$$

in other words, we can simply obtain deviations of block means from the overall means, square them, add them together, and multiply the total by the number of observations in each block. Thus we have

$$2^2\left[\left(\frac{50}{4}-\frac{234}{16}\right)^2+\cdots+\left(\frac{121}{4}-\frac{234}{16}\right)^2\right]=1512.25.$$

The degrees of freedom is 1 less than the number of blocks, $2r-1=3$. When 1512.25 is subtracted from 1558.25 in Table 8.9, we obtain 46.00 as the new residual sum of squares.

We now obtain $s_e^2=\mathrm{SS}_r/(2^k-2)(r-1)=46.00/6=7.67$. Confidence intervals for parameters and tests follow just as in the earlier cases, with r complete blocks or with n observations and no blocks. The variance of any one of the treatment effects is $\sigma_e^2/2^{k-2}r$, which is estimated by $s_e^2/2^{k-2}r$, where $s_e^2=\mathrm{SS}_r/(2^k-2)(r-1)$. In our example, a $1-\alpha$ level confidence interval for $\mathbf{C}$ is

$$C\pm t[1-\alpha/2;(2^k-2)(r-1)]\sqrt{s_e^2/2^{k-2}r} \text{ or } -7.25\pm t[1-\alpha/2;6]\sqrt{7.67/4},$$

or, as a 95% interval, $-10.64<\mathbf{C}<-3.86$.

Similarly, a test of $H_0:\mathbf{C}=0$ is made by comparing

$$F=\frac{210.25/1}{46.00/6}=\frac{210.25}{7.67}=27.41$$

with

$$F[1-\alpha;1,6]=5.99 \qquad \text{for} \quad \alpha=.05.$$

For a somewhat more sophisticated analysis, see Bennett and Franklin [8.1]. In the 2^3 situation, if the block size is taken to be 2 rather than 4, four different types of block could be used and three effects would be confounded. The analysis of designs with very small blocks tends to be complicated. Davies [8.3] discusses confounding different interactions in each pair of blocks, permitting all the interactions to be estimated.

FRACTIONAL REPLICATE DESIGNS

Sometimes the investigator cannot take many observations, but he nevertheless wishes to consider several factors simultaneously. To do an experiment involving just two factors with only four observations, he can make observations as follows:

	B	
A	Low	High
Low	(1)	b
High	a	ab

The effects are then estimated from the Table 8.10 and, as before we have

$$A = \tfrac{1}{2}(a-1)(b+1) = \tfrac{1}{2}[-(1)+a-b+ab]$$

$$B = \tfrac{1}{2}(a+1)(b-1) = \tfrac{1}{2}[-(1)-a+b+ab]$$

$$AB = \tfrac{1}{2}(a-1)(b-1) = \tfrac{1}{2}[+(1)-a-b+ab].$$

If it is known that there is no interaction between the A and B factors ($AB=0$), the size of AB is due simply to sampling variation. In this case, a

TABLE 8.10. TREATMENTS FOR 2^2 DESIGN

	Effects		
Treatment	A	B	AB
$(1) = Y_{11}$	-	-	+
$a = Y_{21}$	+	-	-
$b = Y_{12}$	-	+	-
$ab = Y_{22}$	+	+	+

TABLE 8.11. TREATMENTS FOR $\frac{1}{2}$ REPLICATE OF 2^3 DESIGN

	Effects		
Observation	A	B	C = AB
$c = Y_{112}$	-	-	+
$a = Y_{211}$	+	-	-
$b = Y_{121}$	-	+	-
$abc = Y_{222}$	+	+	+

third factor C can be introduced into the design, provided the new factor does not interact with factors A and B. We use one level of factor C for those treatment combinations which enter AB with plus signs, the other level for those with minus signs. Equating C with AB, we have Table 8.11. Such a design is called a *fractional factorial* design and can be displayed as follows:

	C Low		C High	
	B		B	
A	Low	High	Low	High
Low	—	b	c	—
High	a	—	—	abc

The appropriate model, in the notation of previous chapters, is that each observation is of the form

$$Y_{ijk} = \mu + \alpha_i + \beta_j + \gamma_k + \varepsilon_{ijk}, \tag{8.10}$$

where

$$\sum_{i=1}^{2} \alpha_i = \sum_{j=1}^{2} \beta_j = \sum_{k=1}^{2} \gamma_k = 0 \qquad \text{and} \qquad \varepsilon_{ijk}\ \text{IND}(0, \sigma_e^2).$$

Using $\mathsf{A} = \alpha_2 - \alpha_1$, $\mathsf{B} = \beta_2 - \beta_1$, $\mathsf{C} = \gamma_2 - \gamma_1$ to conform to the notation introduced in this chapter, we calculate A, B, and $AB = C$ as their estimates:

$$A = \tfrac{1}{2}(-c + a - b + abc)$$

$$B = \tfrac{1}{2}(-c - a + b + abc)$$

$$C = \tfrac{1}{2}(c - a - b + abc).$$

The variance of each of these estimates of treatment effects is

$$\left(\tfrac{1}{2}\right)^2(\sigma_e^2 + \sigma_e^2 + \sigma_e^2 + \sigma_e^2) = \sigma_e^2.$$

An investigator who wishes to study three factors with just four observations might be inclined to take observations on the treatment combinations (1), a, b, and c. If he did this, his estimates of A, B, and C would be $a-(1)$, $b-(1)$, and $c-(1)$, respectively. Such estimates are inferior to A, B, and C because of their higher variability. Each of the variances of $a-(1)$, $b-(1)$, and $c-(1)$ equals $2\sigma_e^2$; comparing $2\sigma_e^2$ with σ_e^2 (the variance of A, B, and C), we see that the estimates $a-(1)$, $b-(1)$, and $c-(1)$ are more variable than are A, B, and C. The lower variance for the estimates of treatment effects is the essential reason for taking observations on c, a, b, and abc.

We might have used the high level of factor C with the minus signs in AB and the low level with the plus signs; in this case, our observations would have been (1), ab, ac, and bc.

The treatment combinations a, b, c, and abc are those used in one of the blocks in the design in which the ABC effect was confounded with blocks; the treatment combinations (1), ab, ac, and bc form the other block. In the half-replicate design, just one type of block is used, and we must assume that all interactions are zero. When both types of block are used, we can estimate all interactions except ABC.

If in the fractional replicate the investigator is wrong in his assumption that the $AB = 0$, then $C = \tfrac{1}{2}(c - a - b + abc)$ is an estimate of $C + AB$ rather than an estimate of C. The word *alias* is used to describe the relationship of C and AB; we say that C and AB are aliased in this design.

In looking at the signs for A, B, and C in Table 8.3, we see that the design is symmetric in the three factors. It follows that A and BC are aliased in the design, and also B and AC. Thus A is an estimate of $A + BC$ if the interaction BC is not zero, and the B estimates $B + AC$ if AC is not zero.

As a numerical example, let $c = 9$, $a = 10$, $b = 15$, and $abc = 30$. Calculations with Yates's method appear in Table 8.12. From the table, we have

TABLE 8.12. YATES'S COMPUTATIONAL METHOD

Treatment	Observation	(1)	(2)	Sums of Squares $(2)^2/2^2$
c	9	19	64	
a	10	45	16	64
b	15	1	26	169
abc	30	15	14	49

$A=16/2=8$ as an estimate of A or of $\mathsf{A}+\mathsf{BC}$, $B=13$ as an estimate of B or of $\mathsf{B}+\mathsf{AC}$, and $C=7$ as an estimate of C or of $\mathsf{C}+\mathsf{AB}$. The analysis of variance appears in Table 8.13.

As Table 8.13 indicates, the data provide no sum of squares for estimating the within treatment variance σ_e^2. Sometimes the investigator knows an approximate value of the residual variance from previous experiments. In the example, if he knows that σ_e^2 is about 30, and assumes that the interactions are zero, he would decide that B is not zero; his values of A and C, respectively, are not high enough to reject $H_0:\mathsf{A}=0$ or $H_0:\mathsf{C}=0$.

The foregoing example for a fractional replicate is called a half-replicate design because half the treatment combinations are used. In higher-order fractional replicate designs, it is possible to estimate some of the interactions. Lists of designs are given in Ref. 8.5. Further discussion of the method can be found in Kempthorne [8.4], Davies [8.3], and Bennett and Franklin [8.1].

The half-replicate of the 2^3 design is equivalent to the 2×2 Latin square design of Chapter 6. Table 8.14 shows the correspondence; here treatment I consists of a high level of factor C and treatment II consists of a low level of C. The name of the treatment combination appears below the treatment

TABLE 8.13. ANALYSIS OF VARIANCE TABLE

Source of Variation (1)	Sums of Squares (2)	d.f. (3)
A + BC	64	1
B + AC	169	1
C + AB	49	1
Total	282	3

TABLE 8.14. CORRESPONDENCE BETWEEN 2×2 LATIN SQUARE AND HALF-REPLICATE OF A 2^3 DESIGN

		B	
		Low	High
A	Low	I (c)	II (b)
	High	II (a)	I (abc)

number, and we see that the same observations are taken. As stressed in Chapter 6, the Latin square design is used on the assumption that the interactions are zero.

These designs can be particularly helpful to the investigator who is limited in his ability to study many factors simultaneously because of restrictions imposed on the number of observations he can take (by lack of time, money, ability to keep conditions constant, etc.). The designs can also be used as a preliminary design to determine whether certain factors should be studied further at the levels used. When the investigator is trying to use the lowest possible number of observations, he sometimes uses estimates of σ_e^2 from previous experiments to help in evaluating the present one.

The design and analysis techniques are quite complicated, and the investigator must consider carefully whether their advantages outweigh the difficulty of interpretation. They require especially close cooperation between the statistician and the investigator because it is necessary to know which interactions are expected to be zero and which are not. Moreover, simple errors such as missing values or observations taken at the "wrong" levels make it extremely difficult to analyze the data. In general, these designs are best suited to professional investigators who are willing to devote the necessary time and effort to these rather complex analyses in order to save observations. Examples of use of this technique are often found in agriculture or chemistry among investigators who have worked in their particular area for a long time, have good estimates of σ_e^2, and are sure that they are not likely to obtain wild outliers. Each observation yields a great deal of information and a mistaken observation (outlier) can lead to erroneous conclusions.

We omit illustrative examples for these designs because we feel that they tend to apply well in very specific situations and are dangerous to imitate. Without knowing a precise situation, it is difficult to determine whether a given design is a sensible choice. Examples of the use of these designs in chemistry are found in Bennett and Franklin [8.1] and Davies [8.3]; examples from agriculture appear in Cochran [8.2].

SUMMARY

A brief introduction to factorial designs at two levels was given. Symbolic notation and Yates's computational method, blocking, and fractional replicates were presented.

REFERENCES

8.1. Bennett, C. A., and N. L. Franklin, *Statistical Analysis in Chemistry and the Chemical Industry*, Wiley, New York, 1954.

8.2. Cochran, W. G., and G. M. Cox, *Experimental Designs*, Wiley, New York, 1957.

8.3. Davies, O. L. (Ed.), *The Design and Analysis of Industrial Experiments*, 2nd ed., Oliver & Boyd, Edinburgh, 1956.

8.4. Kempthorne, O., *The Design and Analysis of Experiments*, Wiley, New York, 1952.

8.5. U.S. National Bureau of Standards, *Fractional Factorial Experimental Designs for Factors at Two Levels*, Applied Mathematics Series, Vol. 48 (1957).

PROBLEMS

8.1. Analyze the following 2^3 factorial design on yield of soybeans by the Yates method.

Yield of soybeans ($n=2$)
Planting date—early, late
Fertilizer—yes, no
Seed variety—1, 2

	Variety 1			Variety 2	
	Planting Date			Planting Date	
Fertilizer	Early	Late	Fertilizer	Early	Late
Yes	6	5	Yes	4	2
	7	4		5	4
No	6	3	No	3	1
	5	3		2	1

8.2. Three factors—A (nitrogen), B (potash), and C (phosphate)—are to be varied in a fertilizer. Each of these is set at a high level and at a low level. The following output, in bushels of wheat per acre (a constant has been subtracted to simplify computations), has been obtained:

	Phosphate			
	Low		High	
	Potash		Potash	
Nitrogen	Low	High	Low	High
Low	5	4	10	10
High	8	7	14	15

(*a*) Test for main effects at an $\alpha = .10$ level of significance using the Yates method to obtain the sum of squares.

(*b*) You wish to perform the experiment in blocks (plots) of four observations each confounding the I(AB) interaction with blocks. State which observations will go into each block.

8.3. After an airplane has been deliberately stalled by the instructor during training, the time to recover is measured in seconds. Three factors that were varied were A, type of aircraft; B, recovery method; and C, power setting (in rpm).

A—Aircraft	M				P			
B—Method	1		2		1		2	
C—Power setting (rpm)	1200	1500	1200	1500	1200	1500	1200	1500
Time (sec)	6.8	15.9	3.5	11.2	4.9	13.3	3.1	11.0
Treatment combination	(1)	c	b	bc	a	ac	ab	abc

(*a*) Perform an analysis of variance using Yates's technique.

(*b*) If a short recovery time from a stall is desirable, decide which type of plane and method you prefer.

(*c*) If you are willing to give up information on the AC interaction and can take blocks of four observations apiece, decide which observations go into the two blocks.

**CHAPTER 9

OTHER MODELS IN THE ANALYSIS OF VARIANCE

We have been primarily concerned in the previous chapters with *fixed effects* models, which are frequently referred to as Model I. In the present chapter, we introduce two other models, designated Models II and III; these models are *variable effects models*. We also discuss mixed models.

Model I is characterized by the fact that the experimenter's objective is to study the particular treatments (or particular populations) being considered. In the corn yield problem, for example, Model I was appropriate because the experimenter wished to draw conclusions concerning the four fertilizers studied and no others. Usually Model I is the most reasonable model, but occasionally Model II or III seems to be more appropriate.

In Models II and III, the treatments studied have been sampled from a larger population of treatments, and we wish to draw inferences concerning all treatments. For example, in a one-way analysis of variance in which four laboratory technicians each use the same new method to make 10 determinations of vitamin C content, we may be interested in studying the new method as applied by laboratory technicians in general. If we consider that the four technicians form a random sample from a large (infinite) population of laboratory technicians, Model II is appropriate. In some situations, the four technicians may be a sample from a *finite* population; then Model III can be applied. In a large laboratory, there could be a total of 12 laboratory technicians, and data obtained on just four of them could be used to make inferences concerning the new method as applied by the entire group of 12 technicians.

When in using a treatments (technicians), we wish to make inferences concerning A treatments; if $A = a$, we use Model I; if $A = \infty$, we use Model

II; $a<A<\infty$ implies Model III. Thus Models I and II are limiting cases of Model III. Similarly, in a two-way analysis of variance, the first and second factors might be the machine used and the machine operator, respectively. Model III is appropriate for drawing inferences concerning A machines and B operators using a machines and b operators, provided A and B are finite numbers larger than a and b; Model II is suitable if A and B are large enough to be considered infinite; Model I is the proper choice when $A=a$ and $B=b$.

A *mixed model* is also possible. For example, the study might be concerned with exactly a machines, but it might use just b out of B operators. The model chosen depends on the way the experiment was conducted and also on the scope of the inferences that we wish to make.

A convenient feature of Model III is that the formulae for expected mean squares and many of the formulae for variances can be made in terms of Model III and specialized to the other models. Before discussing this, however, let us think a little about underlying assumptions and about the quantities we wish to estimate in the different models.

THE MATHEMATICAL MODELS

The same expression used for Model I for an individual observation serves for the other models. For example, in the two-way cross classification with a levels of the first factor and b levels of the second factor, we still use for all models

$$Y_{ijk}=\mu+\alpha_i+\beta_j+(\alpha\beta)_{ij}+\varepsilon_{ijk}, \qquad \varepsilon_{ijk}\ \text{IND}(0,\sigma_e^2),$$

$$i=1,\ldots,a, \qquad j=1,\ldots,b, \qquad k=1,\ldots,n. \tag{9.1}$$

The models differ, however, in their statements concerning the α_i, β_j, and $(\alpha\beta)_{ij}$.

MODEL III

In Model III, we say that there are A levels of the first factor (A machines) and B levels of the second factor (B operators). For each machine–operator combination, there exists a population of output measurements. There is thus a set of AB population means μ_{ij}, $i=1,\ldots,A$, $j=1,\ldots,B$, and these population means, without loss of generality, can be expressed as follows:

$$\mu_{ij} = \mu + \alpha_i + \beta_j + (\alpha\beta)_{ij}, \tag{9.2}$$

where

$$\sum_{i=1}^{A} \alpha_i = \sum_{j=1}^{B} \beta_j = \sum_{i=1}^{A} (\alpha\beta)_{ij} = \sum_{j=1}^{B} (\alpha\beta)_{ij} = 0. \tag{9.3}$$

We pick a machines at random and we pick b operators at random, thus obtaining the ab populations to be used in the study. Each of the a machines is used by each of the b operators, thus there are ab treatment combinations studied. For each machine–operator combination there is a population of possible output measurements; from each of the ab populations, we obtain a sample of size n.

In Model I we were able to set

$$\sum_{i=1}^{a} \alpha_i = \sum_{j=1}^{b} \beta_j = \sum_{i=1}^{a} (\alpha\beta)_{ij} = \sum_{j=1}^{b} (\alpha\beta)_{ij} = 0.$$

In Model III the machine and operator were chosen at random, and we cannot say, as in Model I, that the parameters in the *sample* sum to zero; their sums are random variables, for they depend on the machines and the operators drawn. The expected values can be calculated, however, and averaging over all possible sets of a operators and b machines and using the fact that the parameters in the *population* sum to zero as given in (9.3), we have

$$E\alpha_i = E\beta_j = E(\alpha\beta)_{ij} = 0.$$

We shall define σ_a^2 by the equation $\sigma_a^2 = [A/(A-1)]$ Var α_i; similarly, $\sigma_b^2 = [B/(B-1)]$ Var β_j, and $\sigma_{ab}^2 = [AB/(A-1)(B-1)]$ Var $(\alpha\beta)_{ij}$; again we are taking the variances over all possible samples of a machines and b operators.

With some algebraic effort, we can calculate the covariances of the various quantities; and again using (9.3), we find

$$\operatorname{Cov} \alpha_i, \alpha_{i'} = -\frac{\sigma_a^2}{A}$$

$$\operatorname{Cov} \beta_j, \beta_{j'} = -\frac{\sigma_b^2}{B}$$

$$\operatorname{Cov}(\alpha\beta)_{ij},(\alpha\beta)_{i'j'}=+\frac{\sigma_{ab}^2}{AB}$$

$$\operatorname{Cov}(\alpha\beta)_{ij},(\alpha\beta)_{i'j}=-\frac{(1-1/B)\sigma_{ab}^2}{A}$$

$$\operatorname{Cov}(\alpha\beta)_{ij},(\alpha\beta)_{ij'}=-\frac{(1-1/A)\sigma_{ab}^2}{B}. \tag{9.4}$$

Because we selected the ith machine and jth operator independently, the α_i and the β_j are independent of each other. The values of the $(\alpha\beta)_{ij}$, on the other hand, in general depend on the operator and machine chosen, thus the $(\alpha\beta)_{ij}$ are *not* independent of the various α's and β's in the sample. However, it can be demonstrated that the covariance between α_i, and $(\alpha\beta)_{i'j}$ is zero regardless of whether $i=i'$; similarly, the covariance between β_j and $(\alpha\beta)_{ij'}$ is zero. Thus, we have

$$\operatorname{Cov}\alpha_i,\beta_j=0$$

$$\operatorname{Cov}\alpha_i,(\alpha\beta)_{ij}=0$$

$$\operatorname{Cov}\alpha_i,(\alpha\beta)_{i'j}=0$$

$$\operatorname{Cov}\beta_j,(\alpha\beta)_{ij}=0$$

$$\operatorname{Cov}\beta_j,(\alpha\beta)_{ij'}=0. \tag{9.5}$$

The expressions in (9.4) and (9.5) are used in finding the variance of the sample mean and the expected mean squares for the analysis of variance table.

MODEL II

If we allow A and B to approach infinity in Model III, we see that all the covariances approach zero; now we are in a Model II situation. In Model II, we assume that the α_i, β_j, and $(\alpha\beta)_{ij}$ are *independently normally distributed* with zero means and variances σ_a^2, σ_b^2, and σ_{ab}^2, respectively.

The assumption that the main effects and interactions are statistically independent is difficult to justify. It represents an effort to make the mathematics more tractable. The assumption implies that there is no tendency for good machines used by good operators to be better (or worse) than would be expected from adding the "good machine" effect to the "good operator" effect. This is not true in general. Conceivably bad operators working with bad machines might perform poorly, whereas three other combinations (bad operators with good machines, good operators with bad machines, or good operators with good machines) might all perform fairly satisfactorily. In such a case, knowledge of α_i and β_j clearly gives some information on the size of $(\alpha\beta)_{ij}$.

Purely for the sake of mathematical convenience, however, we are making this unpalatable assumption for Model II. We assume, then, that

$$\alpha_i \,\text{IND}(0,\sigma_a^2), \quad \beta_j \,\text{IND}(0,\sigma_b^2), \quad (\alpha\beta)_{ij}\,\text{IND}(0,\sigma_{ab}^2), \quad \varepsilon_{ijk}\,\text{IND}(0,\sigma_e^2),$$

and that all α_i, β_j, $(\alpha\beta)_{ij}$, and ε_{ijk} are mutually independent.

PARAMETERS TO BE ESTIMATED

In Model I problems, we usually estimate the linear contrasts among the means and also the variance σ_e^2. For example, if only a types of fertilizers and b planting times are being studied, we are usually interested in the mean differences in yield between pairs of fertilizers $(\alpha_i - \alpha_{i'})$, and between planting times $(\beta_j - \beta_{j'})$, the effect of a certain fertilizer (α_i), and the interaction of a certain fertilizer with a particular planting time, $(\alpha\beta)_{ij}$. The mean μ (the average of the ab population means) is of secondary interest.

In Models II and III, on the other hand, the parameter μ is of primary interest. Now, picking a machines and b operators at random, μ is the mean response for all machines and all operators. Parameters such as $\mu + \alpha_i$ (the mean response for the ith machine averaged over all operators) and $\alpha_i - \alpha_{i'}$ (the difference in mean response between the ith and i'th machine) are of little interest, for they were drawn randomly. Instead, we are concerned with variability among machines using the same operator; with this in mind, we estimate σ_a^2. Similarly, we estimate σ_b^2 for variability among operators. To obtain an idea of the size of the machine–operator interactions, we estimate σ_{ab}^2.

In Models II and III we also wish to make F tests corresponding to those of Model I. A test of whether machines make any difference in the response has the null hypothesis $H_0: \sigma_a^2 = 0$; a test for machine–operator interaction is $H_0: \sigma_{ab}^2 = 0$.

An unfortunate feature of Models II and III is that many of the tests and confidence intervals used involve either approximations or additional assumptions.

From the foregoing discussion it is clear that the variable effects models are not our favorite. We know little about how well the approximations work out in practice, and even less about the results of the assumption that $(\alpha\beta)_{ij}$ is statistically independent of α_i and β_j. On the other hand, when the treatment levels are chosen at random and we wish to draw inferences to a larger population, we must resort to Models II or III or mixed models. We therefore treat them briefly here, beginning with the analysis of variance table, followed by a discussion of F tests, and finally a little about variances.

Throughout the chapter, the output data in Table 9.1 serve as a numerical example. Here each of two machines was used by four operators, and three observations were made on each of the eight machine–operator combinations. Thus we have a two-way classification problem, which can be used to illustrate various procedures for the different models.

THE ANALYSIS OF VARIANCE TABLE

In a simple one-way analysis of variance, the entire analysis of variance table and the F test are precisely the same for all three models. For more complicated designs, the analysis of variance table is made exactly the same for all models, except for the EMS column. The EMS columns differ, however, and therefore the F tests are frequently different for the various models.

Tables 9.2 and 9.3 are the analysis of variance tables for the one-way hierarchical classification of Chapter 6 and for the two-way cross classification of Chapter 7, respectively. For the three-way cross classification, Table 9.4 presents only the source of variation and the EMS columns of

TABLE 9.1. OUTPUT DATA FOR TWO MACHINES AND FOUR OPERATORS

Machine	Operator				
	1	2	3	4	
1	16.0	29.5	19.2	23.0	
	12.9	23.6	25.2	23.2	
	15.5	23.2	26.4	29.1	
	$\bar{Y}_{11.} = 14.80$	$\bar{Y}_{12.} = 25.43$	$\bar{Y}_{12.} = 23.60$	$\bar{Y}_{14.} = 25.10$	$\bar{Y}_{1..} = 22.23$
2	19.6	24.5	19.1	14.3	
	19.4	15.8	21.6	20.1	
	20.0	17.1	16.9	21.8	
	$\bar{Y}_{21.} = 19.67$	$\bar{Y}_{22.} = 19.13$	$\bar{Y}_{23.} = 19.20$	$\bar{Y}_{24.} = 18.73$	$\bar{Y}_{2..} = 19.18$
	$\bar{Y}_{.1.} = 17.23$	$\bar{Y}_{.2.} = 22.28$	$\bar{Y}_{.3.} = 21.40$	$\bar{Y}_{.4.} = 21.92$	$\bar{Y}_{...} = 20.71$

TABLE 9.2. ANALYSIS OF VARIANCE TABLE FOR ONE-WAY CLASSIFICATION NESTED DESIGN, EQUAL SAMPLE SIZES, ALL MODELS

Source of Variation (1)	Sum of Squares (2)	d.f. (3)	MS (4)	EMS (5)
Due A	$SS_a = nb \sum_{i=1}^{a} (\bar{Y}_{i..} - \bar{Y}_{...})^2$	a-1	MS_a	$\sigma_e^2 + n(1-b/B)\sigma_b^2 + nb\sigma_a^2$
Due B	$SS_b = n \sum_{i=1}^{a} \sum_{j=1}^{b} (\bar{Y}_{ij.} - \bar{Y}_{i..})^2$	a(b-1)	MS_b	$\sigma_e^2 + n\sigma_b^2$
Residual	$SS_r = \sum_{i=1}^{a} \sum_{j=1}^{b} \sum_{k=1}^{n} (Y_{ijk} - \bar{Y}_{ij.})^2$	ab(n-1)	s_e^2	σ_e^2
Total	$SS_t = \sum_{i=1}^{a} \sum_{j=1}^{b} \sum_{k=1}^{n} (Y_{ijk} - \bar{Y}_{...})^2$	abn-1		

the analysis of variance table; the sums of squares, degrees of freedom, and mean square columns are as in Table 7.10.

The EMS columns as given in the tables are for Model III. For other models, we use the same formulae by setting A, B, C, and so on, equal to their proper values. For example, if a machines are chosen from an infinite

TABLE 9.3. ANALYSIS OF VARIANCE TABLE FOR TWO-WAY CROSS CLASSIFICATION, ALL MODELS

Source of Variation (1)	Sum of Squares (2)	d.f. (3)	MS (4)	EMS (5)
Due A	$SS_a = nb \sum_{i=1}^{a} (\bar{Y}_{i..} - \bar{Y}_{...})^2$	a-1	MS_a	$\sigma_e^2 + n(1-\frac{b}{B})\sigma_{ab}^2 + nb\sigma_a^2$
Due B	$SS_b = na \sum_{j=1}^{b} (\bar{Y}_{.j.} - \bar{Y}_{...})^2$	b-1	MS_b	$\sigma_e^2 + n(1-\frac{a}{A})\sigma_{ab}^2 + na\sigma_b^2$
Due AB	$SS_{ab} = n \sum_{i=1}^{a} \sum_{j=1}^{b} (\bar{Y}_{ij.} - \bar{Y}_{i..} - \bar{Y}_{.j.} + \bar{Y}_{...})^2$	(a-1)(b-1)	MS_{ab}	$\sigma_e^2 + n\sigma_{ab}^2$
Residual	$SS_r = \sum_{i=1}^{a} \sum_{j=1}^{b} \sum_{k=1}^{n} (Y_{ijk} - \bar{Y}_{ij.})^2$	ab(n-1)	s_e^2	σ_e^2
Total	$SS_t = \sum_{i=1}^{a} \sum_{j=1}^{b} \sum_{k=1}^{n} (Y_{ijk} - \bar{Y}_{...})^2$	abn-1		

TABLE 9.4. EXPECTED MEAN SQUARE COLUMN FOR THE THREE-WAY CROSS CLASSIFICATION DESIGN, ALL MODELS

Source of Variation	EMS
A	$\sigma_e^2+n(1-\frac{b}{B})(1-\frac{c}{C})\sigma_{abc}^2+nc(1-\frac{b}{B})\sigma_{ab}^2+nb(1-\frac{c}{C})\sigma_{ac}^2+nbc\ \sigma_a^2$
B	$\sigma_e^2+n(1-\frac{a}{A})(1-\frac{c}{C})\sigma_{abc}^2+na(1-\frac{c}{C})\sigma_{bc}^2+nc(1-\frac{a}{A})\sigma_{ab}^2+nac\ \sigma_b^2$
C	$\sigma_e^2+n(1-\frac{a}{A})(1-\frac{b}{B})\sigma_{abc}^2+na(1-\frac{b}{B})\sigma_{bc}^2+nb(1-\frac{a}{A})\sigma_{ac}^2+nab\ \sigma_c^2$
AB	$\sigma_e^2+n(1-\frac{c}{C})\sigma_{abc}^2+nc\ \sigma_{ab}^2$
AC	$\sigma_e^2+n(1-\frac{b}{B})\sigma_{abc}^2+nb\ \sigma_{ac}^2$
BC	$\sigma_e^2+n(1-\frac{a}{A})\sigma_{abc}^2+na\ \sigma_{bc}^2$
ABC	$\sigma_e^2+n\sigma_{abc}^2$
Residual	σ_e^2

population of machines, if b machine settings are used, and if c operators are chosen from a population of size C, with n determinations made with every machine–setting–operator combination, we set $A=\infty$, $B=b$, and $C=C$ in Table 9.4.

Expected mean squares such as those given in the tables can be obtained by first expressing the mean squares in terms of the $\alpha,\ldots,\varepsilon$'s and then taking expected values, using the definitions of $\sigma_a^2,\ldots$, and the expressions for the covariances given in (9.4) and (9.5). The algebraic effort in obtaining the EMS is considerable; therefore, at the end of this chapter we give a set of rules for writing down expected mean squares.

Table 9.5 is the analysis of variance calculated from the data of Table 9.1. We proceed now to consider F tests.

F TESTS

As an example of the complications that occur in making F tests, we consider the EMS columns of Tables 9.3 and 9.5. Our two-way classifica-

TABLE 9.5. ANALYSIS OF VARIANCE FOR MACHINE–OPERATOR EXAMPLE, TWO-WAY CLASSIFICATION

Source of Variation (1)	SS (2)	d.f. (3)	MS (4)	EMS (5)
Due Machines	55.70	1	55.70	$\sigma_e^2+3(1-4/4)\sigma_{ab}^2+12\sigma_a^2$
Due Operators	99.03	3	33.01	$\sigma_e^2+3(1-2/\infty)\sigma_{ab}^2+6\sigma_b^2$
Machine-Operator Interaction	129.10	3	43.03	$\sigma_e^2+3\sigma_{ab}^2$
Residual	170.43	16	10.65	σ_e^2
Total	454.26	23		

tion problem involves $a=2$ out of A machines and $b=4$ operators out of B operators, with $n=3$ observations made at each treatment combination. If we are interested only in these four operators, we can use a mixed model in which factor A is a variable effect and factor B is fixed. If A is infinite, then in table 9.3 we set $A=\infty$ and $B=b=4$, giving $\text{EMS}_a=\sigma_e^2+nb\sigma_a^2=\sigma_e^2+12\sigma_a^2$, $\text{EMS}_b=\sigma_e^2+n\sigma_{ab}^2+na\sigma_b^2=\sigma_e^2+3\sigma_{ab}^2+6\sigma_b^2$, $\text{EMS}_{ab}=\sigma_e^2+n\sigma_{ab}^2=\sigma_e^2+3\sigma_{ab}^2$, and $\text{EMS}_r=\sigma_e^2$. From these, we see that the test for machine effects $(H_0:\sigma_a^2=0)$ is just as in Model I, and we compare $F=\text{MS}_a/s_e^2=55.70/10.65$ with $F[1-\alpha;(a-1),ab(n-1)]=F[.95;1,16]=4.49$ (for a test with $\alpha=.05$). The test of $H_0:\sigma_{ab}^2=0$ also is as in Model I. Here $F=\text{MS}_{ab}/s_e^2=43.03/10.65=4.04$, which, for a test with significance level .05, is compared with $F[1-\alpha;(a-1)(b-1),ab(n-1)]=F[.95;3,16]=3.24$.

For testing whether there are operator effects $(H_0:\sigma_b^2=0)$, we can no longer use the residual mean square s_e^2 in the denominator of the F statistic; when H_0 is true, MS_b and s_e^2 do not estimate the same quantity unless σ_{ab}^2 is zero. We see, however, that MS_b and MS_{ab} are both estimates of $\sigma_e^2+n\sigma_{ab}^2$, and therefore we use the interaction mean square in the denominator of the test statistic. Thus the test of $H_0:\sigma_b^2=0$ consists of comparing $F=\text{MS}_b/\text{MS}_{ab}$ with $F[1-\alpha;b-1,(a-1)(b-1)]$. In the example, $F=33.01/43.03=0.77$, compared with $F[.95;3,3]=9.28$.

Thus for our data there seem to be differences among machines and interactions between machine and operator, but there is no evidence that the four operators differ in performance averaged over machines.

If the $a=2$ machines are drawn from a finite number of machines, say $A=8$, then factor A is Model III and factor B is still Model I. The expected mean squares EMS_a, EMS_{ab}, and EMS_r are the same as when A was Model II; now, however, $\mathrm{EMS}_b=\sigma_e^2+n(1-a/A)\sigma_{ab}^2+na\sigma_b^2=\sigma_e^2+3(1-2/8)\sigma_{ab}^2+6\sigma_b^2$. The tests of $H_0:\sigma_a^2=0$ and $H_0:\sigma_{ab}^2=0$ are as before, but we face some difficulty in testing $H_0:\sigma_b^2=0$. There is no exact F test of the latter hypothesis, for if $\sigma_b^2=0$, MS_b estimates $\sigma_e^2+n(1-a/A)\sigma_{ab}^2$ and MS_{ab} estimates $\sigma_e^2+n\sigma_{ab}^2$. Possible tests for this hypothesis appear later in the chapter. We turn first to the estimation of parameters.

ESTIMATION OF VARIANCE COMPONENTS

The estimation of the variance components $\sigma_a^2,\ldots$ is of more interest in the variable effects model than in the fixed effect model. Point estimates are found in all the models from the MS and EMS columns of the analysis of variance table. As a first example, we take the one-way nested design of Table 9.2. This might be a situation in which a operators were drawn from a population of size A using a sample of size b from the output of each operator, where each operator produced B units and n determinations were made on each unit of each sample.

As usual, we estimate σ_e^2 by s_e^2. To obtain an estimate of the variance among the different units, we use

$$\sigma_e^2+n\sigma_b^2\sim\mathrm{MS}_b$$

$$\sigma_e^2\sim s_e^2$$

Subtracting these approximations and dividing by n, we have

$$\sigma_b^2\sim\frac{\mathrm{MS}_b-s_e^2}{n}. \qquad (9.6)$$

We estimate σ_b^2, then, by $(\mathrm{MS}_b-s_e^2)/n$, except that whenever the estimate is negative we replace it by zero. Such an estimate is slightly biased (i.e., slightly higher on the average than σ_b^2) because we have replaced any negative values by zero. Here we have an example of a biased statistic that is clearly preferable to an unbiased statistic.

In the two-way analysis (the numerical example of Table 9.1) with $A=8$, $B=10$ (Model III), our estimates are

$$\sigma_e^2\sim s_e^2=10.65$$

$$\sigma_e^2+n\sigma_{ab}^2\sim\mathrm{MS}_{ab}=43.03$$

$$\sigma_{ab}^2\sim\frac{\mathrm{MS}_{ab}-s_e^2}{n}=10.79. \qquad (9.7)$$

Then, from

$$\sigma_e^2 + n\left(1 - \frac{a}{A}\right)\sigma_{ab}^2 + na\sigma_b^2 \sim \mathrm{MS}_b$$

we subtract

$$n\left(1 - \frac{a}{A}\right)\sigma_{ab}^2 \sim n\left(1 - \frac{a}{A}\right)\frac{\mathrm{MS}_{ab} - s_e^2}{n}$$

and

$$\sigma_e^2 \sim s_e^2$$

to obtain

$$na\sigma_b^2 \sim \mathrm{MS}_b - \left(1 - \frac{a}{A}\right)(\mathrm{MS}_{ab} - s_e^2) - s_e^2$$

and finally

$$\sigma_b^2 \sim \left[\mathrm{MS}_b - \left(1 - \frac{a}{A}\right)\mathrm{MS}_{ab} - \frac{a}{A}s_e^2\right]/na = -0.32. \qquad (9.8)$$

Because a variance component cannot be negative, we replace -0.32 by zero, yielding $\sigma_b^2 \sim 0$.

In a similar fashion, we obtain

$$\sigma_a^2 \sim \left[\mathrm{MS}_a - \left(1 - \frac{b}{B}\right)\mathrm{MS}_{ab} - \frac{b}{B}s_e^2\right]/nb = 2.13. \qquad (9.9)$$

CONFIDENCE INTERVALS FOR VARIANCE COMPONENTS

Inspection of the MS and EMS columns of an analysis of variance table tells us which quantities can easily be estimated by a confidence interval. Each one of the entries of the MS column can be converted into a variable that has a χ^2 distribution by multiplying by the corresponding degrees of freedom and dividing by the corresponding EMS; that variable can then be used to obtain a confidence interval for the EMS. In the two-way classification, for example, for any model in Table 9.3 we see s_e^2 in the MS column, $ab(n-1)$ degrees of freedom, and $\mathrm{EMS}_r = \sigma_e^2$. Therefore, $ab(n-1)$ s_e^2/σ_e^2 is a χ^2 variable with $ab(n-1)$ degrees of freedom, and we can easily write down a $1-\alpha$ level confidence interval for σ_e^2:

$$0 < \sigma_e^2 < \frac{ab(n-1)s_e^2}{\chi^2[\alpha; ab(n-1)]}. \qquad (9.10)$$

Similarly, in the due B row we have MS_b with $b-1$ degrees of freedom estimating $\sigma_e^2+n(1-a/A)\sigma_{ab}^2+na\sigma_b^2$. Therefore, $(b-1)\text{MS}_b/[\sigma_e^2+n(1-a/A)\sigma_{ab}^2+nb\sigma_b^2]$ has a χ^2 distribution with $b-1$ degrees of freedom; thus a $1-\alpha$ level confidence interval is

$$0<\sigma_e^2+n\left(1-\frac{a}{A}\right)\sigma_{ab}^2+na\sigma_b^2<\frac{(b-1)\text{MS}_b}{\chi^2[\alpha;b-1]}. \tag{9.11}$$

The expressions in the EMS column are the *only* ones for which we can find confidence intervals in this simple manner.

When simple confidence intervals are not available, we can resort to approximations. In general, estimates of the variance components are highly variable, and we do not consider it worthwhile to spend much effort obtaining them. When the intervals are unduly long, however, they may at least be of value in keeping the investigator from having too much faith in his estimate. There are various methods for obtaining approximate intervals; see, for example, Bennett and Franklin [9.1]. Here we give a conservative interval (conservative in the sense that we know that the confidence level is at least $1-\alpha$).

A Conservative Confidence Interval for Variance Components. Because variances are always positive or zero, it is possible to obtain a conservative confidence interval by merely omitting the undesired terms in the ordinary confidence intervals formed with the χ^2 table. For example, from (9.11) we can omit the terms containing σ_e^2 and σ_{ab}^2; dividing by na, we have as a conservative interval whose confidence level is at least 1-α:

$$0<\sigma_b^2<\frac{(b-1)\text{MS}_b}{na\chi^2[\alpha;b-1]}. \tag{9.12}$$

In the Model III numerical example, the conservative .05 level interval is

$$0<\sigma_b^2<\frac{(3)(33.01)}{na\chi^2[.05;3]}$$

or

$$0<\sigma_b^2<\frac{3(33.01)}{6(0.352)}$$

or

$$0<\sigma_b^2<46.93.$$

ESTIMATION OF LINEAR COMBINATIONS OF MEANS

Since point estimates for linear combinations of means are the same in all models as in Model I, they cause no problem, except that as mentioned earlier, many of the parameters routinely estimated in Model I are not estimated in the variable effects models. The situation with confidence intervals is entirely different, however, in the different models. The usual type of confidence interval based on a t statistic cannot be obtained without an expression for the variance of the estimator, and such expressions differ in the various models. If we are studying the output of a machines and b operators, then in Model I μ is the mean output for just those ab operator–machine combinations. Thus $\bar{Y}_{...}$ varies from one study to the next merely because a particular machine–operator combination does not perform exactly the same each time. In the variable effects models, $\bar{Y}_{...}$ estimates the mean output for *all* machine–operator combinations. From one experiment to the next, $\bar{Y}_{...}$ varies because of different operators and different machines, as well as because of random variation on the part of each machine–operator combination. We must expect the variance of $\bar{Y}_{...}$ to be larger in the variable effects models than in the fixed effect model.

Table 9.6 gives the variances for the point estimate of μ for several designs. The expressions for the variances can be used for Models I, II, and III and for mixed models by substituting $A=a$, ∞, A, and so on. These variances were obtained from the variances and covariances among the α's, β's, ε's and so on, for Model III by expressing the overall sample mean as a linear combination of the α's, ε's,..., and then using the ordinary rule for variances of linear combinations.

To obtain a confidence interval for μ for any model, we first compare the variance of the overall mean with the EMS column of the analysis of variance table. If the variance is found to be identical (except for a constant multiple) to an entry in the EMS column, a χ^2 variable can be formed for estimating the variance and a simple confidence interval can be made. For example, setting $A=\infty$ in the one-way classification in Model II, we find the variance of $\bar{Y}_{..}$ to be $[\sigma_e^2+n\sigma_a^2]/na$. The analysis of variance table, as mentioned earlier, is the same for all models; therefore, we can use $\text{EMS}_a=\sigma_e^2+n\sigma_a^2$ from Table 5.3. The variance of $\bar{Y}_{..}$ is thus estimated by MS_a/na, and a $1-\alpha$ level confidence interval for μ is

$$\bar{Y}_{..}\pm t[1-\alpha/2;a-1]\sqrt{\text{MS}_a/na}\ ,$$

TABLE 9.6. VARIANCE OF OVERALL SAMPLE MEAN AND ITS ESTIMATE FOR SEVERAL DESIGNS, ALL MODELS

Design (1)	Statistic (2)	Variance (3)	Estimate of Variance (4)
1-way	$\bar{Y}_{..}$	$[n(1-\frac{a}{A})\sigma_a^2 + \sigma_e^2]/na$	$[(1-\frac{a}{A})MS_a + \frac{a}{A}\, s_e^2]/na$
1-way nested	$\bar{Y}_{...}$	$[nb(1-\frac{a}{A})\sigma_a^2 + n(1-\frac{b}{B})\sigma_b^2 + \sigma_e^2]/nab$	$[(1-\frac{a}{A})MS_a + \frac{a}{A}(1-\frac{b}{B})MS_b + \frac{ab}{AB}\, s_e^2]/nab$
Randomized complete block	$\bar{Y}_{..}$	$[b(1-\frac{a}{A})\sigma_a^2 + a(1-\frac{b}{B})\sigma_b^2 + \sigma_e^2]/ab$	$[(1-\frac{a}{A})MS_A + (1-\frac{b}{B})MS_b + (-1 + \frac{a}{A} + \frac{b}{B})s_e^2]/ab$
Latin Square	$\bar{Y}_{...}$	$[p(1-\frac{p}{A})\sigma_a^2 + p(1-\frac{p}{B})\sigma_b^2 + p(1-\frac{p}{C})\sigma_c^2 + \sigma_e^2]/p^2$	$[(1-\frac{p}{A})MS_a + (1-\frac{p}{B})MS_b + (1-\frac{p}{C})MS_c + (-1+\frac{p}{A}+\frac{p}{B}+\frac{p}{C})s_e^2]/p^2$
2-way cross	$\bar{Y}_{...}$	$[nb(1-\frac{a}{A})\sigma_a^2 + na(1-\frac{b}{B})\sigma_b^2 + n(1-\frac{a}{A})(1-\frac{b}{B})\sigma_{ab}^2 + \sigma_e^2]/nab$	$[(1-\frac{a}{A})MS_a + (1-\frac{b}{B})MS_b - (1-\frac{a}{A})(1-\frac{b}{B})MS_{ab} + \frac{ab}{AB}\, s_e^2]/nab$
3-way cross	$\bar{Y}_{....}$	$[nbc(1-\frac{a}{A})\sigma_a^2 + nac(1-\frac{b}{B})\sigma_b^2 + nab(1-\frac{c}{C})\sigma_c^2$ $+ nc(1-\frac{a}{A})(1-\frac{b}{B})\sigma_{ab}^2 + nb(1-\frac{a}{A})(1-\frac{c}{C})\sigma_{ac}^2 + na(1-\frac{b}{B})(1-\frac{c}{C})\sigma_{bc}^2$ $+ n(1-\frac{a}{A})(1-\frac{b}{B})(1-\frac{c}{C})\sigma_{abc}^2 + \sigma_e^2]/nabc$	$[(1-\frac{a}{A})MS_a + (1-\frac{b}{B})MS_b + (1-\frac{c}{C})MS_c$ $-(1-\frac{a}{A})(1-\frac{b}{B})MS_{ab} - (1-\frac{a}{A})(1-\frac{c}{C})MS_{ac}$ $-(1-\frac{b}{B})(1-\frac{c}{C})MS_{bc} + (1-\frac{a}{A})(1-\frac{b}{B})(1-\frac{c}{C})MS_{abc}$ $+ \frac{abc}{ABC}\, s_e^2]/nabc$

where the degrees of freedom $a-1$ correspond to MS_a and are read from the analysis of variance table.

A Conservative Confidence Interval. We are seldom fortunate enough to find the variance in the EMS column. As an illustration, we return to the two-way classification with factor A fixed and factor B Model III. The variance of the overall sample mean is from Table 9.6 with $A=a$:

$$\operatorname{Var}\bar{Y}_{...}=\left[na\left(1-\frac{b}{B}\right)\sigma_b^2+\sigma_e^2\right]/nab, \tag{9.13}$$

and we find no entry in the EMS column which is a multiple of the variance. For rough work, we can be satisfied with a conservative confidence interval as follows. We note that if we drop the $-b/B$ in the parentheses in (9.13) the expression is made larger; thus $\operatorname{Var}\bar{Y}_{...}$ is less than $(na\sigma_b^2+\sigma_e^2)/nab \sim \mathrm{MS}_b/nab$. If we form a confidence interval for μ using MS_b in the estimate of $\operatorname{Var}\bar{Y}_{...}$, it will be somewhat too long. This gives us as a conservative interval for μ, with confidence level greater than or equal to $1-\alpha$:

$$\bar{Y}_{...}\pm t[1-\alpha/2;b-1]\sqrt{\mathrm{MS}_b/nab}\,. \tag{9.14}$$

For the numerical example, with $a=A=2, b=4, B=10$, the confidence interval with a conservative 95% level is

$$20.71\pm 3.18\sqrt{33.01/24} \qquad \text{or} \qquad 17.0<\mu<24.4.$$

An Approximate Confidence Interval. Where there is no simple estimate of the variance, it is also possible, with a little more effort, to find an approximate confidence interval using an approximation suggested by Satterthwaite [9.13]. From the estimates of the variance components as found in the preceding section, we estimate the variance of $\bar{Y}_{...}$ and use Satterthwaite's approximate χ^2 variable in a confidence interval. In the same example ($a=A=2, b=4, B=10, n=3$), we have

$$\sigma_e^2 \sim s_e^2$$

$$\sigma_b^2 \sim \left(\mathrm{MS}_b-\frac{a}{A}s_e^2\right)/na,$$

giving

$$\operatorname{Var} \overline{Y}_{\dots} = \left[na\left(1 - \frac{b}{B}\right)\sigma_b^2 + \sigma_e^2 \right] / nab$$

$$\sim \left[\left(1 - \frac{b}{B}\right)\mathrm{MS}_b + \left(1 - \frac{a}{A} + \frac{ab}{AB}\right)s_e^2 \right] / nab$$

$$= \frac{\widetilde{\mathrm{MS}}}{nab}.$$

Satterthwaite's Approximation. Satterthwaite has suggested that such linear combinations of mean squares can be used to form variables that are distributed approximately as χ^2 variables. We form

$$\nu\widetilde{\mathrm{MS}} / \left[na\left(1 - \frac{b}{B}\right)\sigma_b^2 + \sigma_e^2 \right] = \nu\left[\left(1 - \frac{b}{B}\right)\mathrm{MS}_b + \left(1 - \frac{a}{A} + \frac{ab}{AB}\right)s_e^2 \right] / nab \operatorname{Var} \overline{Y}_{\dots} \tag{9.15}$$

and treat it as if it were a χ^2 variable with ν degrees of freedom. The statistic in (9.15) does *not* have a χ^2 distribution, but we hope that its distribution can be approximated by a χ^2 distribution. The degrees of freedom ν are chosen in such a manner that the variance of the statistic approximates that of the χ^2 distribution with ν degrees of freedom.

Here ν is calculated from

$$\nu = \frac{\left[(1 - b/B)\mathrm{MS}_b + (1 - a/A + ab/AB)s_e^2 \right]^2}{\left[(1 - b/B)^2 \mathrm{MS}_b^2 \right] / (b-1) + \left[(1 - a/A + ab/AB)^2 (s_e^2)^2 \right] / ab(n-1)}. \tag{9.16}$$

In general, if the linear combination of the mean squares $\mathrm{MS}_1, \mathrm{MS}_2, \dots, \mathrm{MS}_k$ is $\widetilde{\mathrm{MS}} = C_1(\mathrm{MS}_1) + C_2(\mathrm{MS}_2) + \cdots + C_k(\mathrm{MS}_k)$, and if the degrees of freedom for MS_i are ν_i, $i = 1, \dots, k$, the degrees of freedom are chosen as

$$\nu = \frac{(\widetilde{\mathrm{MS}})^2}{C_1^2(\mathrm{MS}_1)^2/\nu_1 + C_2^2(\mathrm{MS}_2)^2/\nu_2 + \cdots + C_k^2(\mathrm{MS}_k)^2/\nu_k}. \tag{9.17}$$

In the example, the two mean squares are $\mathrm{MS}_1 = \mathrm{MS}_b$, $\mathrm{MS}_2 = s_e^2$, and we recall that $[ab(n-1)s_e^2]/\sigma_e^2$ is a χ^2 variable with $ab(n-1)$ degrees freedom and that $[(b-1)\mathrm{MS}_b]/[\sigma_e^2 + n(1-a/A)\sigma_{ab}^2 + na\sigma_b^2]$ is a χ^2 variable with $b-1$ degrees of freedom. Therefore, $\nu_1 = b-1 = 3$ and $\nu_2 = ab(n-1) = 16$. We calculate $\widetilde{\mathrm{MS}} = 24.07$, $C_1 = (1-b/B) = 1-4/10 = 0.6$, $C_2 = (1-a/A + ab/AB) = 1-2/2+8/20 = 0.4$, $\nu = 4.39$, and a $1-\alpha$ level confidence interval for μ is

$$\bar{Y}_{\dots} \pm t[1-\alpha/2;\nu]\sqrt{\widetilde{\mathrm{MS}}/nab} \qquad \text{or} \qquad 18.0 < \mu < 23.4 \qquad \text{for} \quad \alpha = .05. \tag{9.18}$$

If factor B is II rather than III, then the $-b/B$ term in (9.13) is zero. The confidence interval used for μ is then (9.14); now the confidence level is equal to $1-\alpha$ and the interval is not a conservative one as it was for Model III. In the last column of Table 9.6 there are unbiased estimates $\widetilde{\mathrm{MS}}$ for the variance of the overall mean; these are useful in working with Satterthwaite intervals.

The methods given can always be used to find confidence intervals for the overall mean, the parameter most important in the variable models. For mixed models, the investigator may wish to estimate other parameters. With factor A fixed and B variable in the two-way classification, it would be appropriate to estimate $\mu + \alpha_i, \alpha_i, \alpha_i - \alpha_{i'}, \dots$. Point estimates follow as in Model I, but there are complications in obtaining confidence intervals when factor B is Model III. The expressions for the variances now contain parameters other than σ_e^2, σ_a^2, σ_b^2, and σ_{ab}^2, and the situation calls for either additional assumptions or other methods. These difficulties do not arise when factor B is Model II, and we can write the necessary expression for the variance with little difficulty. To estimate $\mu + \alpha_i$ in the same example, averaging Y_{ijk} over j and k, we have

$$\bar{Y}_{i..} = \mu + \alpha_i + \bar{\beta}_. + (\overline{\alpha\beta})_{i.} + \bar{\varepsilon}_{i..}. \tag{9.19}$$

Because $\bar{\beta}_.$ is the mean of b values of β_j, its variance is σ_b^2/b; similarly, the variance of $(\overline{\alpha\beta})_{i.}$ is σ_{ab}^2/b, and the variance of $\bar{\varepsilon}_{i..}$ is σ_e^2/nb. Thus we set

$$\operatorname{Var} \bar{Y}_{i..} = \frac{\sigma_e^2 + n\sigma_b^2 + n\sigma_{ab}^2}{nb}.$$

To find an approximate confidence interval, we could proceed to find $\widetilde{\mathrm{MS}}$ for Satterthwaite's approximation.

Satterthwaite's approximate χ^2 variable also serves in making approximate F tests. It is considered to be a good approximation when all the coefficients are positive. We do not recommend the use of this method when some coefficients are negative, and thus we do not introduce it for obtaining confidence intervals for variance components.

A CONSERVATIVE F TEST

We now return to the problem mentioned earlier, when there is no exact F test and we wish conservative or approximate F tests. We assume that the two-way cross classification has factor A Model III and factor B Model III, and wish to test $H_0:\sigma_b^2=0$. A conservative test can be obtained by neglecting $-a/A$ as follows.

When $\sigma_b^2=0$, the quantity

$$F=\frac{\text{MS}_b/[\sigma_e^2+n(1-a/A)\sigma_{ab}^2]}{\text{MS}_{ab}/(\sigma_e^2+n\sigma_{ab}^2)}$$

has an F distribution with degrees of freedom $b-1$ and $(a-1)(b-1)$. It cannot be used for testing $H_0:\sigma_b^2=0$, however, because the expressions involving unknown parameters do not cancel and therefore cannot be evaluated. But if we change the coefficient of σ_{ab}^2 from $n(1-a/A)$ to n, $\sigma_e^2+n\sigma_{ab}^2$ cancels, and we can calculate the statistic $\text{MS}_b/\text{MS}_{ab}$. We have decreased the value of the statistic; thus if we compare it with $F[1-\alpha;$ $b-1,(a-1)(b-1)]$ we have made it more difficult to reject the null hypothesis. Such a test is conservative; if we try to have a critical region of $\alpha=.05$, we actually have one with $\alpha\leqslant.05$. When A is fairly large compared with a, this test is satisfactory.

In the numerical example, with $a=2$, $A=8$, $b=4$, $B=10$, and $n=3$, the computed F statistic equals 0.767, which is less than $F(.95;3,3)=9.28$ for a conservative test with $\alpha\leqslant.05$.

AN APPROXIMATE F TEST

To test $H_0:\sigma_b^2=0$, we need a mean square with which to compare MS_b. Its expected value should be $\sigma_e^2+n(1-a/A)\sigma_{ab}^2$. We do not have one, but we can find a linear combination of the other mean squares whose expected value equals $\sigma_e^2+n(1-a/A)\sigma_{ab}^2$. This mean square can be used in an F statistic. It is easy to see that $\widetilde{\text{MS}}=(1-a/A)\text{MS}_{ab}+(a/A)s_e^2$ has the expected value we need. With ν calculated from (9.17), the expression

$\nu\widetilde{\text{MS}}/[\sigma_e^2+n(1-a/A)\sigma_{ab}^2]$ has approximately a χ^2 distribution with ν degrees of freedom. Therefore we can form a statistic $F=\text{MS}_b/\widetilde{\text{MS}}$ that will have approximately an F distribution with $b-1$ and ν degrees of freedom. This statistic can be compared with $F[1-\alpha;b-1,\nu]$ to give a test whose level of significance is approximately α.

In the example, $\widetilde{\text{MS}}=34.9$ and $\nu=3.50$; the F statistic is 0.946, which is less than $F[.95;3,3.50]=7.94$.

RULE FOR OBTAINING THE EMS COLUMN

As we have seen, the entries in the EMS column of any analysis of variance table are basic to the analysis. They can be obtained (with considerable algebraic manipulation) by writing out each sum of squares in terms of the parameters μ, α_i, and so on, and using the usual rules for taking variances and covariances of linear combinations. As designs increase in complication, it becomes exceedingly tedious to derive these expected mean squares. We proceed now to give a cookbook method for writing them down; the method is applicable to both crossed and nested classifications.

We first make an auxiliary table—Tables 9.7 and 9.8, respectively, are the auxiliary tables for the three-way classification and for $Y_{ijk}=\mu+\alpha_i+\beta_{(i)j}+\varepsilon_{(ij)k}$, the nested design.

The first column of the auxiliary table consists of the various terms contained in $Y-\mu$. In parentheses beside each parameter are given the names of the corresponding effect and of the corresponding variance component. For example, in Table 9.7 in the second line of column 1, we have $\beta_j(B,\sigma_b^2)$; the second line in Table 9.8 has $\beta_{(i)j}(B,\sigma_b^2)$ in the first column.

As headings for the other columns of the table we use each of the various subscripts (i, j, etc.), and in parentheses following each subscript we write the corresponding symbols (a, A, etc.). In Table 9.7 the headings are $i(a,A), j(b,B), k(c,C), l(n,N=\infty)$; in Table 9.8 they are $i(a,A), j(b,B)$, and $k(n,N=\infty)$.

To complete the body of the auxiliary table, we must distinguish between crossed and nested classifications. Whenever a factor is nested within one or more other factors, the subscripts of factors within which it is nested are enclosed in parentheses. In the design given in Table 9.8, for example, the parameters are α_i, $\beta_{(i)j}$, and $\varepsilon_{(ij)k}$, where the j is nested within i and the k is nested within ij. In the three-way classification, the subscript l is nested within the ijkth treatment, and we write $\varepsilon_{(ijk)l}$. Because only the error term in this design is nested, these parentheses are frequently omitted. Here we must include them, however.

TABLE 9.7. AUXILIARY TABLE FOR OBTAINING EMS COLUMN, THREE-WAY CLASSIFICATION

(1) Column	(2) $i(a,A)$	(3) $j(b,B)$	(4) $k(c,C)$	(5) $\ell(n,N=\infty)$
$\alpha_i(A,\sigma_a^2)$	$1-a/A$	b	c	n
$\beta_j(B,\sigma_b^2)$	a	$1-b/B$	c	n
$\gamma_k(C,\sigma_c^2)$	a	b	$1-c/C$	n
$(\alpha\beta)_{ij}\ (AB,\sigma_{ab}^2)$	$1-a/A$	$1-b/B$	c	n
$(\alpha\gamma)_{ik}\ (AC,\sigma_{ac}^2)$	$1-a/A$	b	$1-c/C$	n
$(\beta\gamma)_{jk}\ (BC,\sigma_{bc}^2)$	a	$1-b/B$	$1-c/C$	n
$(\alpha\beta\gamma)_{ijk}\ (ABC,\sigma_{abc}^2)$	$1-a/A$	$1-b/B$	$1-c/C$	n
$\varepsilon_{(ijk)\ell}\ (Res;\sigma_e^2)$	1	1	1	$1-n/N=1$

To fill in a cell in the table, we look at the subscript appearing in the heading above the cell and also look at the parameter listed in the first column of the same row as the cell. We then write a number in the cell as follows:

1. If the subscript listed in the heading is *missing* as a subscript of the parameter in column 1, we write down the letter to which that subscript runs (a for i, b for j, etc.; these letters can be read from the headings).
2. If the subscript listed in the heading is *present* enclosed in parentheses as a subscript of the parameter in column 1, we write down "1."

TABLE 9.8. AUXILIARY TABLE FOR OBTAINING EMS COLUMN, NESTED CLASSIFICATION

(1) Column	(2) $i,(a,A)$	(3) $j,(b,B)$	(4) $k,(n,N=\infty)$
$\alpha_i(A;\sigma_a^2)$	$1-a/A$	b	n
$\beta_{(i)j}(B;\sigma_b^2)$	1	$1-b/B$	n
$\varepsilon_{(ij)k}\ (Res;\sigma_e^2)$	1	1	$1-n/N$

3. If the subscript listed is *present* and *not* enclosed in parentheses as a subscript of the parameter in column 1, we write down $1-p/P$ where (p,P) are from the column heading. For example, for the column headed $i(a,A)$, we write $1-a/A$.

Application of these rules in filling in the auxiliary table should be clear from Tables 9.7 and 9.8. In the α_i row of Table 9.8, for example, i is present and outside parentheses; thus $1-a/A$ is written in the first column. The subscript j is missing in α_i; therefore, b is written in the second column. Similarly, n is written in the third column.

In the $\beta_{(i)j}$ row, i is present in parentheses, and 1 appears in the first column of that row. The subscript j is present and outside parentheses, and thus we write $1-b/B$ in the second column. The subscript k is missing, and so n is written in the last column.

In the $\varepsilon_{(ij)k}$ row, i and j are both within parentheses; therefore, 1 appears in each of these cells. Because k is present outside parentheses, we write $1-n/N$ in the third column. We then take N to be infinite and replace $1-n/N$ by 1.

After completion of the auxiliary table, we use it in order to find the expected mean squares. First, we must decide which variance components to include in each expected mean square. Second, we must obtain coefficients that multiply those components which are included. In the two steps, parentheses around the subscripts have no effect.

1. An expected mean square must be found for each of the effects listed in column 1 of the auxiliary table. For each effect whose EMS we seek, we look at the subscripts; we include every variance component listed in column 1 whose effect contains all those subscripts.

To find the EMS for the AC effect in Table 9.7, for example, we see the subscripts i and k on $(\alpha\gamma)_{ik}$, $(\alpha\beta\gamma)_{ijk}$, and $\varepsilon_{(ijk)l}$; therefore,the expected mean square for the AC effect must contain σ_{ac}^2, σ_{abc}^2, and σ_e^2. As a second example, for the B effect in Table 9.8 we see the subscripts i and j on $\beta_{(i)j}$. These subscripts occur in $\beta_{(i)j}$ and in $\varepsilon_{(ij)k}$, and the components σ_b^2 and σ_e^2 must be included in the expected mean square for the "due B" row.

2. The coefficient of a variance component is also found from the auxiliary table. For any particular variance component, we begin with the product of all the cell entries in that component's row in the auxiliary table. We then cancel the cell entries occurring in columns whose headings contain letters in the name of the particular effect whose EMS we are obtaining.

For an example, for the AC effect, we need the coefficients of σ_{ac}^2, σ_{abc}^2, and σ_e^2. For σ_{ac}^2, we find $1-a/A$, b, $1-c/C$, and n from the σ_{ac}^2 row in the auxiliary table. Because we are working with the AC effect, $1-a/A$, and $1-c/C$ are canceled from the product, and the coefficient of σ_{ac}^2 is bn. For σ_{abc}^2 we find $1-a/A$, $1-b/B$, $1-c/C$, and n from the σ_{abc}^2 row of the auxiliary table. Because we are finding the EMS for the AC effect, we cancel $1-a/A$ and $1-c/C$, and the coefficient of σ_{abc}^2 is $(1-b/B)n$. The coefficient of σ_e^2 is always 1; this can be obtained using the foregoing rules by calling the residual term the $ABCN$ effect.

SUMMARY

Variable effects models, both infinite and finite, and mixed models have been described briefly and contrasted with the fixed effects model. Satterthwaite's approximation was introduced and used for F tests and for confidence intervals for means. Conservative methods were also described for F tests, for confidence intervals for means, and for confidence intervals for variance components. A rule for writing down the EMS column of the analysis of variance table was given.

REFERENCES

APPLIED STATISTICS

9.1. Bennett, C. A., and N. L. Franklin, *Statistical Analysis in Chemistry and the Chemical Industry*, Wiley, New York, 1954.

9.2. Brownlee, K. A., *Statistical Theory and Methodology in Science and Engineering*, 2nd ed., Wiley, New York, 1965.

MATHEMATICAL STATISTICS

9.3. Bross, I., "Fiducial Intervals for Variance Components," *Biometrics*, Vol. 6 (1950), p. 136.

9.4. Bush, N., and R. L. Anderson, "A Comparison of Three Different Procedures for Estimating Variance Components," *Technometrics*, Vol. 5, No. 4 (1963), pp. 421–440.

9.5. Cornfield, Jerome, and John W. Tukey, "Average Values of Mean Squares in Factorials," *Annals of Mathematical Statistics*, Vol. 27 (1956), pp. 907–949.

9.6. Eisenhart, C., "The Assumptions Underlying the Analysis of Variance," *Biometrics*, Vol. 3 (1947), pp. 1–21.

9.7. Fleiss, J. L., "Distribution of a Linear Combination of Independent Chi-Squares," *Journal of the American Statistical Association*, Vol. 66 (1971), pp. 142–144.

9.8. Scheffé, H., *The Analysis of Variance*, Wiley, New York, 1967.

9.9. Searle, S. R., and R. F. Fawcett, "Expected Mean Squares in Variance Component Models Having Finite Populations," *Biometrics*, Vol. 26 (1970), pp. 243-254.

9.10. Thompson, W. A., and J. R. Moore, "Non-negative Estimates of Variance Components," *Technometrics*, Vol. 5, No. 4 (1963), pp. 441–450.

MEAN SQUARE ESTIMATION

9.11. Gaylor, D. W., and F. N. Happer, "Estimating Degrees of Freedom for Linear Combinations of Mean Squares by Satterthwaite's Formula," *Technometrics*, Vol. 11 (1969), pp. 691–706.

9.12. Howe, R. B., and R. H. Myers, "An Alternative to Satterthwaite's Test Involving Positive Linear Combinations of Variance Components," *Journal of the American Statistical Association*, Vol. 65 (1970), pp. 404–412.

9.13. Satterthwaite, F. E., "An Approximate Distribution of Estimates of Variance Components," *Biometrics, Bulletin 2*, Vol. 6 (1946), pp. 110–114.

PROBLEMS

9.1. A study to compare two teaching methods involved three teachers. Each teacher taught four students by each of the two methods. The data represent scores from a final examination. The three teachers were picked at random from a large group of teachers.

	Teaching Method							
Teacher	1				2			
1	67	73	59	84	75	61	67	58
2	92	84	94	83	54	78	61	70
3	74	72	76	64	42	44	80	83

Preliminary calculations were as follows:

$\sum_k Y_{11k} = 283.$ $\quad \sum_k Y_{12k} = 261.$ $\quad \sum_{jk} Y_{1jk} = 544.$

$\sum_k Y_{11k}^2 = 20355.$ $\quad \sum_k Y_{12k}^2 = 17199.$ $\quad \sum_{jk} Y_{1jk}^2 = 37554.$

$\sum_k Y_{21k} = 353.$ $\quad \sum_k Y_{22k} = 263.$ $\quad \sum_{jk} Y_{2jk} = 616.$

$\sum_k Y_{21k}^2 = 31245.$ $\quad \sum_k Y_{22k}^2 = 17621.$ $\quad \sum_{jk} Y_{2jk}^2 = 48866.$

$\sum_k Y_{31k} = 286.$ $\quad \sum_k Y_{32k} = 249.$ $\quad \sum_{jk} Y_{3jk} = 535.$

$\sum_k Y_{31k}^2 = 20532.$ $\quad \sum_k Y_{32k}^2 = 16989.$ $\quad \sum_{jk} Y_{3jk}^2 = 37521.$

$\sum_{ik} Y_{i1k} = 922.$ $\quad \sum_{ik} Y_{i2k} = 773.$ $\quad \sum_{ijk} Y_{ijk} = 1695.$

$\sum_{ik} Y_{i1k}^2 = 72132.$ $\quad \sum_{ik} Y_{i2k}^2 = 51809.$ $\quad \sum_{ijk} Y_{ijk}^2 = 123941.$

$i = 1,2,3$ (teacher)
$j = 1,2$ (method)
$k = 1,\ldots,4$

(*a*) State an appropriate model and assumptions in terms of the problem.
(*b*) Fill in the analysis of variance table, including the EMS column.
(*c*) Test $H_0:\sigma_a^2=0$ and $H_0:\beta_1=\beta_2=0$.
(*d*) Estimate by point estimates and confidence intervals μ, β_1, β_2, and $\beta_1-\beta_2$.

9.2. In Chapter 6, we described randomized block designs in which both treatment and block effect were assumed to be Model I. Frequently it is appropriate to make the block effect Model II. For treatment effect Model I and block effect Model II, obtain the EMS column by first writing out an auxiliary table such as Table 9.7. Describe how you would test $H_0:\alpha_1=0$ and $H_0:\sigma_b^2=0$.

9.3. In a three-way cross-classification model $A=a$ (Model I), $B=\infty$ (Model II) and C is finite and larger than c (Model III). Find the EMS from Table 9.4 and describe how you would test for the three main effects.

CHAPTER 10

LINEAR REGRESSION AND CORRELATION

When two measurements are made on each individual, it is often desirable to study them simultaneously. In Chapter 1 we introduced the scatter diagram and the correlation coefficient, which are both widely used in working with two variables. In the first part of the present chapter, we introduce another important method, regression analysis. In the second part of the chapter we return to the population correlation coefficient and consider methods for making statistical inferences regarding it.

LINEAR REGRESSION

After plotting a scatter diagram, the investigator often wishes to draw a curve that in some sense fits the points as well as possible. There are several possible objectives in fitting a curve, including the following:

1. *General study*. The investigator may wish to study a general underlying pattern connecting the two variables, say age and weight.
2. *Prediction*. He may wish a curve that shows the relation between age and weight, in order to be able to predict a person's weight from his age.
3. *Removal of an unwanted factor*. He may feel that he can study weight better if he "removes" the effect of age. If he is able to obtain a curve giving a usual relation between age and weight, he can replace the weight for each individual by its deviation from the value predicted for his age, thus eliminating age as an unwanted factor.

FITTING A STRAIGHT LINE

Of the various types of curve that are sometimes fitted to data, the one most commonly used is the straight line. Frequently the decision to fit a straight line is made simply because the line can be fitted easily. On the other hand, the straight line is often an excellent choice for the data. When a straight line does not fit the data well, it is sometimes possible to make a transformation on the data in such a way that a straight line fits reasonably well. For example, a straight line fitting the logarithms of weight to the square root of age seems to fit better than one fitting weight to age.

In this section we present the most usual method for fitting a straight line; in a later section we discuss briefly other types of regression. The regression line given in this section is called the least squares line. The least squares line has the property that the sum of the squared vertical distances from it to points on the scatter diagram is smaller than the similar sum for any other line.

If X_i and Y_i denote the measurements on each of n individuals, our notation is $x_i = X_i - \bar{X}$, $y_i = Y_i - \bar{Y}$; in general, $x = X - \bar{X}$, $y = Y - \bar{Y}$, where $\bar{X}$ and $\bar{Y}$ are the two means, each based on the n measurements. The Y's are measured along the vertical axis, the X's along the horizontal axis. The equation of any straight line except a vertical one can be written in the form $Y = a + b(X - \bar{X}) = a + bx$. (It could also be written as $Y = a' + bX$, but the formulae for regression are simpler if we work with $x = X - \bar{X}$ instead of with X.) We use Y' to denote any point on the least squares regression line and Y_i' to denote the ordinate of the point on the regression line corresponding to X_i. We have said that the sum of squared vertical deviations is as small as possible when the least squares line is used. This is equivalent to saying that a and b must be chosen so that the expression $Q = \sum_{i=1}^{n} (Y_i - a - bx_i)^2$ is as small as possible. The appropriate a and b can be found from (10.1) and (10.2).*

$$\sum_{i=1}^{n} Y_i = na + b \sum_{i=1}^{n} x_i \tag{10.1}$$

* Finding a and b to satisfy this condition is a problem in calculus. The two partial derivatives of Q with respect to a and b are set equal to zero. We have

$$\frac{\partial Q}{\partial a} = -2 \sum_{i=1}^{n} (Y_i - a - bx_i) = 0$$

$$\frac{\partial Q}{\partial b} = -2 \sum_{i=1}^{n} x_i (Y_i - a - bx_i) = 0,$$

which are equivalent to (10.1) and (10.2).

$$\sum_{i=1}^{n} x_i Y_i = a \sum_{i=1}^{n} x_i + b \sum_{i=1}^{n} x_i^2. \tag{10.2}$$

Because $\sum_{i=1}^{n} x_i = 0$, it can be shown that $\sum_{i=1}^{n} x_i Y_i = \sum_{i=1}^{n} x_i y_i$ and that the equations become

$$\sum_{i=1}^{n} Y_i = na \tag{10.3}$$

$$\sum_{i=1}^{n} x_i y_i = b \sum_{i=1}^{n} x_1^2.$$

Therefore the solution for a and b is

$$a = \bar{Y}$$

$$b = \frac{\sum_{i=1}^{n} x_i y_i}{\sum_{i=1}^{n} x_i^2}.$$

When there is no ambiguity, we often write $\sum x^2$ for $\sum_{i=1}^{n} x^2$, $\sum xy$ for $\sum_{i=1}^{n} x_i y_i$, and $\sum y^2$ for $\sum_{i=1}^{n} y_i^2$. The well-known least squares line then becomes

$$Y' = \bar{Y} + b(X - \bar{X}). \tag{10.4}$$

where (X, Y') denotes any point on the line and $b = \sum xy / \sum x^2$. Here $\bar{Y}$ is the ordinate or height of the line at the point $\bar{X}$. The quantity b is the *slope* of the line; when X increases by one unit, the height of the line changes by b units. The line can alternately be written as $y' = bx$, where $y' = Y' - \bar{Y}$ and $x = X - \bar{X}$.

AN EXAMPLE OF THE LEAST SQUARES LINE

As an example of the computation of the least squares line, we use the data from Table 10.1 on fiber diameter (microns) and the log to the base 10 of breaking strength; calculations and formulae appear in the last column of the table.

Preliminary calculations include the summations $\sum X$ and $\sum Y$ and the sums of squares and cross products—$\sum X^2$, $\sum XY$, and $\sum Y^2$. From them, the so-called corrected sums of squares and cross products are obtained—$\sum x^2$, $\sum xy$, and $\sum y^2$—and $\bar{X}$, a, and b are calculated. From $\bar{X}$, a, and b,

TABLE 10.1. FIBER DIAMETER AND LOG BREAKING STRENGH OF 15 FIBER

	X Fiber Diameter (microns)	Y Log_{10} Breaking Strength
1	22.5	.19
2	28.0	.62
3	27.5	.51
4	25.5	.53
5	22.0	.24
6	30.5	.87
7	23.0	.25
8	25.0	.25
9	23.5	.37
10	27.0	.32
11	21.5	.13
12	22.0	.35
13	29.0	.53
14	20.5	.22
15	27.0	.65

Preliminary Calculations

$$\Sigma X = 374.5$$
$$\Sigma Y = 6.03$$
$$\Sigma X^2 = 9{,}482.75$$
$$\Sigma XY = 158.250$$
$$\Sigma Y^2 = 3.0315$$
$$\Sigma x^2 = \frac{n\Sigma X^2 - (\Sigma X)^2}{n} = 132.733$$
$$\Sigma xy = \frac{n\Sigma XY - \Sigma X\Sigma Y}{n} = 7.701$$
$$\Sigma y^2 = \frac{n\Sigma Y^2 - (\Sigma Y)^2}{n} = .6074$$

Calculation of Regression Line

$$\bar{X} = 24.97$$
$$\bar{Y} = a = 0.402$$
$$b = \Sigma xy/\Sigma x^2 = +\ .058$$
$$Y' = .402 + .058(X-24.97)$$
$$\text{or} \quad Y' = -1.046 + .058X$$
$$\text{or} \quad y' = .058x$$
$$\sum_{i=1}^{n} (Y_i - Y_i')^2 = \Sigma y^2 - b\Sigma xy$$
$$= .6074 - .058(7.701)$$
$$= .1608$$
$$s_e^2 = \sum_{i=1}^{n} (Y_i - Y_i')^2/(n-2)$$
$$= .1608/13 = .01236$$

the regression line is written down. The calculations also include $\sum_{i=1}^{n}(Y_i - Y_i')^2$, which is used to compute s_e^2, as discussed later. The formulae given in the table for calculating $\sum x^2$, $\sum xy$, $\sum y^2$, and $\sum_{i=1}^{n}(Y_i - Y'_i)^2$ were chosen for ease in desk calculator operations; because of round-off errors, they should not be used for computations with an electronic computer (see Table 1.4 for other formulae).

Figure 10.1 presents a scatter diagram of the data and the least squares line.

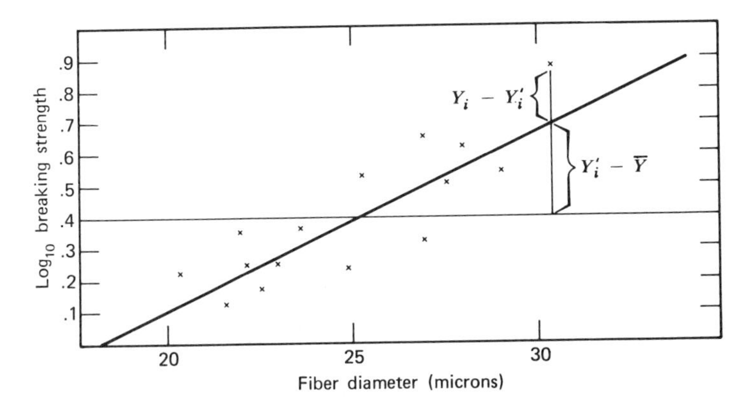

Figure 10.1. Log_{10} breaking strength versus fiber diameter.

If we wish to compute a line of the form $Y' = a' + bX$, the slope of the line is the same as that given in (10.4). The intercept a' is

$$a' = \bar{Y} - b\bar{X}. \tag{10.5}$$

The line in either form passes through the point $\bar{X}, \bar{Y}$. By substituting an $\bar{X}$ for X in $Y' = \bar{Y} + b(X - \bar{X})$ we have $Y' = \bar{Y} + b(\bar{X} - \bar{X}) = \bar{Y}$. Substituting $X = 0$ for X, we obtain the Y intercept: $Y' = \bar{Y} + b(0 - \bar{X}) = \bar{Y} - b\bar{X}$, given in (10.5).

REGRESSION WITH FIXED X'S

A straight line is often fitted to a set of data in the hopes of making inferences concerning the underlying populations. We now state some assumptions about these underlying populations; under these assumptions we shall be able to estimate population parameters with confidence intervals and to make tests.

For each *fixed* X of a certain set, we assume that there is a normally distributed population of Y's. All the means of these populations lie on a straight line, which is called the *population regression line*, and all the variances are equal.

The population regression line is estimated by the least squares line, $Y' = \bar{Y} + b(X - \bar{X})$, which is called the *sample regression line*, and any parameter associated with the model can be estimated by the corresponding sample statistic.

For example, we consider a population of fibers all having the same diameter, say, $X = 30$ microns. The log breaking strengths of fibers with $X = 30$ form a population of Y's; this population has a mean denoted by

$E(Y|X=30)$, the expected value of Y given that the fiber diameter equals 30 microns. The population has a variance that is called $\text{Var}(Y|X=30)$. For each fiber strength under consideration, such a population of Y's exists.

In the model we are describing, we make three assumptions concerning these populations.

1. They are normally distributed—more specifically, log breaking strengths for fibers with a particular diameter are normally distributed.
2. Their means all lie on a straight line. In symbols, this assumption is $E(Y|X)=\alpha+\beta(X-\bar{X})$.
3. Their variances are all equal. This can be expressed by writing $\text{Var}(Y|X)=\sigma_e^2$ [i.e., we can use a single symbol σ_e^2 for $\text{Var}(Y|X)$, no matter what value X has].

These three assumptions concerning the underlying population can be summarized by writing

$$Y|X \text{ ND}\left(\alpha+\beta(X-\bar{X}),\sigma_e^2\right).$$

The assumptions that the population means lie on a straight line and that the population variances are equal may be quite unrealistic. Often, however, if X is confined to a moderate range, they are reasonable assumptions.

An alternate expression for the model is

$$Y_i=\alpha+\beta\left(X_i-\bar{X}\right)+\varepsilon_i, \qquad i=1,\ldots,n \tag{10.6}$$

where $\varepsilon_i \text{ IND}(0,\sigma_e^2)$.

We note that the ε's are independently distributed. If we knew that the first fiber was unusually strong for its diameter, this knowledge would tell us nothing about the strength of the other 14 fibers.

Point Estimates of Parameters. The parameter α can be estimated by $a=\bar{Y}$. Just as a is the ordinate of the sample regression line at the point $\bar{X}$, α is the ordinate of the population regression line at the point $\bar{X}$. It is also the mean for the population of Y's with $X=\bar{X}$.

The parameter β (the *population regression coefficient*) is estimated by $b=\sum xy/\sum x^2$ (the *sample regression coefficient*). The regression coefficient is the *slope* of the regression line. When X increases by one unit, the height of the population regression line changes by β units.

The variance σ_e^2 is estimated by $s_e^2=\sum_{i=1}^n(Y_i-Y_i')^2/(n-2)=\sum_{i=1}^n(Y_i-\bar{Y}-b(X_i-\bar{X}))^2/(n-2)$. It is natural to use the sum of squared deviations around the sample regression line to estimate the variance of the population. If we had known the true parameters α and β, we could have used $\sum_{i=1}^{n}(Y_i - \alpha - \beta(X_i - \overline{X}))^2/n$ to estimate σ_e^2. On the average, the use of $\overline{Y}$ and b underestimates σ_e^2 in repeated sampling, and to take account of this we subtract 2 from n and divide by $n-2$. We say that 2 degrees of freedom have been used in estimating α and β; $n-2$ remain.

Before obtaining estimates for various quantities of interest, confidence intervals for them, and tests of hypotheses, we give an analysis of variance table.

The Analysis of Variance Table and F Test. In the analysis of variance models of the earlier chapters, each deviation of an observation from the overall mean of the data was divided into two or more parts; this can also be done in regression. The ith deviation from $\overline{Y}$, $Y_i - \overline{Y}$, can be expressed as follows:

$$Y_i - \overline{Y} = \left(Y_i' - \overline{Y}\right) + \left(Y_i - Y_i'\right). \tag{10.7}$$

Just as in the analysis of variance problems, each term in (10.7) estimates a similar term from the model. In our regression model, from (10.6) we can write

$$Y_i - \alpha = \beta\left(X_i - \overline{X}\right) + \varepsilon_i. \tag{10.8}$$

It is clear that $Y_i - \overline{Y}$ corresponds to $Y_i - \alpha$ and that $Y_i - Y_i'$ corresponds to ε_i. That $Y_i' - \overline{Y}$ corresponds to $\beta(X_i - \overline{X})$ can be seen from

$$Y_i' - \overline{Y} = \overline{Y} + b\left(X_i - \overline{X}\right) - \overline{Y} = b\left(X_i - \overline{X}\right),$$

where $\overline{Y} + b(X_i - \overline{X})$ has been substituted for Y_i'.

In Figure 10.1 a deviation $Y_i - \overline{Y}$ is divided into two parts. One part $Y_i' - \overline{Y}$ shows how far the sample regression line is from the horizontal line $Y = \overline{Y}$ at $X = X_i$; the other part is the vertical distance of Y_i from the sample regression line.

When all n deviations are broken into two parts in this way, squared, and added, the cross-product terms add to zero, and we have

$$\sum_{i=1}^{n}\left(Y_i - \overline{Y}\right)^2 = \sum_{i=1}^{n}\left(Y_i' - \overline{Y}\right)^2 + \sum_{i=1}^{n}\left(Y_i - Y_i'\right)^2. \tag{10.9}$$

For convenience, we often write the total sum of squares $\sum_{i=1}^{n}(Y_i - \overline{Y})^2$, as $\sum y^2$. On the right-hand side of (10.9), $\sum_{i=1}^{n}(Y_i' - \overline{Y})^2$ is called the "due regression on X" sum of squares and is denoted by SS_x; $\sum_{i=1}^{n}(Y_i - Y_i')^2$ is the residual sum of squares SS_r—it is sometimes called the "deviations from regression on X" sum of squares. The computational formula for SS_x is obtained as follows:

$$\sum_{i=1}^{n}\left(Y_i' - \bar{Y}\right)^2 = \sum_{i=1}^{n} b\left(X_i - \bar{X}\right)^2 = b^2 \sum x^2$$

$$= b\frac{\sum xy}{\sum x^2}\left(\sum x^2\right)$$

$$= b\sum xy.$$

By subtraction, SS_r is $\sum y^2 - b\sum xy$.

Tables 10.2 and 10.3 give the analysis of variance table for linear regression in formulae and for the data on fiber diameter and log breaking strengths, respectively.

The calculated F and tabled F in the last two columns of the tables are for testing the hypothesis $H_0: \beta = 0$. If β is zero, the analysis of variance table provides two independent estimates of σ_e^2—MS_x and s_e^2; their ratio MS_x/s_e^2 is used as a test statistic.

In the example, $MS_x/s_e^2 = 36.14$, which is greater than $F[.95; 1, 13]$. Our decision, therefore, is that β is not equal to zero.

The test of $H_0: \beta = 0$ is sometimes called a test for independence. If β equals zero, then $E(Y|X) = \alpha$; thus at each value of X there is a normally distributed population of Y's, all with the same mean and the same variance. Knowledge of X can give us no information concerning Y, and X and Y are statistically independent.

Estimation of Parameters and Confidence Intervals. The point estimates for α and β have already been obtained; $a = \bar{Y}$ and $b = \sum xy/\sum x^2$. For σ_e^2, the point estimate s_e^2, can now be read from the analysis of variance table. A $1-\alpha$ level confidence interval for σ_e^2 is

$$\frac{(n-2)s_e^2}{\chi^2[1-\alpha/2; n-2]} < \sigma_e^2 < \frac{(n-2)s_e^2}{\chi^2[\alpha/2; n-2]}. \tag{10.10}$$

There are other quantities to estimate in addition to α, β, and σ_e^2. We may wish to estimate the *population mean* for individuals having a particular X value—say, $X = X^*$. This is the height of the population regression curve at $X = X^*$, $E(Y|X^*) = \alpha + \beta(X^* - \bar{X})$, and it is estimated by the

TABLE 10.2. ANALYSIS OF VARIANCE TABLE, LINEAR REGRESSION

Source of Variation (1)	SS (2)	df (3)	MS (4)	EMS (5)	F Calc. (6)	F Tabled (7)
Due Regression on X	$SS_x = \sum_{i=1}^{n} (Y_i' - \overline{Y})^2 = b\Sigma xy$	1	$MS_x = SS_x/1$	$\sigma_e^2 + \beta^2 \Sigma x^2$	MS_x/s_ϵ^2	$F[1-\alpha;1,n-2]$
Residual	$SS_r = \sum_{i=1}^{n} (Y_i - Y_i')^2 = \Sigma y^2 - b\Sigma xy$	n-2	$s_e^2 = SS_r/(n-2)$	σ_e^2		
Total	$SS_t = \sum_{n=1}^{n} (Y_i - \overline{Y})^2 = \Sigma y^2$	n-1				

TABLE 10.3. ANALYSIS OF VARIANCE TABLE FOR REGRESSION OF LOG BREAKING STRENGTHS OF FIBERS ON FIBER DIAMETERS

Source of Variation (1)	SS (2)	df (3)	MS (4)	F Calc. (5)	F Tabled (6)
Due Regression on X	.4467	1	.4467	36.14	4.68
Residual	.1607	13	.01236		
Total	.6074	14			

height of the sample regression line at X^*, $\bar{Y}+b(X^*-\bar{X})$.

We may also wish to estimate the Y value for *an individual* whose X value is known. If we have a fiber whose measured diameter is X^* and whose breaking strength (say, Y^*) is unknown, we naturally use the sample regression line to estimate Y^*. Our estimate for Y^* is $\bar{Y}+b(X^*-\bar{X})$, the same number as the estimate of the population mean breaking strength for fibers with diameter X^*.

In order to be able to construct confidence intervals, we must now obtain estimates of the variances of these statistics.

The statistics a and b are simply linear combinations of the n independent observations $Y_1, Y_2, \ldots, Y_n$. We have

$$a=\bar{Y}=\frac{1}{n}(Y_1+\cdots+Y_n). \tag{10.11}$$

Then, using (2.4) for the variance of linear combinations, we set

$$\operatorname{Var} a=\frac{1}{n^2}\operatorname{Var} Y_1+\frac{1}{n^2}\operatorname{Var} Y_2+\cdots+\frac{1}{n^2}\operatorname{Var} Y_n=\frac{1}{n^2}(n\sigma_e^2)=\frac{\sigma_e^2}{n}. \tag{10.12}$$

The expression for b can be written in an algebraically equivalent form:

$$b=\frac{x_1Y_1+x_2Y_2+\cdots+x_nY_n}{\sum_{i=1}^{n}x_i^2}. \tag{10.13}$$

Using (2.4) again, and assuming that the numbers $x_1,\ldots,x_n$ are fixed in repeated sampling, we have

$$\operatorname{Var} b = \frac{(x_1^2 + \cdots + x_n^2)\sigma_e^2}{\left(\sum_{i=1}^{n} x_i^2\right)^2} = \frac{\sigma_e^2}{\sum x^2}. \tag{10.14}$$

For finding the variance of $a + b(X^* - \bar{X})$, we need the covariance between a, the height of the sample line at $\bar{X}$, and the slope of the sample line b. From (2.5)

$$\operatorname{Cov}(a,b) = \frac{(x_1 + \cdots + x_n)\sigma_e^2}{n\sum_{i=1}^{n} x_i^2} = 0 \tag{10.15}$$

because $\sum x_i = \sum (X_i - \bar{X}) = 0$. Therefore, we have

$$\operatorname{Var}\left(a + b\left(X^* - \bar{X}\right)\right) = \frac{\sigma_e^2}{n} + \frac{\left(X^* - \bar{X}\right)^2 \sigma_e^2}{\sum x^2} = \sigma_e^2\left(\frac{1}{n} + \frac{\left(X^* - \bar{X}\right)^2}{\sum\left(X_i - \bar{X}\right)^2}\right)$$

$$= \sigma_e^2\left(\frac{1}{n} + \frac{x^{*2}}{\sum_{i=1}^{n} x_i^2}\right), \tag{10.16}$$

where $x^* = X^* - \bar{X}$. This variance is used in a confidence interval for $E(Y|X^*)$, the mean of the Y population at $X = X^*$. Note that in obtaining all these formulae for variances, we treat the X's as constants.

It is also convenient to have the variance of the difference between a fiber's breaking strength and its breaking strength as predicted from its diameter; that is, $Y^* - [a + b(X^* - \bar{X})]$. Here, as before, Y^*, X^* are the log breaking strength and diameter of a new fiber—one not included in the original sample of n fibers. Because Y^* is independent of a and b, we can write

$$\operatorname{Var}\left(Y^* - (a + bx^*)\right) = \sigma_e^2\left(1 + \frac{1}{n} + \frac{x^{*2}}{\sum_{i=1}^{n} x_i^2}\right). \tag{10.17}$$

Each of the variances just obtained enables us to write down a confidence interval for a particular parameter. Table 10.4 summarizes confidence intervals for linear regression.

TABLE 10.4. CONFIDENCE INTERVALS IN LINEAR REGRESSION

Parameter (1)	Statistic (2)	Variance (3)	Confidence Interval* (4)
α	$\overline{Y}$	σ_e^2/n	$a \pm t[1-\alpha/2m;n-2] \quad s_e/\sqrt{n}$
β	b	$\frac{\sigma_e^2}{\Sigma x^2}$	$b \pm t[1-\alpha/2m;n-2] \quad s_e/\sqrt{\Sigma x^2}$
$E(Y \mid x^*) = \alpha + \beta x^*$	$Y^{*\prime} = a + bx^*$	$\sigma_e^2 \left(\frac{1}{n} + \frac{x^{*2}}{\Sigma x^2}\right)$	$(a + b\,x^*) \pm t[1-\alpha/2m;n-2] \quad s_e \sqrt{\frac{1}{n} + \frac{x^{*2}}{\Sigma x^2}}$
Y^*	$Y^{*\prime} = a + bx^*$	$\sigma_e^2 \left(1 + \frac{1}{n} + \frac{x^{*2}}{\Sigma x^2}\right)^{**}$	$(a + b\,x^*) \pm t[1-\alpha/2m;n-2] \quad s_e \sqrt{1 + \frac{1}{n} + \frac{x^{*2}}{\Sigma x^2}}$
σ_e^2	s_e^2		$\frac{(n-2)\,s_e^2}{\chi^2[1-\alpha/2;n-2]} < \sigma_e^2 < \frac{(n-2)\,s_e^2}{\chi^2[\alpha/2;n-2]}$

* m is the number of intervals to be made with a simultaneous confidence level $1-\alpha$.

** $\sigma_e^2 \left(1 + \frac{1}{n} + \frac{x^{*2}}{\Sigma x^2}\right)$ is the variance of $Y^* - Y^{*\prime} = Y^* - (a + bx^*)$.

We notice that the same statistic, $a+bx^*$, is used to estimate two quantities—$\alpha+\beta x^*$ and Y^*. If we wish to estimate the population mean for log breaking strength for fibers whose diameter is X^*, we use $Y^{*\prime}=a+bx^*$, the corresponding point on the sample regression line. On the other hand, if we wish to estimate the log breaking strength Y^*, for a particular fiber whose diameter is X^*, we also use $Y^{*\prime}$ as the point estimate.

The intervals in these two situations differ, however, as shown in Table 10.4. The confidence interval for $E(Y|X^*)=\alpha+\beta x^*$ is straightforward. The interval for Y^*, on the other hand, is formed from a confidence interval for the population mean of the statistic $Y^*-Y^{*\prime}$. In repeated sampling, the mean of $Y^*-Y^{*\prime}$ is zero; thus we write down a confidence interval for zero using the variance of $Y^*-Y^{*\prime}$, and then transform it into the interval for Y^* as given in the table.

The confidence interval for σ_e^2 has the usual form. For the example, the variance of $\bar{Y}$ is estimated by $s_e^2/n=0.01236/15=0.00082$, and the 95% confidence interval for α—the population mean of Y's at $\bar{X}=24.97$—is $0.4020\pm(2.16)\sqrt{0.00082}$ or $0.340<\alpha<0.464$. The variance of b is estimated by $0.01236/132.733=0.0000931$, and a 95% confidence interval for β is $0.0580\pm(2.16)\sqrt{0.0000931}$ or $0.0372<\beta<0.0788$. The population mean log breaking strength for fibers with diameter $X^*=30$ is estimated by 0.694, and the 95% confidence interval is

$$0.694\pm(2.16)\sqrt{0.01236\left[1/15+(30-24.97)^2/132.733\right]}$$

$$\text{or} \qquad 0.680<E(Y|X=30)<0.708.$$

Tests of Hypotheses. As usual, the confidence intervals of the previous section can be used to test hypotheses concerning either the regression coefficient or the height of the population regression line at particular points. Alternatively, we may wish to make tests without calculating confidence intervals. A commonly made test is that of the null hypotheses $H_0:\alpha-\beta\bar{X}=\mu_0$; this is the test of whether the height of the regression line at $X=0$ equals μ_0. The test statistic is $[(a-b\bar{X})-\mu_0]/\sqrt{s_e^2(1/n+\bar{X}^2/\sum x^2)}$, which is compared with $t[1-\alpha/2;n-2]$. Similarly, to test whether the regression coefficient equals any particular value $\beta_0(H_0:\beta=\beta_0)$, we compare the statistic $(b-\beta_0)/\sqrt{s_e^2/\sum x^2}$ with $t[1-\alpha/2;n-2]$. Each of the foregoing tests has a significance level of α; to make m tests having an overall significance level of α, we replace $t[1-\alpha/2;n-2]$ by $t[1-\alpha/2m;n-2]$.

**OTHER TOPICS IN REGRESSION

***The Straight Line through the Origin.* Another regression line is sometimes fitted: a straight line passing through the point $X=0$, $Y=0$. This should be done only when it is known from the problem that the true regression line must pass through the origin. The appropriate model is

$$Y_i=\beta X_i+\varepsilon_i, \qquad i=1,\ldots,n, \tag{10.18}$$

where ε_i IND$(0,\sigma_e^2)$.

From (10.18) the mean of the population of Y's for a given X value is $E(Y|X)=\beta X$; thus the population regression line is a straight line that passes through the origin. The assumptions implied by (10.18) are similar to those for linear regression. We assume that

1. The Y_i's are independently normally distributed.
2. Their population means lie on a line of the form $E(Y|X)=\beta X$.
3. Their population variances are equal.

To obtain the least squares estimate of β from a set of data, we find the line $Y'=bX$ such that the sum of squared vertical distances from the data points to the line is a minimum. The calculus solution of this problem is

$$b=\frac{\sum_{i=1}^{n} X_iY_i}{\sum_{i=1}^{n} X_i^2}.$$

The variance σ_e^2 is of any Y population estimated by using the sum of squared vertical deviations from the regression line:

$$s_e^2=\frac{\sum_{i=1}^{n}(Y_i-bX_i)^2}{n-1}$$

or, for calculation,

$$s_e^2=\frac{\sum Y_i^2-b\sum X_iY_i}{n-1}.$$

Here $n-1$ is appropriate in the denominator, because a single parameter (β) has been estimated from the data.

Table 10.5 shows confidence intervals for the various parameters. Instead of constructing a confidence interval for β, we can test $H_0:\beta=\beta_0$ using the statistic $(b-\beta_0)/\sqrt{s_e^2/\sum_{i=1}^{n}X_i^2}$, which is compared with $t[1-\alpha/2;\ n-1]$.

TABLE 10.5. LINEAR REGRESSION THROUGH THE ORIGIN; ESTIMATES OF PARAMETERS, VARIANCES OF THE ESTIMATES, AND CONFIDENCE INTERVALS

Parameter or Expected Value (1)	Statistic (2)	Variance (3)	Confidence Interval[1] (4)
a) **Variance** $Y\|X_i = \sigma_e^2$			
β	$b = \frac{\sum_{i=1}^{n} X_i Y_i}{\sum X_i^2}$	$\sigma_e^2 / \sum_{i=1}^{n} X_i^2$	$b \pm t[1-\alpha/2m;n-1] \sqrt{s_e^2 / \sum_{i=1}^{n} X_i^2}$
$E(Y\|X^*) = \beta X^*$	$Y' = bX^*$	$\sigma_e^2 X^{*2} / \sum_{i=1}^{n} X_i^2$	$bX^* \pm t[1-\alpha/2m;n-1] \sqrt{s_e^2 X^{*2} / \sum_{i=1}^{n} X_i^2}$
Y^*	$Y'^* = bX^*$	$\sigma_e^2(1+X^{*2}/\sum_{i=1}^{n} X_i^2)$**	$bX^* \pm t[1-\alpha/2m;n-1] \sqrt{s_e^2 (1+X^{*2}/\sum_{i=1}^{m} X_i^2)}$
σ_e^2	$s_e^2 = \frac{\sum_{i=1}^{n} Y_i^2 - b\sum_{i=1}^{n} X_i Y_i}{n-1}$		$(n-1)s_e^2/\chi^2[1-\alpha/2;n-1] < \sigma_e^2 < (n-1)s_e^2/\chi^2[\alpha/2;n-1]$
b) Variance $Y\|X_i = X_i^2 \sigma_e^2$; $Z_i = Y_i/X_i$			
β	$b = \overline{Z}$	σ_e^2/n	$b \pm t[1-\alpha/2m;n-1] \sqrt{s_e^2/n}$
$E(Z\|X^*) = \beta X^*$	$Y^{*\prime} = bX^*$	$X^{*2}\sigma_e^2/n$	$bX^* \pm t[1-\alpha/2m;n-1] \sqrt{X^{*2}s_e^2/n}$
Y^*	$Y^{*\prime} = bX^*$	$X^{*2}\sigma_e^2(1 + 1/n)$†	$bX^* \pm t[1-\alpha/2m;n-1] \sqrt{X^{*2}s_e^2(1+1/n)}$
σ_e^2	$s_e^2 = \sum_{i=1}^{n} (Z_i-\overline{Z})^2/(n-1)$		$(n-1)s_e^2/\chi^2[1-\alpha/2;n-1] < \sigma_e^2 < (n-1)s_e^2/\chi^2[\alpha/2;n-1]$

[1] m is the number of confidence intervals to be made with an overall confidence level of $1-\alpha$.

** $\sigma_e^2(1+X^{*2}/\Sigma X_i^2)$ is the variance of $Y^* - Y^{*\prime} = Y^* - bX^*$.

† $X^{*2}s_e^2(1+1/n)$ is the variance of $Y^* - Y^{*\prime} = X^*(Z^*-\overline{Z})$.

As mentioned earlier, a straight line of the form $Y' = bX$ should not be fitted unless it is certain that the true regression line goes through the origin. It is insufficient to know merely that at $X=0$; Y must also be zero. Often the population regression curve is very close to a straight line for most of the range of X, but as X approaches zero, the regression curve suddenly curves down to zero. In such cases, it is appropriate to use $Y' = a + b(X - \overline{X})$ rather than $Y' = bX$. When X is fiber diameter and Y is breaking strength (rather than $\log_{10}$ breaking strength), it is certainly true that if X equals zero Y must also be zero. But we have no reason to suppose that any straight-line relationship continues to the origin, and therefore a line of the form $Y' = a + b(X - \overline{X})$ should be fitted.

A further possibility should be mentioned. Sometimes when we are convinced that the population regression line is a straight line that passes through the origin, it may be quite unrealistic to expect the variances about the line to be the same for large and for small values of X. It may be more reasonable to assume that the standard deviation varies with the size of X: if this is true, the variance of the population of Y's for a given value of X equals $X^2\sigma_e^2$. The model then becomes

$$Y_i = \beta X_i + \eta_i, \qquad i = 1, \ldots, n, \tag{10.19}$$

where η_i IND$(0, X_i^2\sigma_e^2)$.

This regression model can be handled very easily by dividing each Y value by the corresponding X value. If we let $Z_i = Y_i / X_i$, $i = 1, \ldots, n$, the model in terms of Z_i can be written (denoting η_i / X_i by ε_i) as follows:

$$Z_i = \beta + \varepsilon_i, \qquad i = 1, \ldots, n, \tag{10.20}$$

where ε_i IND$(0, \sigma_e^2)$.

The estimates β and σ_e^2 are easily obtained as $b = \overline{Z}$ and $s_e^2 = \sum_{i=1}^n (Z_i - \overline{Z})^2 / (n-1)$. Estimates and confidence intervals appear in part b of Table 10.5.

** *Weighted Regression.* In fitting a regression line, sometimes certain observations should be weighed more heavily than others. In astronomy, for example, when it is necessary to read distances from photographs, photographs taken on some nights may be clearer than other photographs, and we may wish to weigh measurements from the clearer photographs more heavily. An arbitrary set of weights W_i can be constructed and used. As a second example, the observations might fall in n groups that are so nearly alike that the regression should be performed on the n pairs of means $\overline{X}_i, \overline{Y}_i$. If W_i is the number of observations averaged to obtain each of the ith pair of means, it is appropriate to use the W_i as weights. It is assumed that W_i are fixed.

The regression line for weighted regression is obtained in a similar manner as for ordinary regression. Values of a and b are determined that minimize the quantity $Q=\sum_{i=1}^{n}W_i(Y_i-a-b(X_i-\overline{X}))^2$, where $\overline{X}$ is defined by $\overline{X}=\sum_{i=1}^{n}W_iX_i/\sum_{i=1}^{n}W_i$. If all the weights in a given problem are multiplied by any one constant, it is clear that the values of a and b that minimize Q remain unchanged; it is the relative weights that determine the sample regression line rather than the weights themselves. Therefore, to simplify the computational formulae, we can choose weights in such a way that they add to n. This can be accomplished by using w_i, $i=1,\ldots,n$, as weights, where $w_i=nW_i/\sum_{i=1}^{n}W_i$.

In terms of the weights w_i, with $\sum_{i=1}^{n}w_i=n$, we have

$$a=\overline{Y}=\frac{\sum_{i=1}^{n}w_iY_i}{n} \tag{10.21}$$

and

$$b=\frac{\sum_{i=1}^{n}w_ix_iy_i}{\sum_{i=1}^{n}w_ix_i^2} \tag{10.22}$$

where

$$y_i=Y_i-\overline{Y}, \qquad x_i=X_i-\overline{X}, \qquad \text{and} \qquad \overline{X}=\frac{\sum_{i=1}^{n}w_iX_i}{n}.$$

The estimate of the variance is the weighted sum of squared deviations divided by $n-2$:

$$s_e^2=\frac{\sum_{i=1}^{n}w_i(Y_i-Y_i')^2}{n-2}. \tag{10.23}$$

The variances of a and b are

$$\operatorname{Var}a=\frac{\sigma_e^2}{n} \tag{10.24}$$

$$\operatorname{Var}b=\frac{\sigma_e^2}{\sum_{i=1}^{n}w_ix_i^2}. \tag{10.25}$$

The variances in (10.24) and (10.25) can be used in confidence intervals and tests just as in unweighted linear regression.

****Transformations.*** The research worker may know theoretically or from looking at his data that the three assumptions underlying the linear regression model are not true. For each X, the Y's may not be normally distributed; their variances may be unequal, or their means may lie on a curve rather than on a straight line.

When the means lie on a curve a polynomial in X can be fitted to the data. Fitting a polynomial logically belongs in this chapter, but because it is accomplished by the methods used in multiple regression, we postpone its discussion till Chapter 12. Besides polynomials, there are numerous special curves that can be fitted in special situations.

A method widely used when assumptions are not met is that of transforming the data. Transformations are made with the objective of obtaining new variables that satisfy all three assumptions: normality, equal variances, and means on a straight line. For example, we may have data such that the curve appearing in Figure 10.2 seems to be appropriate. If we use $\log(X+A)$ instead of X, we can fit a straight line $Y' = c + d\log(X+A)$. The appropriate transformation is sometimes based on graphical analysis and sometimes on professional knowledge of the behavior of the variables. It can also be chosen by a trial-and-error approach—various transformations are tried, lines are fitted, and the residuals are examined over the range of X.

Transformations are discussed further in Chapter 14. In regression analysis, the transformation can be made on X, on Y or on both. Frequently used transformations are $\log X, X^c$, where c is a constant, $X^{1/2} + (X+1)^{1/2}, \sin^{-1}\sqrt{X/(N+1)} + \sin^{-1}\sqrt{(X+1)/(N+1)}$, or the same transformations applied to Y. See Hald [10.6] or Draper and Smith [10.4] for a discussion of transformations and nonlinear regression. As an example of transforming the independent variable, log breaking strength was used rather than breaking strength in the fiber example.

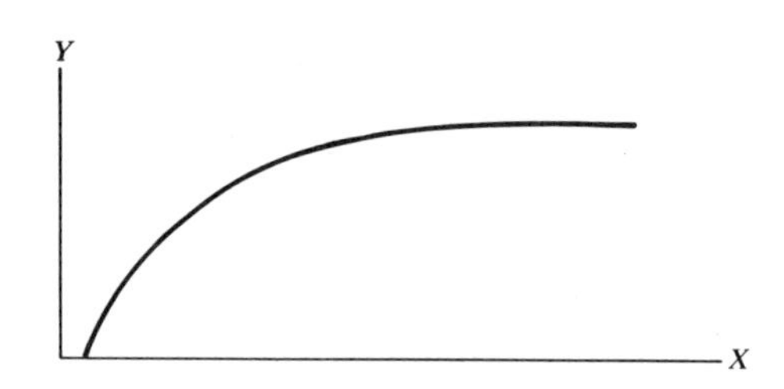

Figure 10.2. Regression curve for which logarithm transformation of X is appropriate.

SAMPLING FROM THE JOINT NORMAL DISTRIBUTION OF X AND Y

THE BIVARIATE MODEL

A second model is often appropriate when considering two variates, X and Y. If X and Y denote two measurements on the same individual—say, his age and his weight—and if each separately is normally distributed, then together they are assumed to be jointly normally distributed.* When sampling is done by selecting n individuals and measuring X and Y on each individual, we have a sample from the joint normal distribution of X and Y.

In the case of a single, normally distributed variate X, the distribution is completely characterized by two parameters—the mean $EX = \mu_x$, and the variance σ_x^2.

For two normally distributed variates X and Y, the distribution is described by five parameters. Four of these are μ_x, μ_y, σ_x^2, and σ_y^2. The fifth can be taken as the covariance of X and Y, $\operatorname{Cov} X, Y = E(X-\mu_x)(Y-\mu_y)$. For many purposes it is more convenient to work with the correlation coefficient of X and Y, $\rho = \operatorname{Cov} X, Y/\sigma_x\sigma_y$ rather than with their covariance.

The correlation coefficient has the advantage of being independent of the units of measurement. If weights are measured in kilograms instead of pounds, the covariance and variance are changed, but the correlation coefficient remains the same. It measures the linear relationship between X and Y. If the entire population of weights and ages lies on a straight line with either positive or negative slope, ρ is ± 1; $+1$ if the slope of the line is positive, -1 if the slope is negative. If X and Y are statistically independent, and an individual's weight bears no relation to his age, ρ equals zero.

We use μ_x, μ_y, σ_x^2, σ_y^2, and ρ as the five parameters characterizing the normal distribution.

If we select a sample of 24 elementary school children and write down their ages and weights, we are effectively sampling from the joint distribution of X and Y. Perhaps instead we decide to pick certain ages—say, 5, 7,

* The joint distribution is

$$f(x,y) = \frac{1}{2\pi\sigma_x\sigma_y\sqrt{1-\rho^2}} \exp\left\{\frac{-1}{2(1-\rho^2)}\left[\frac{(x-\mu_x)^2}{\sigma_x^2} - \frac{2\rho(x-\mu_x)(y-\mu_y)}{\sigma_x\sigma_y} + \frac{(y-\mu_y)^2}{\sigma_y^2}\right]\right\}.$$

Theoretically it is possible that two variables, each of which is normally distributed, have some other joint distribution. We are not concerned with this possibility; in practical situations of the type we are discussing, the situation does not occur.

9, and 11 years—and to use six children of each of these ages. That is, we somehow select six children at random from a population of 5-year-old children and measure their weights. This seems to be the appropriate sampling procedure for the earlier model $[E(Y|X)=\alpha+\beta(X-\bar{X})]$. Let us now examine the estimation and tests appropriate to each of the methods of sampling.

For the earlier model (sampling from the conditional distributions), we are able to estimate: (1) the mean of the Y's for any given X (the mean weight for a given age), (2) the slope of the regression line β, (3) the variance of the Y's for any given X, and (4) Y^* for a new individual whose X value is X^*. This completes the list. It is clear that we cannot hope to estimate the mean age of elementary school children, for we chose the ages. We cannot even estimate the mean weight of school children; the mean weight of our sample would probably turn out to be quite different if we took 12 children of age 9 years and 12 children of age 11 years instead of a half-dozen children, each, at 5, 7, 9, and 11 years. We cannot estimate the correlation coefficient between age and weight of school children, for this too is a parameter of the unconditional distribution.

When we sample from the joint distribution of X and Y, on the other hand, we are able to estimate much more. We can estimate all five parameters, μ_x, μ_y, σ_x^2, σ_y^2, and ρ.

It can be shown mathematically that if X and Y are jointly normally distributed, the populations of Y's corresponding to a given value of X have the following properties: (1) they are normally distributed, (2) they have means that lie on a straight line, and (3) they have equal variances. Thus if X and Y are both normally distributed, they automatically satisfy the conditions for the earlier model: there is a population of Y's associated with each fixed value of X; the populations are normally distributed, they have means that lie on a straight line, and they all have equal variances. There is a population regression line of Y on X, and similarly there is a population regression line of X on Y. These two lines in general do not coincide. Their equations are

$$E(Y|X)=\mu_y+\frac{\rho\sigma_y}{\sigma_x}(X-\mu_x) \tag{10.26}$$

$$E(X|Y)=\mu_x+\frac{\rho\sigma_x}{\sigma_y}(Y-\mu_y). \tag{10.27}$$

For weights and ages, we probably want only the regression of weight on age. Such is often the case, although it is quite possible to estimate both lines. The two lines differ most when the correlation coefficient ρ is zero; in this case, they become two perpendicular lines paralleling the two axes: $E(Y|X)=\mu_y$ and $E(X|Y)=\mu_x$.

As pointed out earlier, the joint normal model implies all the assumptions of the earlier conditional model. This means that if we take our sample of 24 children from the population of elementary school children, we can estimate all parameters, make all tests, and form all confidence intervals, just as we did with the earlier model. That is, we do these things just as if we had selected certain ages and selected the samples from the subpopulations. This point should be clearly understood. It seldom causes difficulty for mathematical statisticians or applied statisticians; it can be most confusing to the student, however, who reads over the fixed X model with all its assumptions and then turns to the examples, which were clearly drawn from the joint distribution.

To repeat, in sampling from the joint normal distribution of X and Y, we can do anything that can be done when sampling from conditional distributions with X fixed. This includes making point and confidence interval estimates and testing hypotheses. Everything is performed in precisely the same way as when sampling from the conditional distribution. In addition, we can estimate and test hypotheses concerning μ_x, μ_y, σ_x^2, σ_y^2, and ρ.

In the fiber example used for fitting a straight line, it seems clear from the data that the sample was selected from the joint distribution. That is, we simply took a sample of 15 fibers and measured their diameters and log breaking strengths.

By comparing (10.26) and (10.6), it can be shown algebraically that $\beta = \rho\sigma_y/\sigma_x$ and $\alpha = \mu_y + \beta(\overline{X} - \mu_x)$. It can also be shown that the variance of the conditional distribution is $\sigma_e^2 = \sigma_y^2(1-\rho^2)$. This completes the correspondence between the two models. Table 10.6 summarizes the two models with their parameters.

The estimates of parameters for sampling from the joint distribution are exactly what we expect. They are the usual sample means and variances for μ_x, μ_y, σ_x^2, and σ_y^2, the sample correlation coefficient r for the population correlation coefficient ρ. The parameters involved in the regression line are precisely as in the fixed X model.

We need learn very little for the joint model, then, in addition to what we know for the fixed X model. We know already how to obtain confidence intervals for μ_x, μ_y, σ_x^2, and $\sigma_y^2, \alpha, \beta$, and so on. There remains only ρ, whose point estimate $r = \sum xy/\sqrt{\sum x^2 \sum y^2}$ was met in Chapter 2.

INTERPRETATION OF THE CORRELATION COEFFICIENT

Before discussing confidence intervals and tests for ρ, we stop to point out a slightly different interpretation for r. Using $b = \sum xy/\sum x^2$ and $r = \sum xy/\sqrt{\sum x^2 \sum y^2}$, the due regression sum of squares of the analysis of

TABLE 10.6. PARAMETERS AND THEIR ESTIMATES IN REGRESSION AND CORRELATION

X, Y Sampled from Joint Normal Distribution		Y Sampled from Conditional Normal Distribution, X Fixed	
Parameter	Estimate	Parameter	Estimate
μ_x	$\overline{X}$	. . . .	. . . .
μ_y	$\overline{Y}$	. . . .	. . . .
σ_x^2	s_x^2	. . . .	. . . .
σ_y^2	s_y^2	. . . .	. . . .
ρ	$r = \frac{\Sigma xy}{\sqrt{\Sigma x^2 \Sigma y^2}}$		
$E(Y\|X^*) = \mu_y + \rho \frac{\sigma_y}{\sigma_x}(X^* - \mu_x)$ $= \alpha + \beta(X^*-\overline{X})$	$\overline{Y} + bx^*$	$E(Y\|X^*) = \alpha + \beta(X^*-\overline{X})$	$\overline{Y} + bx^*$
$\beta = \frac{\rho\sigma_y}{\sigma_x}$	$b = \frac{\Sigma xy}{\Sigma x^2}$	β	$b = \frac{\Sigma xy}{\Sigma x^2}$
$\alpha = \mu_y + \beta(\overline{X} - \mu_x)$	$a = \overline{Y}$	$\alpha = \mu_y + \beta(\overline{X}-\mu_x)$	$a = \overline{Y}$
$Var(Y\|X^*) = \sigma_y^2(1-\rho^2) = \sigma_e^2$	$s_e^2 = (\Sigma y^2 - b\Sigma xy)/(n-2)$	$Var(Y\|X^*) = \sigma_e^2$	$s_e^2 = (\Sigma y^2 - b\Sigma xy)/(n-2)$
$E(X\|Y^*) = \mu_x + \rho \frac{\sigma_x}{\sigma_y}(Y^* - \mu_y)$ $= \alpha' + \beta'(Y^*-\overline{Y})$	$\overline{X} + b'y^*$	. . . .	. . . .
$\beta' = \frac{\rho\sigma_x}{\sigma_y}$	$b' = \frac{\Sigma xy}{\Sigma y^2}$	. . . .	. . . .
$\alpha' = \mu_x + \beta'(\overline{Y} - \mu_y)$	$a' = \overline{X}$	. . . .	. . . .
$Var(X\|Y^*) = \sigma_x^2(1-\rho^2) = \sigma_e^{2'}$	$s_e^{2'} = (\Sigma x^2 - b'\Sigma xy)/(n-2)$	. . . .	. . . .

$x = X-\overline{X}$, $y = Y-\overline{Y}$, $x^* = X^*-\overline{X}$, $y^* = Y^*-\overline{Y}$

TABLE 10.7. ANALYSIS OF VARIANCE TABLE IN TERMS OF r^2

Source of Variation (1)	Sums of Squares (2)	d.f. (3)	MS (4)
Due Regression on X	$r^2\Sigma y^2$	1	$r^2\Sigma y^2$
Residual	$(1-r^2)\Sigma y^2$	n-2	$(1-r^2)\Sigma y^2/(n-2)$
Total	Σy^2	n-1	

variance table (Table 10.2) can be expressed as follows:

$$b\sum xy = \frac{\left(\sum xy\right)^2}{\sum x^2} = r^2\sum y^2.$$

In Table 10.7 the sums of squares column of the analysis of variance table has been written in terms of r^2.

From the sums of squares columns, we see that $1-r^2$ is the proportion of the original variation ($\sum y^2$), which is left after fitting the straight line. The quantity r^2 is the proportion of the original sum of squared deviations, which is removed by fitting the straight line. If $r=.5$, only one-fourth of the sum of squares is removed by the sample regression line.

THE CORRELATION COEFFICIENT: CONFIDENCE INTERVALS AND TESTS

The shape of the distribution of the statistic r depends heavily on the value of the population correlation coefficient ρ, as well as on the sample size. The distribution for $\rho=0$ is symmetric; for ρ large in magnitude, it is highly skewed. Tables for the distribution of r, therefore, are not convenient for obtaining confidence intervals or for making tests concerning ρ. We give here a method for testing $H_0:\rho=0$, and also a method for making confidence intervals for ρ and for testing $H_0:\rho=\rho_0$.

The t Test of $H_0:\rho=0$. For a test of $H_0:\rho=0$, we form the statistic

$$t = \frac{\sqrt{n-2}\,r}{\sqrt{1-r^2}}. \qquad (10.28)$$

If ρ is actually zero, this statistic has a t distribution with $n-2$ degrees of freedom. The t calculated from the data is compared with $t[1-\alpha/2; n-2]$; if its magnitude is greater than $t[1-\alpha/2; n-2]$, we conclude that ρ is different from zero.

The F test (see Table 10.2) of $H_0: \beta=0$ is equivalent to this test. From Table 10.7, the test statistic can be written

$$F = \frac{r^2 \sum y^2}{(1-r^2)\sum y^2/(n-2)} = \frac{(n-2)r^2}{1-r^2}, \qquad (10.29)$$

which is compared with $F[1-\alpha; 1, n-2]$. From (10.28) and (10.29) we see that the calculated F statistic is the square of the calculated t statistic. It can be shown mathematically that the square of a t variable with ν d. f. is an F variable with degrees of freedom 1 and ν. It should not be surprising that this test of $H_0: \rho=0$ is really the same test as the previous test of $H_0: \beta=0$.The relation $\beta=(\sigma_y/\sigma_x)\rho$ implies that if either β or ρ is zero, both are zero.

For the fiber data, we have $r=.8577, t=6.014$ and $t[.975; 13]=2.16$. The hypothesis that ρ is 0 is therefore rejected at a significance level of .05. We note that $t^2=36.17$ equals (except for round-off error) the F value obtained earlier.

Fisher's z Statistic. A simple way of obtaining confidence intervals for ρ or for testing the null hypothesis that ρ equals any particular value ρ_0 makes use of the statistic

$$z = \tfrac{1}{2}\log_e\left(\frac{1+r}{1-r}\right). \qquad (10.30)$$

It has been shown that the statistic z is approximately normally distributed. The mean of the distribution of z's is approximately

$$Ez = \frac{1}{2}\log_e\left(\frac{1+\rho}{1-\rho}\right) = \mathcal{Z} \text{ (say)}$$

and its variance is approximately

$$\operatorname{Var} z = \frac{1}{n-3}.$$

In other words, we have

$$z\text{ND}\left(\mathcal{Z}, \frac{1}{n-3}\right). \qquad (10.31)$$

We can calculate z from (10.30) using Table A.6. For example, in the fiber example, $r = .8577$ and, interpolating linearly in the table, $z = 1.284$. A $1-\alpha$ confidence interval for $\mathcal{Z}$ can be formed in the usual way from (10.31). It is

$$z \pm z[1-\alpha/2]\sqrt{1/(n-3)}\ . \qquad (10.32)$$

The 95% confidence interval for $\mathcal{Z} = \frac{1}{2}\log_e[(1+\rho)/(1-\rho)]$ then becomes

$$1.284 \pm 1.96\sqrt{1/12} \qquad \text{or} \qquad 0.718 < \mathcal{Z} < 1.850.$$

Because the value of $\mathcal{Z}$ increases as ρ increases, if $\mathcal{Z}$ lies between 0.718 and 1.850, ρ lies between the numbers corresponding to 0.718 and 1.850. From Table A.6, these are .615 and .952. The 95% confidence interval for ρ is therefore $.615 < \rho < .952$.

The same z statistic can be used to test $H_0: \rho = \rho_0$, where ρ_0 is any value between -1 and $+1$ (with the foregoing t test we can test only $H_0: \rho = 0$). If we wish to test $H_0: \rho = .5$ using the fiber data, we find that $\mathcal{Z} = 0.549$. Thus under H_0, z is normally distributed with mean 0.549 and standard deviation $1/\sqrt{n-3} = 1/\sqrt{15-3} = 0.289$. To enter the standard normal tables with our value of Fisher's z of 1.284, we calculate

$$\frac{1.284 - 0.549}{0.289} = 2.543.$$

From Table A.1, $P = .0110$, and with $\alpha = .05$ we reject the null hypothesis. Note that we could have used the .95 level confidence interval .602 to .952 to reject $H_0: \rho = .5$. Note also that we have used the same symbol z for $\frac{1}{2}\log_e (1+r)/(1-r)$ that is used for a standard normal variable.

The z statistic can also serve in testing whether two population correlation coefficients are equal. For example, we might wish to test whether the correlation between blood pressure and dosage of a drug is the same in men and in women. If the population correlations are ρ_1 and ρ_2, the null hypothesis is $H_0: \rho_1 = \rho_2$. If the sample correlations r_1 and r_2 are based on independent samples of size n_1 and n_2, we have

$$z_i \text{ND}\left(\mathcal{Z}_i, \frac{1}{n_i - 3}\right), \qquad i = 1,2 \qquad \text{and} \qquad z_1 - z_2 \text{ND}\left(\mathcal{Z}_1 - \mathcal{Z}_2, \frac{1}{n_1 - 3} + \frac{1}{n_2 - 3}\right).$$

under $H_0: \rho_1 = \rho_2$, $\mathcal{Z}_1 - \mathcal{Z}_2 = 0$; therefore, $z_1 - z_2 \text{ND}[0, 1/(n_1 - 3) + 1/(n_2 - 3)]$, and we can use the normal tables for the test.

If ρ_1 is the correlation between IQ and educational level and ρ_2 is the correlation between IQ and socioeconomic status, and if r_1 and r_2 have been calculated from a *single* sample with measurements on the three variables, then r_1 and r_2 are not statistically independent. We no longer have $\mathrm{Var}(z_1 - z_2) = 2/(n-3)$, since z_1 and z_2 are undoubtedly correlated, and the foregoing test does not apply. See Ref. 10.8 or 10.9 for a discussion of dependent correlations.

ILLUSTRATIVE EXAMPLES

As an illustration of a situation in which a fixed X model is reasonable, we consider an investigator who wishes to gather some data on total yield of crop A and amount of water. He hopes to be able to use the data in order to predict yield Y from knowledge of amount of water X. In planning a field trial, he is able to select certain water levels and to assign a water level to each of his n fields.

His first task is to choose levels of X. No doubt he has had some experience that will help him in selecting the water levels; he may know, perhaps, that the crop fails completely below (or above) a certain water level. It can be shown that for a given number of fields the variance of a predicted crop yield $Y^{*\prime}$ can be minimized by using the lowest water level under consideration for half the fields and the highest water level for half the fields. This is a dangerous procedure, however, for it does not provide an opportunity to see whether a straight line fits the data well. A much safer procedure, even though it gives a somewhat larger variance for $Y^{*\prime}$, is to use the lowest water level for one-third of the fields, the highest for one-third, and a water level approximately midway between the lowest and highest for the remaining third.

With this choice of X values, a scatter diagram may appear as in Figure 10.3*a*; here a straight line is clearly an appropriate fit. If the scatter diagram appears as in Figure 10.3*b*, the investigator can either make a transformation on X and/or Y before fitting a straight line, or he can fit a polynomial regression line using the methods of Chapter 12.

Always, before a straight line is fitted, the scatter diagram should be examined to see whether a straight line is appropriate. With a high-speed computer, programs such as BMD02D and EPID41 can be used to plot the scatter diagram; otherwise it can be plotted by hand.

After deciding to fit a straight line, the least squares line (10.4) is calculated by hand or with an appropriate computer program (BMD01R or EPID41). Computer programs often give the line in the form $Y' = a' + bX$ instead of $Y' = a + b(X - \bar{X})$, where a' denotes the height of the regression line at $X = 0$.

The investigator can now use his sample regression line in various ways. He frequently uses it to estimate a yield Y^* when water level X^* is known and to determine whether the slope of the population regression line is positive, zero, or negative.

If the investigator had been simply studying data on yields and water level from various farms (instead of fixing water levels in conducting a field trial), the sample of data would have been drawn from the joint distribution of X and Y, and he could consider using the model of a sample from the joint *normal* distribution. He would let X denote water level and Y denote yield because he believes that yield depends on water level. The scatter diagram would be drawn as before. If water levels and yields were approximately normal, indicating that the sample was from the joint normal distribution, the points should lie as in Figure 10.3*c*—approximately within an ellipse, with the density of points higher in the center than at the outer edges. In such a case, he would proceed as in the fixed X model to fit his straight line. If, on the other hand, the points appeared in a crescent form as in Figure 10.3*d*, a straight line should not be fitted without making a transformation on the data. Programs such as

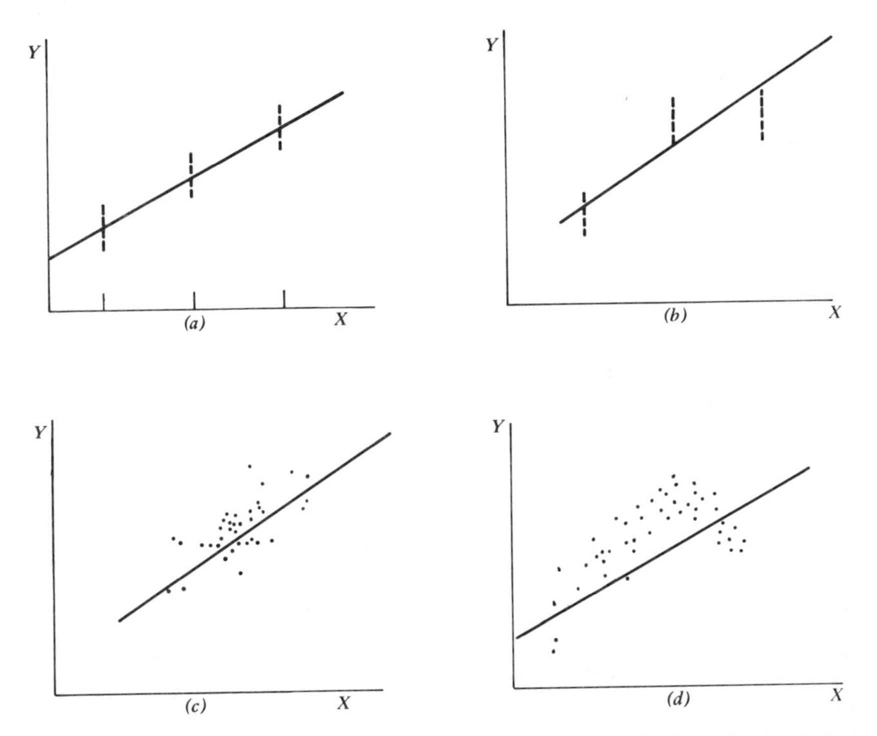

Figure 10.3. Illustrative scatter diagrams with least squares lines: (*a*) fixed X, straight line; (*b*) fixed X, nonstraight line; (*c*) joint distribution, straight line; (*d*) joint distribution, nonstraight line.

BMD01R or EPID41 allow him to try various transformations and make the scatter diagrams after each one. With smaller sample sizes, unfortunately, scatter diagrams are often more difficult to interpret than those in Figure 10.3. In this situation, theory or experience may lead us to expect straight line regression.

After the investigator has decided that a straight line is reasonable and has calculated the least squares line, he can proceed as in the fixed X model. Now, however, interpretation is more difficult. In the fixed X study, the investigator assigned the water levels to the different fields in some random way; presumably all the fields were treated alike except for water level, and any difference in initial fertility among fields was random. Now, sampling from the joint distribution of water levels and yields, his study is a survey rather than an experiment and he must be very careful in his conclusions. In the situation represented in Figure 10.3*c*, it is true that yield seems to increase with water level. However, it is possible that the water level for the various fields depended on the amount of fertilizer used, with more water being used for heavily fertilized fields. In this case, an increase in water alone might not result in any increase in average yield.

Because the investigator sampled from the joint distribution, he can also estimate the overall mean yield, the overall mean water level, and the correlation between water level and yield.

SUMMARY

In this chapter, regression analysis was introduced. First, formulae for the slope and the height of the least squares line were given; these can be used regardless of the distribution of X and Y. Next, a fixed X model was discussed, and confidence intervals and tests of hypothesis were made. Other types of regression on a single variate were mentioned briefly. A bivariate normal distribution model was introduced, along with confidence intervals and tests for the population correlation coefficient.

REFERENCES

APPLIED STATISTICS

10.1. Bennett, C. A., and N. L. Franklin, *Statistical Analysis in Chemistry and the Chemical Industry*, Wiley, New York, 1954.

10.2. Brownlee, K. A., *Statistical Theory and Methodology in Science and Engineering*, 2nd ed., Wiley, New York, 1965.

10.3. Dixon, W. J., and F. J. Massey, Jr., *Introduction to Statistical Analysis*, 3rd ed., McGraw-Hill, New York, 1969.

10.4. Draper, N. R., and H. Smith, *Applied Regression Analysis*, Wiley, New York, 1966.

10.5. Dunn, O. J., *Basic Statistics: A Primer for the Biomedical Sciences*, Wiley, New York, 1964.

10.6. Hald, A., *Statistical Theory with Engineering Applications*, Wiley, New York, 1952.

10.7. Snedecor, G. W., and W. G. Cochran, *Statistical Methods*, 6th ed., Iowa State University Press, Ames, Iowa, 1967.

DEPENDENT CORRELATION

10.8. Dunn, O. J., and V. Clark, "Correlation Coefficients Measured on the Same Individuals," *Journal of the American Statistical Association*, Vol. 64 (March 1969).

10.9. Dunn, O. J., and V. Clark., "Tests of Equality of Dependent Correlation Coefficients," *Journal of the American Statistical Association*, Vol. 66 (December 1971).

PROBLEMS

10.1. Yearly rainfall and yield of cotton were as follows:

X Rainfall (in.)	Y Yield (lb/acre)
7.12	1037
63.54	380
47.38	416
45.92	427
8.68	619
50.86	388
44.46	321

(*a*) How do you think that the data were obtained? Do you think the sampling was done from the joint distribution of rainfall and yield, or from yields for fixed rainfall?
(*b*) What model or models can be used for the data?
(*c*) Write out the least squares regression line to predict yield of cotton from yearly rainfall.
(*d*) Draw a scatter diagram and draw the regression line.
(*e*) Make an analysis of variance table.
(*f*) Give a 95% confidence interval for:
(i) the regression coefficient.
(ii) the mean yield of cotton for a yearly rainfall of 50 in.
(iii) σ_e^2.
(*g*) Predict yield of cotton if rainfall is 30 in. Find an estimate of the variance for the predicted yield.
(*h*) Test whether ρ is zero using a t statistic.
(*i*) Give a 95% confidence interval for ρ.

10.2. Blood glucose level was measured on four individuals 1 hr after glucose load, on seven individuals at 2 hr after glucose load, and on six individuals at 4 hr after load. (There were 17 individuals involved altogether.) The data obtained were as follows:

Interval	Observation Number 1	2	3	4	5	6	7
1	106.5	124.7	123.5	158.5			
2	86.2	59.8	64.5	106.8	95.8	117.5	59.2
4	72.7	62.5	51.8	58.8	44.0	41.0	

(*a*) Calculate the least squares line fitting blood glucose level to interval after load.
(*b*) Plot a scatter diagram and the least squares line.
(*c*) Make an analysis of variance table.
(*d*) Test whether $\beta=0$.
(*e*) Find a 95% prediction interval for the blood glucose level of an individual 3 hr after glucose load.

10.3. In an investigation of the relation of age and physical strength for adults, strength of right-hand grip was measured for a sample of adult men. The data were as follows:

	Observation Number 1	2	3	4	5	6	7	8	9	10	11	12	13	14	15
Age, X (years)	30	23	43	56	29	52	59	42	23	27	59	24	24	37	62
Right-hand grip strength, Y (lb)	86	88	80	83	93	87	71	91	76	82	88	100	92	97	78

(*a*) Fit a straight line $Y'=a+b(X-\overline{X})$ predicting physical strength from age.
(*b*) What is an appropriate model for this problem?
(*c*) Give an estimate of σ_e^2.
(*d*) Test $H_0:\beta=0$.
(*e*) Give a 95% confidence interval for β.

10.4. For 15 selected cities in the United States, average yearly temperature Y and degrees of latitude X were determined. Data were as follows:

Latitude, X	34	32	39	39	41	45	41	33
Temperature, Y (°F)	56.4	51.0	36.7	37.8	36.7	18.2	30.1	55.7

Latitude, X	34	47	44	39	41	32	40
Temperature, Y (°F)	46.6	13.3	34.0	36.3	34.0	49.1	34.5

(*a*) Plot the data.
(*b*) Find the least squares line.
(*c*) State a possible model and discuss whether the model seems to be appropriate.
(*d*) Test $H_0:\rho=0$.

(*e*) Find a 95% confidence interval for ρ.
(*f*) Find a 95% confidence interval for σ_e^2.

10.5. IQ tests were given to children of three different age groups. The scores were as follows:

Age (years)	IQ Scores							
3	105	94	108	101	100	—	—	—
4	96	119	103	107	110	112	106	—
5	120	114	121	116	128	115	118	123

(*a*) State the model.
(*b*) Write out the least squares regression line for predicting IQ from age.
(*c*) Plot the data and the regression line from (*b*).
(*d*) Make an analysis of variance table.
(*e*) Test whether $\beta=0$.
(*f*) Give a 95% confidence interval for β.
(*g*) Give and interpret a 95% confidence interval for the mean IQ for children of age 4.

10.6. The following data represent yields of rye for four preselected levels of fertilizer.

Level of Fertilizer	Yields (bushels per acre)													
50	13.1	28	12.2	10.1	13.8	5.2	8.2	38.3	11.1	14.5	—	—	—	—	—
60	29.6	25.7	28.6	33.8	29.7	9.0	4.4	27.0	16.1	12.4	—	—	—	—	—
70	33.7	42.6	22.1	41.3	23.5	36.4	35.3	26.5	35.4	34.3	15.3	46.4	20.1	46.0	16.5
80	26.8	31.4	25.6	24.2	33.8	32.7	22.0	42.1	40.9	30.1	—	—	—	—	—

(*a*) State an appropriate model.
(*b*) Obtain the estimated regression line.
(*c*) Calculate and interpret confidence intervals for α and β with an overall confidence level of .90.
(*d*) Test the hypothesis $H_0: \beta=0$.
(*e*) Give a prediction interval for the yield of rye for a plot using 75 lb of fertilizer per acre.
(*f*) Give a 95% confidence interval for the population mean yield for 75 lb of fertilizer per acre.
(*g*) Give a 95% confidence interval for the population mean yield with no fertilizer.
(*h*) Do you have as much faith in the confidence interval in (*g*) as in the confidence interval in (*f*)? Why?

CHAPTER 11

MULTIPLE REGRESSION AND CORRELATION

THE REGRESSION OF Y ON SEVERAL VARIABLES

In the previous chapter we studied two variables X and Y simultaneously; two general models were introduced—the fixed X model and the joint distribution model. In this chapter we consider several variables $X_1,\dots,X_k$ used to predict Y, again with the two models.

When an investigator has measurements on, say, $k+1$ variables for a set of n individuals, he frequently selects one variable (denoted by Y) which he wishes to predict from knowledge of the other variables $(X_1,\dots,X_k)$. He wishes to determine from his sample data the least squares plane of the form

$$Y' = a + b_1\left(X_1 - \overline{X}_1\right) + \cdots + b_k\left(X_k - \overline{X}_k\right), \tag{11.1}$$

where $\overline{X}_1,\dots,\overline{X}_k$ are the sample means for the k variates. Setting $x_i = X_i - \overline{X}_i$, $i = 1,\dots,k$, the equation can also be written

$$Y' = a + b_1 x_1 + \cdots + b_k x_k. \tag{11.2}$$

The least squares plane has the property that the sum of the squared differences $\sum_{l=1}^{n}(Y_l - Y_l')^2$ is a minimum among all possible planes of the form (11.1). Here Y_l' is $a + b_1(X_{1l} - \overline{X}_1) + \cdots + b_k(X_{kl} - \overline{X}_k)$, and $Y_l, X_{1l},\dots, X_{kl}$ are the measurements for the lth individual. When there are just two X variables, data can be plotted as points in a three-dimensional space, the vertical axis serving as the Y axis. Equation 11.1 is plotted as a plane in Figure 11.1. The least squares plane is the one that minimizes the sum of squared *vertical* distances, exactly as in the case of a single X variable. For the lth data point, the vertical distance $Y_l - Y_l'$ is illustrated in Figure 11.1 when X_1 and X_2 are measured on the horizontal axes. The least squares regression plane is entirely analogous to the least squares line discussed in Chapter 10. The mechanics of obtaining it are more involved, but the interpretation is similar. Before obtaining the regression equation, we give the underlying model.

THE REGRESSION MODEL

The fixed X model for regression is an extension of the model for a single fixed X. For each particular set of values of $X_1, X_2, \ldots, X_k$, there exists a population of Y's. Each population of Y's is normally distributed; the means of the various Y populations lie on a plane, and the variances of the Y populations are equal.

The plane on which the population means lie is called the population regression plane. It can be written

$$E(Y|X_1,\ldots,X_k) = \alpha + \beta_1(X_1 - \bar{X}_1) + \cdots + \beta_k(X_k - \bar{X}_k)$$
$$= \alpha + \beta_1 x_1 + \cdots + \beta_k x_k \tag{11.3}$$

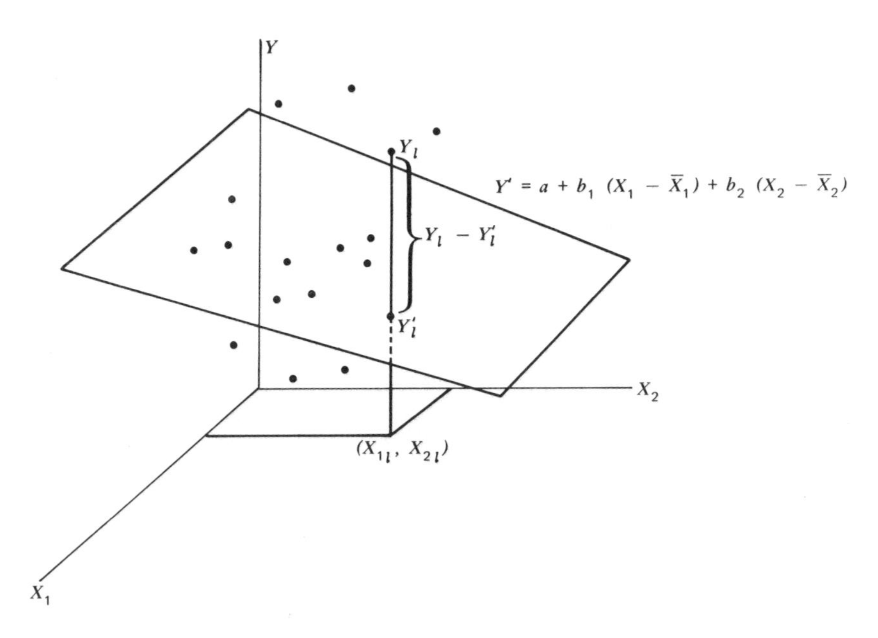

Figure 11.1. The least squares plane in three dimensions.

where α is the height of the regression plane at the mean point $\bar{X}_1,\ldots,\bar{X}_k$. The parameters $\beta_1,\ldots,\beta_k$ are the partial regression coefficients. If $\beta_1=+2$, for example, the height of the regression plane is increased by two units if X_1 is increased by one unit and $X_2,\ldots,X_k$ are kept unchanged.

The assumptions can be summarized by the statement

$$Y_l\,\text{IND}(\alpha+\beta_1x_{1l}+\cdots+\beta_kx_{kl},\sigma_e^2),\qquad l=1,\ldots,n,$$

or, equivalently,

$$Y_l=\alpha+\beta_1x_{1l}+\cdots+\beta_kx_{kl}+\varepsilon_l,\qquad l=1,\ldots,n$$

where $\varepsilon_l\,\text{IND}(0,\sigma_e^2)$.

From a set of n measurements on $Y, X_1,\ldots,X_k$, the population regression plane is estimated by the least squares sample regression plane. We now obtain the least squares plane for kX variables. An example utilizing just two X variables follows.

THE LEAST SQUARES PLANE

The values of $a,b_1,\ldots,b_k$ are such that the sum of squared vertical deviations of sample points from the plane is as small as possible. The quantity to be minimized is thus

$$Q=\sum_{l=1}^{n}(Y_l-Y_l')^2$$

$$=\sum_{l=1}^{n}(Y_l-a-b_1x_{1l}-b_2x_{2l}-\cdots-b_kx_{kl})^2, \tag{11.4}$$

where

$$x_{1l}=X_{1l}-\bar{X}_1,\ x_{2l}=X_{2l}-\bar{X}_2,\ldots,\ x_{kl}=X_{kl}-\bar{X}_k.$$

The solution to this minimization problem is found by solving (11.5) and (11.6). The equations are called the normal equations.*

* To minimize Q, the partial derivatives of Q with respect to $a,b_1,b_2,\ldots,b_k$ are set equal to zero. Thus we have

$$\frac{\partial Q}{\partial a}=-2\sum_{l=1}^{n}[Y_l-a-b_1x_{1l}-b_2x_{2l}-\cdots-b_kx_{kl}]=0$$

$$\frac{\partial Q}{\partial b_1}=-2\sum_{l=1}^{n}[Y_l-a-b_1x_{1l}-b_2x_{2l}-\cdots-b_kx_{kl}]x_{1l}=0$$

$$\sum_{l=1}^{n} Y_l = na \qquad (11.5)$$

$$b_1 \sum x_1^2 + b_2 \sum x_1 x_2 + \cdots + b_k \sum x_1 x_k = \sum x_1 y$$

$$b_1 \sum x_2 x_1 + b_2 \sum x_2^2 + \cdots + b_k \sum x_2 x_k = \sum x_2 y$$

$$\cdots$$

$$b_1 \sum x_k x_1 + b_2 \sum x_k x_2 + \cdots + b_k \sum x_k^2 = \sum x_k y \qquad (11.6)$$

where

$$\sum x_i^2 = \sum_{l=1}^{n} \left(X_{il} - \bar{X}_i\right)^2, \qquad i = 1, \ldots, k,$$

and

$$\sum x_i x_j = \sum_{l=1}^{n} \left(X_{il} - \bar{X}_i\right)\left(X_{jl} - \bar{X}_j\right), \qquad i \neq j.$$

These equations are $k+1$ linear equations in $k+1$ unknowns. The solution of (11.5) is $a = \bar{Y}$. To obtain $b_1, b_2, \ldots, b_k$, the k linear equations (11.6) must be solved.

Calculations. The preliminary calculations in a regression analysis include the calculation of $\bar{Y}, \bar{X}_1, \ldots, \bar{X}_k, \sum y^2, \sum x_1 y, \ldots, \sum x_k y, \sum x_i^2, i = 1, \ldots, k$, and $\sum x_i x_j$, $i \neq j$. After these calculations have been made, the normal equations (11.6) can be written down and solved in any convenient way. Before the advent of the high-speed computer, methods for solving the normal equations using a desk calculator were of importance. For k larger than 3 or 4, such calculations are exceedingly tedious, and today the large computer is generally used except for $k = 2$ and 3. The present-day student of statistics may never have occasion to solve the normal equations by hand, but he will probably need to understand the particular regression program under which his data are analyzed.

An Example. To illustrate the solution of the normal equations, we use some data (Table 11.1) on eight patients who underwent a thyroidectomy. Measurements were made on the following items: Y = change in hemoglo-

---

$$\frac{\partial Q}{\partial b_2} = -2 \sum_{l=1}^{n} [Y_l - a - b_1 x_{1l} - b_2 x_{2l} - \cdots - b_k x_{kl}] x_{2l} = 0$$

$$\cdots$$

$$\frac{\partial Q}{\partial b_k} = -2 \sum_{l=1}^{n} [Y_l - a - b_1 x_{1l} - b_2 x_{2l} - \cdots - b_k x_{kl}] x_{kl} = 0.$$

Algebraic manipulation yields (11.5) and (11.6).

TABLE 11.1. DATA ON DURATION OF OPERATION, BLOOD LOSS, AND HEMOGLOBIN CHANGE

	X_1 Duration of Operation	X_2 Blood Loss	Y % Change in Hemoglobin
1	105	503	-1.7
2	80	490	-4.6
3	86	471	-9.8
4	112	505	-1.1
5	109	482	-4.1
6	100	490	-3.3
7	96	513	+0.4
8	120	464	-2.9

bin (%), X_1 = duration of operation (min), and X_2 = blood loss (ml). We fit a regression plane in order to predict percentage change in hemoglobin from duration of operation and blood loss. With $k=2$ in the example, calculations are relatively simple, but the principles are the same as for higher values of k. The normal equations are

$$8a = -27.10 \tag{11.7}$$

$$1254b_1 - 171b_2 = 138.10 \tag{11.8}$$

$$-171b_1 + 2023.5b_2 = 258.225.$$

Preliminary calculations are recorded in Table 11.2. Here (as in Chapter 1) the formula used for computing $\sum x_1x_2$ was

$$\sum x_1x_2 = \frac{n\sum X_1X_2 - \sum X_1 \sum X_2}{n}.$$

From (11.7) we have $a = \overline{Y} = -3.3875$.

Equations 11.8 are often solved directly for b_1 and b_2. Instead, we obtain them a different way which involves first finding the numbers $c_{11}, c_{12}, c_{21}, c_{22}$. These quantities are solutions of the equations

$$1254c_{11} - 171c_{12} = 1$$

$$-171c_{11} + 2023.5c_{12} = 0 \tag{11.9}$$

$$1254c_{21} - 171c_{22} = 0$$

$$-171c_{21} + 2023.5c_{22} = 1. \tag{11.10}$$

TABLE 11.2. PRELIMINARY CALCULATIONS, REGRESSION OF HEMOGLOBIN CHANGE ON DURATION OF OPERATION AND BLOOD LOSS

n=8	ΣX_1 = 808.00	ΣX_2 = 3918.00	ΣY = -27.10
	$\bar{X}_1$ = 101.00	$\bar{X}_2$ = 489.75	$\bar{Y}$ = - 3.3875
	ΣX_1^2 = 82862.00	$\Sigma X_1 X_2$ = 395547.00	$\Sigma X_1 Y$ = - 2599.00
		ΣX_2^2 = 1920864.00	$\Sigma X_2 Y$ = -13014.00
	Σx_1^2 = 1254.00	$\Sigma x_1 x_2$ = - 171.00	$\Sigma x_1 y$ = 138.10
		Σx_2^2 = 2023.50	$\Sigma x_2 y$ = 258.225

We shall not stop here to solve (11.9) and (11.10). Details of the numerical solution of (11.9) are given in a starred section at the end of this chapter. Equations such as these can be solved by a variety of methods; see Bennett and Franklin [11.1] or Dwyer [11.6].

The solution of (11.9) and (11.10) is

$$c_{11}=0.000807 \qquad c_{12}=0.0000682$$

$$c_{21}=0.0000682 \qquad c_{22}=0.000500\ .$$

Note that $c_{12}=c_{21}$; this is always the case except for possible differences due to rounding. The c_{ij}'s we have obtained are sometimes called Gaussian multipliers.*

To calculate b_1 and b_2 using the c_{ij}'s, we use

$$b_1=c_{11}\sum x_1 y+c_{12}\sum x_2 y$$

$$b_2=c_{21}\sum x_1 y+c_{22}\sum x_2 y$$

or

$$b_1=(0.000807)(138.10)+(0.000068)(258.225)$$

$$=0.1114+0.0176$$

$$=0.129$$

$$b_2=(0.000068)(138.10)+(0.000500)(258.225)$$

$$=0.0094+0.1291$$

$$=0.138.$$

*Here we have not introduced matrix notation; but for readers who are familiar with matrix terminology, the c_{ij} are the elements of the inverse matrix of the matrix of corrected sums of squares and cross products, frequently denoted by S. That is, $C=S^{-1}$.

We have presented the foregoing indirect way of finding b_1 and b_2 because we need $c_{11}, c_{12}=c_{21}$, and c_{22} later in producing confidence intervals. The least squares plane, is

$$\begin{aligned} Y' &= -3.387+0.129x_1+0.138x_2 \\ &= -3.387+0.129(X_1-101.)+0.138(X_2-489.75) \\ &= [-3.387-0.129(101.)-.138(489.75)]+0.129X_1+0.138X_2 \\ &= -84.002+0.129X_1+0.138X_2. \end{aligned} \quad (11.11)$$

The numbers b_1 and b_2 are called *sample partial regression coefficients*. Their interpretation is as follows. For any fixed value of X_2, Y' increases by $b_1=0.129$ unit as X_1 is increased by 1 unit. Similarly, for any fixed value of X_1, Y' increases by $b_2=0.138$ unit as X_2 is increased by 1 unit.

If duration of the operation had not been taken into account, we could have obtained a line $Y''=\overline{Y}+b_2(X_2-\overline{X}_2)$, fitting $Y=$hemoglobin change to $X_2=$blood loss. From the last chapter, such a line can be written down as follows:

$$\begin{aligned} Y'' &= -3.387+0.128(X_2-489.75) \\ Y'' &= -66.075+0.128X_2. \end{aligned} \quad (11.12)$$

We see that the regression coefficient b_2 on X_2 differs in the two equations (11.11) and (11.12).

The regression coefficient on any variable differs according to which other X variables are included in the least squares equation. Regression coefficients should therefore be interpreted with extreme caution. From a regression of Y on X_1 and X_2, we should say "Y tends to increase with X_1 when X_2 is held constant," rather than merely "Y tends to increase with X_1."

Solution of Normal Equations for $k>2$. For higher values of k, (11.5) always yields $a=\overline{Y}$. Equations 11.6 can be solved by first obtaining the Gaussian multipliers:

$$\begin{gathered} c_{11}, c_{12}, \ldots, c_{1k} \\ c_{21}, c_{22}, \ldots, c_{2k} \\ \cdots \\ c_{k1}, c_{k2}, \ldots, c_{kk}. \end{gathered}$$

These are obtained by solving k sets of linear equations. For example, to find the values of $c_{11}, c_{12}, \ldots, c_{1k}$, we solve the first set of equations:

$$c_{11}\sum x_1^2 + c_{12}\sum x_1x_2 + \cdots + c_{1k}\sum x_1x_k = 1$$

$$c_{11}\sum x_2x_1 + c_{12}\sum x_2^2 + \cdots + c_{1k}\sum x_2x_k = 0$$

$$\ldots$$

$$c_{11}\sum x_kx_1 + c_{12}\sum x_kx_2 + \cdots + c_{1k}\sum x_k^2 = 0. \tag{11.13}$$

Similarly, to obtain $c_{21}, c_{22}, \ldots, c_{2k}$, we solve the second set of equations:

$$c_{21}\sum x_1^2 + c_{22}\sum x_1x_2 + \cdots + c_{2k}\sum x_1x_k = 0$$

$$c_{21}\sum x_2x_1 + c_{22}\sum x_2^2 + \cdots + c_{2k}\sum x_2x_k = 1$$

$$\ldots$$

$$c_{21}\sum x_kx_1 + c_{22}\sum x_kx_2 + \cdots + c_{2k}\sum x_k^2 = 0.$$

The other sets of equations are similarly formed. In the equations to be solved for $c_{i1}, c_{i2}, \ldots, c_{ik}$, there is a 1 on the right-hand side of the ith equation; the other right-hand terms are all zeros. Solution of the k sets of equations always yields $c_{12} = c_{21}$, $c_{13} = c_{31}$, and so on, except for differences due to rounding. After calculating the c_{ij}, we find the sample regression coefficients from the following equations:

$$b_1 = c_{11}\sum x_1y + c_{12}\sum x_2y + \cdots + c_{1k}\sum x_ky$$

$$b_2 = c_{21}\sum x_1y + c_{22}\sum x_2y + \cdots + c_{2k}\sum x_ky$$

$$\ldots$$

$$b_k = c_{k1}\sum x_1y + c_{k2}\sum x_2y + \cdots + c_{kk}\sum x_ky. \tag{11.14}$$

Computer programs such as BMD02R or 03R are generally employed. (These programs do not print out the c_{ij} nor the covariances among the regression coefficients; they do print out the variances of the b's. The c_{ij} can be obtained with standard inverse routines such as that available in the IBM Scientific Subroutine Package.)

ANALYSIS OF VARIANCE TABLE AND F TEST OF $H_0: \beta_1 = \beta_2 = \cdots = \beta_k = 0$

Before discussing estimates and confidence intervals, we give an analysis of variance table analogous to that of linear regression. We divide the deviation of each observation from the mean $\bar{Y}$ into parts, just as in the case of a single X variate: one has $Y_l - \bar{Y} = (Y_l' - \bar{Y}) + (Y_l - Y_l')$. Figure 11.2 shows schematically the regression plane $Y' = -84.002 + 0.129X_1 + 0.138X_2$ for the hemoglobin data and also the horizontal plane $\bar{Y} = -3.387$. The vertical deviation $Y_l - \bar{Y}$ appears for just one data point, broken into the two terms—one the "due regression" term, the other the "deviation from regression."

The sum of squares is

$$\sum_{l=1}^{n} \left(Y_l - \bar{Y}\right)^2 = \sum_{l=1}^{n} \left(Y_l' - \bar{Y}\right)^2 + \sum_{l=1}^{n} (Y_l - Y_l')^2, \qquad (11.15)$$

and we make an analysis of variance table as in linear regression. The first sum of squares on the right-hand side of (11.15) is called the "due regression on $X_1,\ldots,X_k$" sum of squares, the second is the residual sum of squares. When k equals 1, the sums of squares due regression was found to be $b\sum xy$. A similar formula holds for k X variates, for it can be shown that

$$\sum_{l=1}^{n} \left(Y_l' - \bar{Y}\right)^2 = b_1 \sum x_1 y + b_2 \sum x_2 y + \cdots + b_k \sum x_k y. \qquad (11.16)$$

The degrees of freedom for the due regression sum of squares is now k, instead of 1, for there are k b_i's.

Tables 11.3 and 11.4 are analysis of variance tables for k X variates and for the blood hemoglobin data, respectively.

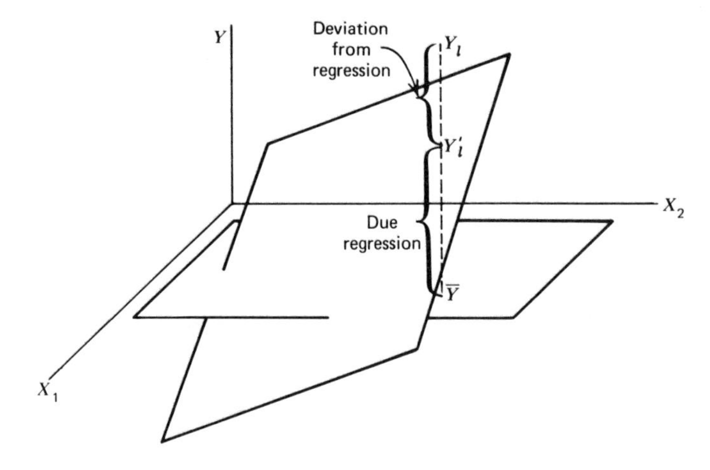

Figure 11.2. The least squares regression plane for regression of percentage change in blood hemoglobin on duration of operation and blood loss.

TABLE 11.3. ANALYSIS OF VARIANCE TABLE FOR MULTIPLE REGRESSION

Source of Variation (1)	Sum of Squares (2)	d.f. (3)	MS (4)	EMS (5)	F Calc. (6)	F Tabled (7)
Due Regression on $X_1, \ldots, X_k$	$\sum_{\ell=1}^{n} (Y'_\ell - \overline{Y})^2 = b_1 \Sigma x_1 y + \ldots + b_k \Sigma x_k y$	k	$MS_{1\ldots k}$	$\sigma_e^2 + \sum_{i=1}^{k} \sum_{j=1}^{k} (\beta_i \beta_j \Sigma x_i x_j)$	$MS_{1\ldots k}/s_e^2$	$F[1-\alpha; k, n-k-1]$
Residual	$\sum_{\ell=1}^{n} (Y_\ell - Y'_\ell)^2 = \Sigma y^2 - b_1 \Sigma x_1 y - \ldots - b_k \Sigma x_k y$	n-k-1	s_e^2	σ_e^2		
Total	$\sum_{\ell=1}^{n} (Y_\ell - \overline{Y})^2 = \Sigma y^2$	n-1				

TABLE 11.4. ANALYSIS OF VARIANCE FOR PERCENTAGE CHANGE IN BLOOD HEMOGLOBIN DATA

Source of Variation (1)	Sum of Squares (2)	d.f. (3)	MS (4)	F Calc. (5)	F Tabled (6)
Due Regression on X_1, X_2	53.450	2	26.725	10.846	5.79
Residual	12.319	5	2.464		
Total	65.769	7			

The F test in Table 11.4 is for the null hypothesis that all regression coefficients are zero—$H_0: \beta_1 = \cdots = \beta_k = 0$. If the null hypothesis is true, both the residual and the due regression mean squares are estimates of σ_e^2, whereas if one or more β_i are nonzero, the due regression sum of squares estimates a quantity larger than σ_e^2.

In the example, we have

$$F = \frac{26.725}{2.464} = 10.846 \qquad \text{and} \qquad F[.95;2,5] = 5.79.$$

The null hypothesis is therefore rejected, and we decide that the population mean hemoglobin change *does* vary with duration and/or with blood loss.

ESTIMATES OF PARAMETERS AND CONFIDENCE INTERVALS

The point estimates of the population regression coefficients β_i are the sample regression coefficients b_i. The height α of the population regression plane at the point $\overline{X}_1, \ldots, \overline{X}_k$ is estimated by $\overline{Y}$. The point estimate of σ_e^2 is $s_e^2 = \sum_{l=1}^{n}(Y_l - Y_l')^2/(n-k-1)$. This quantity can be calculated from the form used in the analysis of variance table:

$$s_e^2 = \frac{\sum y^2 - b_1 \sum x_1 y - \cdots - b_k \sum x_k y}{n-k-1}. \tag{11.17}$$

The regression coefficients can be seen (equations 11.14) to be simply linear combinations of $y_1, \ldots, y_n$, just as was the case in regression on a single X variable. Therefore, expressions for the variances and covariances of the b_i can be written down in terms of σ_e^2. We do not derive these expressions here but simply state them. The variance of the partial regression coefficient b_i is $c_{ii}\sigma_e^2$, which is estimated by $c_{ii}s_e^2$. The covariance of two coefficients b_i and b_j is $c_{ij}\sigma_e^2$, and it is estimated by $c_{ij}s_e^2$. The variance of $\overline{Y}$ is σ_e^2/n, which is estimated by s_e^2/n. The b_i are independent of $\overline{Y}$; thus the covariance of b_i and $\overline{Y}$ is zero.

The importance of obtaining the Gaussian multipliers c_{ij} is now apparent. From them, we can write down expressions for the variances of arbitrary linear combinations of the b_i's and then we can construct confidence intervals.

Table 11.5 lists the various quantities that can be estimated, their estimates, the variances of the estimates, and the confidence intervals. Because two regression coefficients b_i and b_j are not statistically independent, the variance of expressions such as $b_i - b_j$ must include three terms:

$$c_{ii}\sigma_e^2 - 2c_{ij}\sigma_e^2 + c_{jj}\sigma_e^2 = \sigma_e^2(c_{ii} - 2c_{ij} + c_{jj}).$$

TABLE 11.5. ESTIMATION OF PARAMETERS IN MULTIPLE REGRESSION

Parameter (1)	Estimate (2)	Variance of Estimate (3)	Confidence Interval (4)
α	$\overline{Y}$	σ_e^2/n	$\overline{Y} \pm t_{1-\alpha/2m}\sqrt{s_e^2/n}$
β_1	b_1	$c_{11}\sigma_e^2$	$b_1 \pm t_{1-\alpha/2m}\sqrt{c_{11}s_e^2}$
$\vdots$	$\vdots$	$\vdots$	$\vdots$
β_k	b_k	$c_{kk}\sigma_e^2$	$b_k \pm t_{1-\alpha/2m}\sqrt{c_{kk}s_e^2}$
$\beta_i-\beta_j$	b_i-b_j	$\sigma_e^2(c_{ii}+c_{jj}-2c_{ij})$	$(b_i-b_j) \pm t_{1-\alpha/2m}\sqrt{(c_{ii}+c_{jj}-2c_{ij})s_e^2}$
$E(Y^*\mid X_1^* \ldots X_k^*) =$	$Y^{*\prime} = \overline{Y} + b_1x_1^* + \ldots$	$\sigma_e^2(\frac{1}{n} + c_{11}x_1^{*2} + c_{12}x_1^*x_2^* + \ldots + c_{1k}x_1^*x_k^*$	$Y^{*\prime} \pm t_{1-\alpha/2m}\sqrt{(\frac{1}{n} + \ldots)s_e^2}$
$\alpha + \beta_1x_1^* + \ldots + \beta_kx_k^*$	$+ b_kx_k^*$	$+ c_{21}x_2^*x_1^* + c_{22}x_2^{*2} + \ldots + c_{2k}x_2^*x_k^*$	
		$+ \ldots$	
		$+ c_{k1}x_k^*x_1^* + c_{k2}x_2^* + \ldots + c_{kk}x_k^{*2})$	
Y^*	$Y^{*\prime} = \overline{Y} + b_1x_1^* + \ldots$	$\sigma_e^2(1 + \frac{1}{n} + c_{11}x_1^{*2} + c_{12}x_1^*x_2^* + \ldots + c_{1k}x_1^*x_k^*$	$Y^{*\prime} \pm t_{1-\alpha/2m}\sqrt{(1 + \frac{1}{n} + \ldots)s_e^2}$
	$+ b_kx_k^*$	$+ c_{21}x_2^*x_1^* + c_{22}x_2^{*2} + \ldots + c_{2k}x_2^*x_k^*$	
		$+ \ldots$	
		$+ c_{k1}x_k^*x_1^* + c_{k2}x_2^* + \ldots + c_{kk}x_k^{*2})$ [††]	
σ_e^2	s_e^2	$\ldots$	$(n-k-1)s_e^2/\chi^2[1-\alpha/2;n-k-1] < \sigma_e^2 <$
		$x_i^* = X_i^* - \overline{X}_i,\ i=1,\ldots,k$	$(n-k-1)s_e^2/\chi^2[\alpha/2;n-k-1]$

[††] This expression is the variance of $Y^*-Y^{*\prime}$.

In the hemoglobin example, $b_1 = 0.129, b_2 = 0.138$, and $s_e^2 = 2.464$. The variances of b_1 and b_2 are estimated by

$$\widehat{\text{Var}\, b_1} = (0.000807)(2.464) = 0.001988$$

and

$$\widehat{\text{Var}\, b_2} = (0.000500)(2.464) = 0.001232.$$

Their covariance $c_{12}\sigma_e^2$, is estimated by

$$\widehat{\text{Cov}\, b_1, b_2} = (0.0000682)(2.464) = 0.000168$$

and

$$\widehat{\text{Var}(b_1 - b_2)} = 0.002884.$$

We may wish confidence intervals for β_1, β_2 and $\beta_1 - \beta_2$ although $\beta_1 - \beta_2$ is usually of minor interest. From the last column of Table 11.5, confidence intervals with a simultaneous 95% level are:

$$\beta_1 : 0.129 \pm t[.99167; 5]\sqrt{.001988}$$

or

$$0.129 \pm (3.588)(.0446)$$

or

$$-0.031 < \beta_1 < 0.289$$

$$\beta_2 : 0.138 \pm t[.99167; 5]\sqrt{.001232}$$

or

$$-0.012 < \beta_2 < 0.264.$$

$$\beta_1 - \beta_2 : -.009 \pm t[.99167; 5]\sqrt{0.002884}$$

or

$$-0.202 < \beta_1 - \beta_2 < 0.184.$$

As usual, a confidence interval for β_i can serve to test the null hypothesis $H_0 : \beta_i = \beta_{i0}$. The corresponding test statistic is

$$t = \frac{b_i - \beta_{i0}}{\sqrt{c_{ii} s_e^2}}. \tag{11.18}$$

If m tests are to be made with an overall significance level of α, the test statistic is compared with $t[1-\alpha/2m; n-k-1]$.

Confidence intervals for $E(Y^*|X_1^* \cdots X_k^*)$, the population mean of Y's for $X_1 = X_1^*, \ldots, X_k = X_k^*$ also appear in Table 11.5. Such intervals clearly take the same form as those in the case of a single X variable, although the expressions are longer.

If $X_1^* = 100$ and $X_2^* = 500$ in the hemoglobin example, then $x_1^* = -1, x_2^* = 10.25$. We calculate the height of the regression plane as

$$Y^{*\prime} = -3.387 + 0.129x_1^* + 0.138x_2^*$$

$$= -3.387 + 0.129(-1.) + 0.138(10.25)$$

$$= -2.101.$$

The variance of $Y^{*\prime}$ is estimated by

$$s_e^2\left(\frac{1}{n} + c_{11}x_1^{*2} + c_{12}x_1^*x_2^* + c_{21}x_2^*x_1^* + c_{22}x_2^{*2}\right)$$

$$= 2.464\left[\tfrac{1}{8} + 0.000807(-1.)^2 + 2(0.0000682)(-1)(10.25) + 0.0005(10.25)^2\right]$$

$$= 2.464(0.177)$$

$$= 0.436.$$

The 95% level confidence interval for $E(Y|X_1^* = 100., X_2^* = 500.)$ is

$$-2.101 \pm t[.975; 5]\sqrt{0.436}$$

$$-2.101 \pm 2.571(.660)$$

or

$$-3.801 < E(Y|X_1^* = 100, X_2^* = 500) < -0.401.$$

We also can give an interval within which we expect an individual's percentage change in hemoglobin to lie if we know that the duration was $X_1^* = 100.$ and his blood loss was $X_2^* = 500$. The 95% confidence interval for Y^* is

$$-2.101 \pm t[.975; 5]\sqrt{(1 + 0.177)2.464} \qquad \text{or} \qquad -6.479 < Y^* < 2.277.$$

The confidence interval for σ_e^2 is of the usual form:

$$\frac{(n-k-1)s_e^2}{\chi^2[1-\alpha/2;n-k-1]}<\sigma_e^2<\frac{(n-k-1)s_e^2}{\chi^2[\alpha/2;n-k-1]}. \qquad (11.19)$$

In the example, $s_e^2=2.464$ and $n-k-1=5$ can be read from the analysis of variance table; the $1-\alpha$ level interval is

$$\frac{5(2.464)}{12.83} \text{ to } \frac{5(2.464)}{.831} \qquad \text{or} \qquad .960<\sigma_e^2<14.826.$$

TESTS ON SUBSETS OF THE REGRESSION COEFFICIENTS

Earlier we gave a method for testing the hypothesis that *all* the population regression coefficients are zero. We also learned to make confidence intervals for the β_i, and to test the null hypothesis $H_0:\beta_i=0$. We now wish to introduce F tests for testing the null hypothesis that any subset of β_i's are zero. For example, with $k=5$, we might wish to test $H_0:\beta_1=\beta_2=\beta_3=0$.

Such F tests can be helpful in deciding which variables to include in a regression equation—a difficult problem when many variates have been measured on each individual and when there is little theoretical basis for a decision on which variates to include. Often many subsets of the variates are tried and the regression planes are compared. Before giving the F tests for testing that any subsets of the regression coefficients are zero, we give a general method for determining the proper statistic to use in an F test.

An Alternate Way of Obtaining F Tests. In all the examples of exact F tests presented thus far, we decided on the appropriate F statistic by looking at the EMS column of the analysis of variance table. An alternate rule, which we present now, is useful in more complicated problems. As a simple example, we use the linear regression model of Chapter 10.

The procedure consists in first forming a table such as the upper section of Table 11.6. The first column lists various hypotheses that might be made concerning the population means of the Y's for a given value of X. Here we have listed just three among many possible hypotheses. The means for a given X may lie on a straight line $\alpha+\beta x$ (H_1). They may all be equal, so that $\beta=0$ (H_2). Or all the means may equal some particular value α_0 (H_3).

In the second column we write the least squares estimate of the population means under each hypothesis. We have already noted that the sample regression line is the least squares estimate of the population regression line. That is, under H_1, Y' is the least squares estimate of the mean. It is

TABLE 11.6. *F* TESTS FOR LINEAR REGRESSION

$E(Y\|X)$ (1)	Least Squares Estimate (2)	Minimum Sum of Squared Deviations (3)	No. of Independent Parameters to Estimate (4)	d.f. (5)
$H_1: \alpha + \beta x$	$Y' = \overline{Y} + bx$	$SS_r = \sum_{i=1}^{n} (Y_i - Y_i')^2$	2	n-2
$H_2: \alpha$	$\overline{Y}$	$SS_t = \sum_{i=1}^{n} (Y_i - \overline{Y})^2$	1	n-1
$H_3: \alpha_0$	α_0	$SS_t + n(\overline{Y} - \alpha_0)^2 = \sum_{i=1}^{n} (Y_i - \alpha_0)^2$	0	n

Possible F Tests:

$E(Y\|X)$ Assumption (1)	H_0 (2)	F Statistic (3)	F Tabled
$\alpha + \beta x$	α	$\frac{(SS_t - SS_r)/[(n-1) - (n-2)]}{SS_r/(n-2)} = \frac{SS_x/1}{SS_r/(n-2)}$	$F[1-\alpha;\ 1,\ n-2]$
α	α_0	$\frac{[SS_t + n(\overline{Y}-\alpha_0)^2 - SS_t]/[n - (n-1)]}{SS_t/(n-1)} = \frac{n(\overline{Y}-\alpha_0)^2/1}{SS_t/(n-1)}$	$F[1-\alpha;\ 1,\ n-1]$
$\alpha + \beta x$	α_0	$\frac{[SS_t + n(\overline{Y}-\alpha_0)^2 - SS_r]/[n - (n-2)]}{SS_r/(n-2)} = \frac{[SS_x + n(\overline{Y}-\alpha_0)^2]/2}{SS_r/(n-2)}$	$F[1-\alpha;\ 2,\ n-2]$

also true that the mean calculated from a single sample is the least squares estimate of the population mean. If we assume H_2, with all n observations coming from populations with the same mean, the least squares estimate of $E(Y|X)$ is $\bar{Y}$. Column 3 is the minimum sum of squared deviations; that is, it is the sum of squared deviations of each Y_i from Y_i', the least squares estimate of $E(Y_i|X_i)$.

Column 4 gives the number of independent parameters in the population means to be estimated from the data. Sometimes there are relations among the parameters, and we cannot say that they are all independent parameters. For example, in an analysis of variance problem with means $\mu+\alpha_i$, $i=1,\ldots,a$, there are $a+1$ parameters but only a independent parameters, because $\sum_{i=1}^{a}\alpha_i=0$. Column 5 tells the degrees of freedom of the sum of squared deviations; they are always the number of observations minus the number of independent parameters estimated from the sample.

The lower section of Table 11.6 lists F tests that can be made from the upper section of the table. In making any test concerning the means $E(Y|X)$, we need two hypotheses concerning them—the one we wish to assume and the one we wish to test. (Of course in making the test we also make the other assumptions of normal distributions and equal variances.) We can perform several tests from the three hypotheses listed in Table 11.6. We can assume that H_1 is true $(E(Y|X)=\alpha+\beta x)$ and test the null hypothesis that H_2 is true $(E(Y|X)=\alpha)$. Or we can assume H_1 and test H_3. Or assume H_2 and test H_3. The tests made by the researcher are determined by the situation and by the particular questions he hopes to answer. (Note that the hypothesis that is tested is always a particular case of the hypothesis that has been assumed.)

It can be shown mathematically that assuming H_i, H_0 can be tested with the following statistic:

$$F=\frac{(\text{min. SS under } H_0-\text{min SS under } H_i)/(\text{d.f. under } H_0-\text{d.f. under } H_i)}{\text{min SS under } H_i/\text{d.f. under } H_i}$$

This statistic for an α level test is compared with

$$F[1-\alpha;\ \text{d.f. under } H_0-\text{d.f. under } H_i,\ \text{d.f. under } H_i].$$

To assume that the population means lie on a straight line and to test whether the line is horizontal (i.e., to assume H_1 and test H_2), we have

$$F=\frac{(\text{SS}_t-\text{SS}_r)/[(n-1)-(n-2)]}{\text{SS}_r/(n-2)}$$

$$=\frac{\text{SS}_x/1}{\text{SS}_r/(n-2)}.$$

We note that this is our usual F test of $H_0:\beta=0$ as given in Chapter 10. Two other possible F tests are presented in Table 11.6. Regardless of whether appropriate F tests are read directly from the analysis of variance table or obtained from it with the help of a table such as Table 11.6, it is important to realize in every instance exactly what is assumed and exactly what is tested.

We return now to finding an F test for the null hypothesis that the members of a subset of the regression coefficients are all zero.

An Example. Our example involves the data on blood hemoglobin, although with only two X variables, a proper subset of the two regression coefficients contains only one regression coefficient. We wish to test the null hypothesis that $\beta_1=0$. Our basic assumption is that $E(Y|X_1,X_2)=\alpha+\beta_1x_1+\beta_2x_2$, and we are testing whether $E(Y|X_1,X_2)=\alpha+\beta_2x_2$. Table 11.7 shows the F test of this hypothesis. The procedure is general. If the underlying assumption is that $E(Y|X_1,\ldots,X_k)=\alpha+\beta_1x_1+\cdots+\beta_{k'}x_{k'}+\cdots+\beta_kx_k$ and we wish to test that a set of $k-k'$ of the β's are all equal to zero (say, $H_0:\beta_{k'+1}=\cdots=\beta_k=0$, where the variables have been numbered so that the last $k-k'$ variables are the ones in question), we fit least squares planes

$$Y'=\bar{Y}+b_1x_1+\cdots+b_{k'}x_{k'}+\cdots+b_kx_k$$

and

$$Y''=\bar{Y}+b_1^*x_1+\cdots+b_{k'}^*x_{k'}.$$

The F statistic then becomes

$$F=\frac{\left[\left(\sum y^2-b_1^*\sum x_1y-\cdots-b_{k'}^*\sum x_{k'}y\right)-\left(\sum y^2-b_1\sum x_1y-\cdots-b_k\sum x_ky\right)\right]/(k-k')}{\left(\sum y^2-b_1\sum x_1y-\cdots-b_k\sum x_ky\right)/(n-k-1)} \tag{11.20}$$

and is compared with

$$F[1-\alpha;k-k',n-k-1].$$

Analysis of Variance Table for Testing $H_0:\beta_{k'+1}=\cdots=\beta_k=0$. The foregoing tests can be performed by breaking the basic sum of squares into three parts in forming an analysis of variance table. In the blood he-

TABLE 11.7. F TEST FOR $H_0: \beta = 0$, BLOOD HEMOGLOBIN DATA

$E(Y\|X_1, X_2)$ (1)	Least Squares Estimate (2)	Minimum Sum of Squared Deviations (3)	Number of Parameters Estimated (4)	d.f. (5)
$H_1: \alpha + \beta_1 x_1 + \beta_2 x_2$	$Y' = \overline{Y} + b_1 x_1 + b_2 x_2$	$\Sigma y^2 - b_1 \Sigma x_1 y - b_2 \Sigma x_2 y$		
	$= -3.387 + .129 x_1 + .138 x_2$	$= 12.322$	3	$n-3 = 5$
$H_2: \alpha + \beta_2 x_2$	$Y'' = \overline{Y} + b_2^* x_2$	$\Sigma y^2 - b_2^* \Sigma x_2 y$		
	$= -3.387 + .128 x_2$	$= 32.816$	2	$n-2 = 6$

moglobin example, in order to test $H_0:\beta_1=0$, we consider the two best fitting regression planes

$$Y' = -3.387 + 0.129x_1 + 0.138x_2$$

and

$$Y'' = -3.387 + 0.128x_2.$$

Figure 11.3 schematically represents these two regression planes, as well as a third plane, which is a horizontal plane at a height $\bar{Y}$ from the X_1X_2 plane. Each vertical deviation $Y_l - \bar{Y}$ can be broken into three parts as illustrated:

$$Y_l - \bar{Y} = \left(Y_l'' - \bar{Y}\right) + \left(Y_l' - Y_l''\right) + \left(Y_l - Y_l'\right). \tag{11.21}$$

Squaring both sides and summing over the observations, we have

$$\sum_{l=1}^{n}\left(Y_l - \bar{Y}\right)^2 = \sum_{l=1}^{n}\left(Y_l'' - \bar{Y}\right)^2 + \sum_{l=1}^{n}\left(Y_l' - Y_l''\right)^2 + \sum_{l=1}^{n}\left(Y_l - Y_l'\right)^2, \tag{11.22}$$

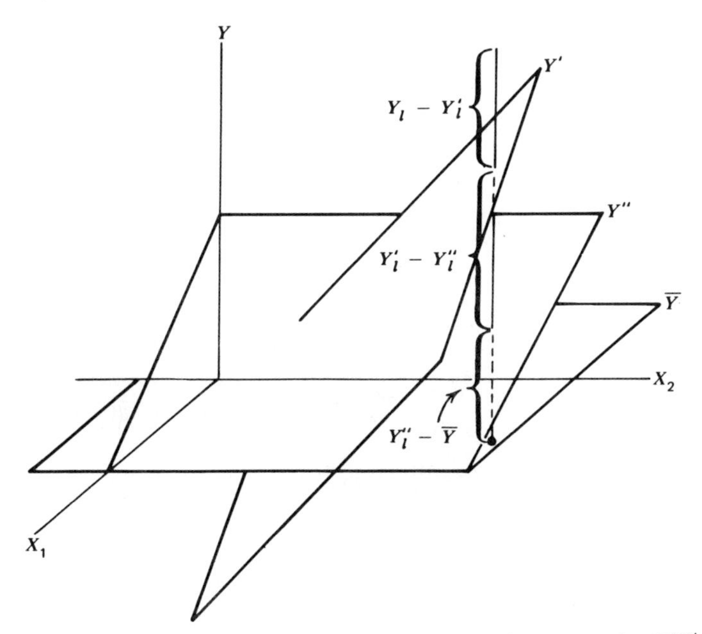

Figure 11.3. The least squares planes of Y under three assumptions concerning $E(Y|X_1, X_2)$.

where the cross-product terms have summed to zero, as usual. The term $\sum_{i=1}^{n}(Y_i'' - \bar{Y})^2$ is said to be due regression on X_2; $\sum_{i=1}^{n}(Y_i' - Y_i'')^2$ is said to be due regression on X_1 *after* fitting X_2. The third sum of squares $\sum_{i=1}^{n}(Y_i - Y_i')^2$ is the residual sum of squares. Table 11.8 is the complete analysis of variance table for the example. The due regression on X_2 sum of squares is $b_2^*\sum x_2 y = (0.128)(258.225) = 33.053$. From Table 11.4 the "due regression on X_1 and X_2 sum of squares" is 53.450. By subtraction, the "due regression on X_1 after X_2 sum of squares" is 20.397. The table allows us to assume that $E(Y|X_1,X_2) = \alpha + \beta_1 x_1 + \beta_2 x_2$ and test the null hypothesis $H_0: E(Y|X_1,X_2) = \alpha + \beta_2 x_2$ (or $H_0: \beta_1 = 0$). The F statistic is

$$F = \frac{20.397}{2.464} = 8.278.$$

Because $F[.95;1,5] = 6.61$, we reject H_0 and decide that $E(Y|X_1,X_2)$ is of the form $\alpha + \beta_1 x_1 + \beta_2 x_2$.

If the test had accepted the null hypothesis, we might have assumed that $E(Y|X_1,X_2) = \alpha + \beta_2 x_2$ and tested $H_0: E(Y|X_1,X_2) = \alpha$ (i.e., $H_0: \beta_2 = 0$). This test is

$$F = \frac{33.053/1}{(20.397 + 12.319)/(1+5)}$$

$$= 6.062,$$

which is compared with $F[.95;1,6] = 5.99$. At a 5% level, the decision is borderline.

TABLE 11.8. REGRESSION OF BLOOD HEMOGLOBIN ON BLOOD LOSS AND DURATION

Source of Variation (1)	SS (2)	d.f. (3)	MS (4)	F Calculated (5)	F Tabled (6)
Due Regression on X_2	33.053	1	33.053	6.062	F[.95;1,6] = 5.99
Due Regression on X_1 after X_2	20.397	1	20.397	8.278	F[.95;1,5] = 6.61
Residual	12.319	5	2.464		
Total	65.769	7			

With k variables, a wide variety of tests can be made and a wide variety of analysis of variance tables could be formed. Even for $k=2$ there are five possible tables. Each one is suitable for testing some hypothesis concerning $E(Y|X_1,X_2)$. Table 11.9 is the analysis of variance for testing $H_0:\beta_{k'+1} = \cdots = \beta_k = 0$. The mean square due regression on $X_1,\ldots,X_{k'}$ is denoted by $\text{MS}_{1,\ldots,k'}$; the mean square due regression on $X_{k'+1},\ldots,X_k$ after $X_1,\ldots,X_{k'}$ by $\text{MS}_{k'+1,\ldots,k|1,\ldots,k'}$.

The problem of deciding which X variates to include in a regression equation is seldom very difficult when the X levels are fixed in advance; in this case, the number of possible X variates tends to be small. With a small number of X variates (say, 4 or less) all possible regression equations can be quickly computed on a high-speed computer. Then the investigator can observe how the partial regression coefficient on any particular variable varies from one equation to another, depending on the other variates used in the regression.

STANDARD REGRESSION COEFFICIENTS

In trying to interpret a regression equation, research workers should not draw conclusions directly from the relative size of the regression coefficients. Besides depending on the set of variates included in the regression, the size of the coefficients depends on the units in which the variables are measured. To remove this second difficulty, *standard regression coefficients* are sometimes used.

Standard regression coefficients can be obtained by first standardizing the data in order to make all means equal to zero and all variances equal to one. Standardization of the data is accomplished by subtracting the mean from each observation and dividing by the standard deviation. In other words, we make the transformation $y'=(Y-\bar{Y})/s_y, x_i'=(X_i-\bar{X}_i)/s_i$, $i=1,\ldots,k$, where

$$s_y^2=\sum_{l=1}^{n}\left(Y_l-\bar{Y}\right)^2/(n-1), \qquad s_i^2=\sum_{l=1}^{n}\left(X_{il}-\bar{X}_i\right)^2/(n-1), \qquad i=1,\ldots,k.$$

With the new data, $\sum y'^2=\sum x_i'^2=n-1$, $i=1,\ldots,k$. Also, each $\sum x_i'x_j' =(n-1)r'_{ij}$, where r'_{ij} is the correlation between x_i' and x_j'. Similarly, $\sum x_i'y'=(n-1)r'_{iy}$, where r'_{iy} is the correlation between y' and x_i'. From the formula for a correlation coefficient, it can be demonstrated that the correlation coefficient is independent of the units of measurement. This means that the correlation between x_i' and x_j' is the same as the correlation between x_i and x_j; that is, $r'_{ij}=r_{ij}$. Therefore, when the normal equations

TABLE 11.9. ANALYSIS OF VARIANCE TABLE

Source of Variation (1)	Sum of Squares (2)	d.f. (3)	Mean Square (4)	F Calculated (5)	F Tabled (6)
Due Regression on $X_1,\ldots,X_{k'}$	$b_1^*\Sigma x_1 y + \ldots + b_{k'}^*\Sigma x_{k'} y$	k'	$MS_{1,\ldots,k'}$		
Due Regression on $X_{k'+1},\ldots,X_k$ after $X_1,\ldots,X_{k'}$	$(b_1 x_1 y + \ldots + b_k x_k y) - (b_1^*\Sigma x_1 y + \ldots + b_{k'}^*\Sigma x_{k'} y)$	$k-k'$	$MS_{k'+1,\ldots,k\mid 1,\ldots,k'}$	$\frac{MS_{k'+1,\ldots,k\mid 1,\ldots,k'}}{s_e^2}$	$F[1-\alpha;k-k',n-k-1]$
Residual	$\Sigma y^2 - b_1\Sigma x_1 y - \ldots - b_k\Sigma x_k y$	$n-k-1$	s_e^2		
Total	Σy^2	$n-1$			

are written for the standardized data, they become, after division by $n-1$:

$$b_1' + r_{12}b_2' + \cdots + r_{1k}b_k' = r_{1y}$$

$$r_{21}b_1' + b_2' + \cdots + r_{2k}b_k' = r_{2y}$$

$$\cdots$$

$$r_{k1}b_1' + r_{k2}b_2' + \cdots + b_k' = r_{ky}. \tag{11.23}$$

The relation between the standardized regression coefficients b_i' and the unstandardized coefficients b_i is

$$b_i' = b_i\sqrt{\sum x_i^2 / \sum y^2}\ . \tag{11.24}$$

We can solve (11.6) directly for $b_1, \ldots, b_k$ and calculate the standardized regression coefficients from (11.24), or we can solve (11.23) for $b_1', \ldots, b_k'$ and calculate the $b_1, \ldots, b_k$. Finding the standardized coefficients first presents a computational advantage when the variances and covariances differ widely in magnitude; the correlations may be closer in magnitude and calculations can be made more accurately.

In the example, we calculate

$$b_1' = 0.129\sqrt{1254./65.769} = 0.129\sqrt{19.0667} = 0.563$$

$$b_2' = 0.138\sqrt{2023.5/65.769} = 0.138\sqrt{30.7668} = 0.765.$$

The relative sizes of b_1' and b_2' now reflect the effect of variables 1 and 2 on the regression plane.

SAMPLING FROM THE JOINT NORMAL DISTRIBUTION

In Chapter 10 we considered drawing a sample of n individuals and measuring X and Y on each individual. Under the assumption that X and Y are jointly normally distributed, we found that the means of the Y's for a given value of X lay on a straight line

$$E(Y|X) = \mu_y + \frac{\rho\sigma_y}{\sigma_x}(X - \mu_x)$$

and that the variance of these populations was $\sigma_e^2 = \sigma_y^2(1-\rho^2)$. We estimated the five parameters μ_x, μ_y, σ_x^2, σ_y^2, and ρ. Then, because the joint normal model implies all the assumptions of the earlier fixed X model, we also estimated the parameters and made the tests appropriate for the fixed X model.

Whenever $X_1, X_2, \ldots, X_k, Y$ are jointly normally distributed, for each fixed set of values of $X_1, \ldots, X_k$ a population of Y's exists whose mean is given by (11.3). All such Y populations have equal variances σ_e^2, and their means lie on a plane. Thus when $X_1, X_2, \ldots, X_k$, Y are jointly normally distributed, we can find the least squares regression plane, making the same estimates and tests as in the fixed $X_1, \ldots, X_k$ model given earlier in this chapter. In addition to these estimates and tests, we can estimate the population means, variances, and correlations.

The population mean for Y, μ_y, is estimated by $\bar{Y}$; the population variance for Y, σ_y^2, is estimated by the usual sample variance s_y^2. The mean and variance of the X_i population, denoted by μ_i and σ_i^2, are estimated by the sample mean and sample variance $\bar{X}_i$ and s_i^2. The population correlation coefficient between Y and X_i is denoted by ρ_{iy} and is estimated by the sample correlation coefficient

$$r_{iy} = \frac{\sum x_i y}{\sqrt{\sum x_i^2 \sum y^2}}$$

Similarly, the population correlation coefficient between X_i and X_j is denoted by ρ_{ij} and is estimated by

$$r_{ij} = \frac{\sum x_i x_j}{\sqrt{\sum x_i^2 \sum x_j^2}}.$$

For formulae used in calculating the sample correlation coefficients, refer to Chapter 2; for tests and confidence intervals, see Chapter 10. In multiple regression problems, the foregoing correlation coefficients are often referred to as the *simple correlation coefficients*. This is done in order to distinguish them from the *multiple correlation coefficient* and from *partial correlation coefficients*.

THE MULTIPLE CORRELATION COEFFICIENT

We now wish to generalize the concept of correlation coefficient to the situation with measurements $X_1, \ldots, X_k$, and Y. To obtain an idea of the strength of the linear relationship between Y and $X_1, \ldots, X_k$, we might calculate the simple correlation coefficients $r_{1y}, \ldots, r_{ky}$; as a set, however, these seem to be difficult to interpret and we look instead for a single statistic.

It is rather easy to see that the correlation coefficient r_{xy} calculated from n observations on X and Y has the same magnitude as the correlation between the n values of m_1+m_2X and Y where m_1 and m_2 are any numbers, provided m_2 is not zero.

In particular, for the least squares line $Y'=a+b(X-\overline{X})$, the correlation coefficient calculated between Y (actual Y) and Y' (Y predicted for that X from the sample regression line) is the same as the correlation between Y and X except for a possible sign difference; that is, $r_{y'y}=\pm r_{xy}$. Table 11.10 gives a small example of this.

Thus the correlation coefficient between Y and X has the same magnitude as the correlation coefficient between Y and its predicted value Y'. The sample multiple correlation coefficient of Y with $X_1,\ldots,X_k$ is defined as the simple correlation coefficient between Y and its predicted value Y'. It is denoted by $R_{Y.1\ldots k}$ or sometimes simply by R. Thus we have

$$R_{Y.1\ldots k}=r_{yy'} \text{ where } Y'=\overline{Y}+b_1x_1+\cdots+b_kx_k.$$

That is, after fitting a least squares plane using a sample of size n, we find the simple correlation coefficient between the two columns of figures

$$\begin{array}{ll} Y_1 & Y_1'=a+b_1x_{11}+\cdots+b_kx_{1k} \\ & \vdots \\ Y_n & Y_n'=a+b_1x_{1n}+\cdots+b_kx_{kn}. \end{array}$$

Because $R_{Y.1\ldots k}=r_{yy'}$, the multiple correlation coefficient reduces to the simple correlation coefficient between X_1 and Y (except for a possible sign difference), when k is 1.

POPULATION MULTIPLE CORRELATION COEFFICIENT

The population multiple correlation coefficient corresponding to $R_{Y.1\ldots k}$ is denoted by $\mathcal{R}_{Y.1\ldots k}$, or sometimes more simply by $\mathcal{R}$. It is defined by

$$\mathcal{R}^2_{Y.1\ldots k}=1-\frac{\sigma_e^2}{\sigma_y^2}, \tag{11.25}$$

where σ_e^2 is the variance of a population of Y's belonging to a certain $X_1,\ldots,X_k$, and σ_y^2 is the variance of the population of all possible Y's. If σ_e^2, the variance of the Y distributions about the population regression plane, is small relative to σ_y^2, the variance of the entire Y distribution, $\mathcal{R}^2_{Y.1\ldots k}$ (and $\mathcal{R}_{Y.1\ldots k}$) is close to 1. In that case, fitting a sample plane is likely to result in a considerably smaller variance of the data around the fitted

TABLE 11.10. CORRELATIONS BETWEEN X AND Y, Y' AND Y

	Y	X	$y=Y-\bar{Y}$	$x=X-\bar{X}$	y^2	xy	x^2	Y'	$y'=Y'-\bar{Y}$	yy'	y'^2
	3	2	-2	-1	4	2	1	4.5	- .5	1	.25
	6	4	1	1	1	1	1	5.5	.5	.5	.25
	5	1	0	-2	0	0	4	4	-1	0	1.00
	6	5	1	2	1	2	4	6	1	1	1.00
Σ	20	12	0	0	6	5	10	20	0	2.5	2.50
	$\bar{Y} = 5$	$\bar{X} = 3$						$\bar{Y}'=5=\bar{Y}$			

$$b = 5/10 = .5$$

$$Y' = 5 + .5(X-3)$$

$$= 3.5 + .5X$$

$$r_{xy} = \frac{\Sigma xy}{\sqrt{\Sigma x^2 \Sigma y^2}} = \frac{5}{\sqrt{10(6)}} = \frac{5\sqrt{60}}{60} = \frac{\sqrt{15}}{6}$$

$$r_{yy'} = \frac{\Sigma yy'}{\sqrt{\Sigma y^2 \Sigma y'^2}} = \frac{2.5}{\sqrt{6(2.5)}} = \frac{2.5\sqrt{15}}{15} = \frac{\sqrt{15}}{6}$$

plane $Y' = \bar{Y} + b_1 x_1 + \cdots + b_k x_k$ than about the plane $\bar{Y}$. Note that (11.25) can be rewritten as $\sigma_e^2 = \sigma_y^2(1 - \mathcal{R}^2_{Y.1\ldots k})$. If $\mathcal{R}^2$ is small, σ_e^2 is nearly as large as σ_y^2, and fitting a plane to $X_1,\ldots,X_k$ is probably not warranted. If $\mathcal{R}^2$ equals zero, we should just use the plane $\bar{Y}$. If $\mathcal{R}^2$ equals 1, σ_e^2 is zero, and all the observations lie exactly on a plane.

We do not know $\mathcal{R}^2$ and must be content with the sample estimate R^2. From the definitions given for R and $\mathcal{R}$, it is not apparent that R estimates $\mathcal{R}$; to show this, we now put R^2 into a form analagous to that given by (11.25) for $\mathcal{R}^2$. The sample multiple correlation coefficient can be introduced into the analysis of variance table. Just as we demonstrated earlier that $b\sum xy = r^2 \sum y^2$, it can be shown that the "due regression on $X_1,\ldots,X_k$" sum of squares $b_1\sum x_1 y + \cdots + b_k \sum x_k y = R^2_{Y.1\ldots k}\sum y^2$. The analysis of variance table then can take the form of Table 11.11.

We can divide the residual sum of squares from Table 11.11 by the total sum of squares:

$$\frac{(1 - R^2_{Y.1\ldots k})\sum y^2}{\sum y^2} = \frac{(n-k-1)s_e^2}{(n-1)s_y^2}.$$

After a little manipulation, we obtain

$$R^2_{Y.1\ldots k} = 1 - \frac{(n-k-1)s_e^2}{(n-1)s_y^2}. \tag{11.26}$$

Comparison of (11.25) and (11.26) reveals that they are similar, except for the factor $(n-k-1)/(n-1)$. This factor is close to 1 for large sample sizes, and thus $R^2_{Y.1\ldots k}$ is actually an estimate of $\mathcal{R}^2_{Y.1\ldots k}$.

Table 11.11 indicates that the square of the sample multiple correlation coefficient $R^2_{Y.1\ldots k}$ is equal to the proportion of the basic sum of squares explained by regression on $X_1,\ldots,X_k$. If $R^2_{Y.1\ldots k}$ is small, there is little advantage in using $X_1,\ldots,X_k$ to predict Y.

Another property of $R_{Y.1\ldots k}$ is that it is the highest correlation between Y and any linear combination of $X_1,\ldots,X_k$. Thus the regression $Y' = \bar{Y} + b_1 x_1 + \cdots + b_k x_k$ can be regarded as simply a linear index based on $X_1,\ldots,X_k$ which has the highest correlation with Y of all such indices. This means that by omitting any subset of the X variables from the regression, we cannot possibly increase the multiple correlation coefficient.

Often an investigator finds that $r_{y1},\ldots,r_{yk}$ are all low. He still hopes, however, that a combination of the X variates can be useful in predicting Y; $R_{Y.1\ldots k}$ helps him decide whether this is true.

A simple procedure for calculating $R_{Y.1\ldots k}$ is taken from Table 11.11.

TABLE 11.11. ANALYSIS OF VARIANCE FOR Y ON $X_1,\ldots,X_k$

Source of Variation (1)	Sum of Squares (2)	d.f. (3)	MS (4)	F Calculated (5)	F Tabled (6)
Due Regression on $X_1,\ldots,X_k$	$R^2_{Y.1\ldots k}\Sigma y^2$	k	$MS_{1\ldots k}$	$\frac{R^2/k}{(1-R^2)/(n-k-1)}$	$F(1-\alpha;k,n-k-1)$
Residual	$(1-R^2_{Y.1\ldots k})\Sigma y^2$	n-k-1	s_e^2		
Total	Σy^2	n-1	s_y^2		

Dividing the "due regression on $X_1,\ldots,X_k$" sum of squares by the total sum of squares, we obtain R^2; that is,

$$R^2_{Y.1\ldots k} = \frac{k\text{MS}_{1\ldots k}}{(n-1)s_y^2}. \tag{11.27}$$

TEST FOR $H_0 : \mathrm{R}^2 = 0$

The test statistic for the null hypothesis that the population correlation coefficient equals zero is

$$F = \frac{R^2/k}{(1-R^2)/(n-k-1)}, \tag{11.28}$$

to be compared with

$$F[1-\alpha; k, n-k-1]$$

for an α level test. This test statistic is the same one given earlier for $H_0 : \beta_1 = \cdots = \beta_k = 0$. In fact, the two null hypotheses are really the same. If the population regression plane is horizontal (all $\beta_i = 0$), the variance of Y's about the regression plane (σ_e^2) is the same as the variance of the entire Y population (σ_y^2); thus $\mathrm{R} = 0$. Conversely, if $\mathrm{R} = 0, \sigma_e^2 = \sigma_y^2$, and the population regression plane must be horizontal.

For the blood hemoglobin data, we have

$$R^2_{Y.12} = \frac{53.450}{65.769} = .813.$$

We can compare $R_{Y.12} = .901$ with $r_{y1} = .481$ and $r_{y2} = .708$.

PARTIAL CORRELATION COEFFICIENTS

When considering a correlation coefficient that has been calculated between two variates, say X_1 and X_2, we often wonder whether $r_{X_1X_2}$ can be explained in terms of another variable X_3. This is a natural question if we are trying to explain an apparent relationship. For example, if we obtained a correlation of .8 between mental age and height, an early thought in interpreting this correlation (rather than plunging into discussions of whether height affects intelligence or intelligence affects height) would be that both variables may be correlated with chronological age. Would there be any correlation between mental age and height if we took children of exactly the same chronological age?

To examine the correlation between mental age and height in the absence of the *linear* effect of chronological age, we remove chronological age as well as possible from our data on mental age and on height and correlate what is left over. That is, we obtain two regression lines, X_1 on X_3 (mental age on chronological age) and X_2 on X_3 (height on chronological age). These are, say,

$$X_1' = \overline{X}_1 + b_3 x_3$$

$$X_2' = \overline{X}_2 + b_3^* x_3.$$

Then we replace each individual's mental age X_{1l} and height X_{2l} by $X_{1l} - X'_{1l}$ and $X_{2l} - X'_{2l}$. Having attempted to remove the linear effect of chronological age, we correlate the new sets of data. Figure 11.4 shows the two regression lines.

The procedure appears to involve a considerable amount of calculation. There is, however, a formula that simplifies calculations. If $r_{12.3}$ denotes the partial correlation, we can write

$$r_{12.3} = \frac{r_{12} - r_{13} r_{23}}{\sqrt{(1 - r_{13}^2)(1 - r_{23}^2)}}. \tag{11.29}$$

For the example of Y = blood hemoglobin, X_1 = duration of operation, and X_2 = blood loss, $r_{y1} = .481, r_{y2} = .708, r_{12} = -.107$. Formula 11.29 becomes

$$r_{y1.2} = \frac{r_{y1} - r_{y2} r_{12}}{\sqrt{(1 - r_{y2}^2)(1 - r_{12}^2)}}$$

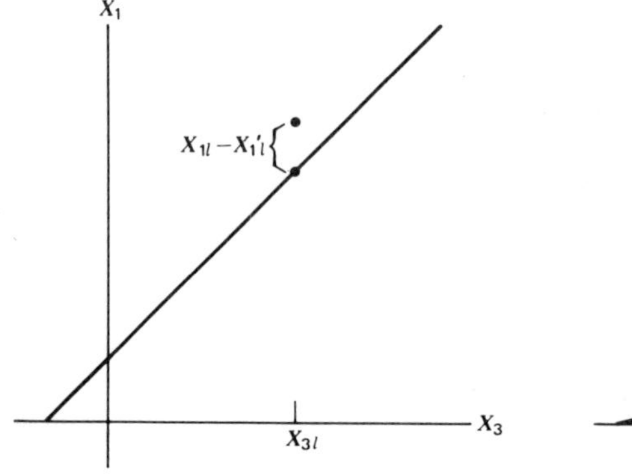

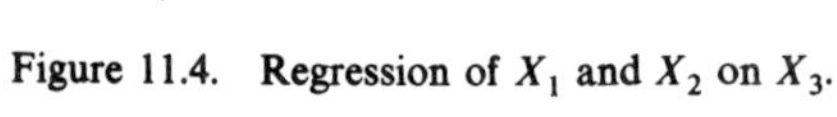

Figure 11.4. Regression of X_1 and X_2 on X_3.

and we calculate

$$r_{y1.2} = .793.$$

The sample partial correlation coefficient $r_{ij.k}$ is an estimate of the population partial correlation coefficient $\rho_{ij.k}$.

For normal populations, $\rho_{ij.k}$ is the population correlation coefficient in the populations of X_i and X_j at a particular value of X_k. It also has a formula similar to (11.29):

$$\rho_{ij.k} = \frac{\rho_{ij} - \rho_{ik}\rho_{jk}}{\sqrt{(1-\rho_{ik}^2)(1-\rho_{jk}^2)}}.$$

It is possible to obtain a value of $r_{12.3} < 0$ when $r_{12} > 0$. For example, X_1 could be height, X_2 waist circumference, and X_3 weight. All the simple correlations are positive for measurements taken on adults. But let us consider the partial correlation of height and waist circumference with the linear effect of weight removed. Persons with the same weight could be either short with a large waist or tall with a small waist. Hence the partial correlation would be negative $r_{12.3} < 0$. As seen from (11.29), $r_{12.3}$ is negative whenever $r_{13}r_{23} > r_{12}$.

Frequently the partial correlation coefficient is somewhat less than the simple correlation coefficient because the effect of a common variate has been removed in a linear fashion. There are many exceptions, however. The denominator of $r_{12.3}$ is less than 1 and thus has a tendency to increase the size $r_{12.3}$. Also, if one of r_{13} or r_{23} is negative and the other two simple correlations are positive, $r_{12.3}$ is larger than r_{12}.

Sometimes it is necessary to transform the variates or to fit a more complicated function in order to remove the effect of a third variate effectively.

The concept of partial correlation coefficient can be extended to eliminate the effect of two or more variates. For example, we may find $r_{12.34}$ the correlation coefficient of X_1 and X_2, with the effect of X_3 and X_4 removed. In terms of partial correlations $r_{12.3}$, $r_{14.3}$, and $r_{24.3}$, the formula for $r_{12.34}$ is

$$r_{12.34} = \frac{r_{12.3} - r_{14.3}r_{24.3}}{\sqrt{(1-r_{14.3}^2)(1-r_{24.3}^2)}}. \qquad (11.30)$$

The extension of (11.30) to the removal of several variables is straightforward.

TESTS AND CONFIDENCE INTERVALS FOR PARTIAL CORRELATION COEFFICIENTS

Tests and confidence intervals for $\rho_{ij.kl\ldots}$ are made using the same methods as those given in Chapter 10 for ρ. The rule is that we decrease the sample size (or degrees of freedom) by 1 for each variate that is removed and proceed as before.

For a confidence interval for $\rho_{1y.2}$ in the blood hemoglobin example, we have

$$z \text{ ND}\left(Z, \frac{1}{n-4}\right).$$

From Table A.6, corresponding to $r_{1y.2}=.793$, we obtain a z value of 1.0794; therefore, a 95% confidence interval for Z is

$$1.0794 \pm (1.96)\left(1/\sqrt{8-4}\,\right) \qquad \text{or} \qquad 0.10 < Z < 2.06.$$

Corresponding to 0.10 and 2.06 in Table A.6, we find r values of .10 and .996, respectively. The 95% confidence interval for $\rho_{1y.2}$ is then .09 to .996, a wide interval indeed!

Besides helping us evaluate associations with the effect of various variates removed, partial correlations can also help us decide on variates to include in a regression program. Efroymson [11.7] gives a stepwise regression procedure that successively includes variates one at a time in a regression. At each step, the variate is added that has the highest partial correlation with Y. This procedure is available in BMD02R. At each step, the program also checks whether the variable being added reduces the residual sums of squares significantly and whether with the new variable included, any of the previously added variates can be discarded. In the printout from such programs, a simple criterion for deciding at which point to stop adding variables is the size of the multiple correlation coefficient between Y and each set of X variates. As each variate is added by the program, the multiple correlation coefficient increases; but when the increase is unimportant (say less than .02), the value of adding additional variables is questionable. Usually no more than four or five variables are included in the final regression equation. Selecting X variates to use in a regression is particularly important when sampling is done from the joint distribution; investigators often consider 20 or more variables as X variates. Stepwise regression programs are not a complete answer to the problem of obtaining the best set of X's for a regression equation, but they can be of help.

Note that it is always possible to decrease the sum of squared deviations by adding more variates. This does not say that the maximum possible number of variates should be included in the analysis. Added variates may improve the fit to the sample data; but they may make the approximation worse for the population as a whole. This problem is exceedingly difficult, but we can obtain some protection against including too many variates by using a stepwise program that tests at each step whether the population regression coefficient of the variable being added is zero. For discussion of variable selection in regression, see Draper and Smith [11.5].

EXTENSIONS OF MULTIPLE REGRESSION

We can extend multiple regression methods just as we did the methods of linear regression. Weights can be assigned to the observations; the formulae obtained are comparable to those given in linear regression.

Transformations also can be used in multiple regression. The same considerations hold as in linear regression; the transformations are selected with a view to making the assumptions more realistic, just as in the case with a single X.

**AN EXAMPLE OF A DESK CALCULATOR SOLUTION OF LINEAR EQUATIONS

As an example of the solution of k linear equations in k unknowns, we solve the two equations (11.9) in the two unknowns c_{11} and c_{12}:

$$1254c_{11} - 171c_{12} = 1$$

$$-171c_{11} + 2023.5c_{12} = 0.$$

A general procedure is to reduce k equations in k unknowns to $k-1$ equations in $k-1$ unknowns, then to reduce the $k-1$ equations in $k-1$ unknowns to $k-2$ equations in $k-2$ unknowns, and so on, until finally there is just one equation in one unknown. This single equation is solved, and the value obtained is substituted in one of the two equations in two unknowns. From this equation, we find the value of a second unknown, and the two values are substituted in one of the three equations in three unknowns, to obtain the value of a third unknown. This backward solution process continues until the values of all k unknowns have been found.

The two equations (11.9) are reduced to a single equation by first manipulating them until the coefficient of c_{11} is the same for both equations and then subtracting the second equation from the first to obtain a new equation in c_{12} alone. We divide the first equation by 1254 to obtain a 1 as the coefficient of c_{11}. The (two) equations are then:

$$1c_{11} - \frac{171}{1254}c_{12} = \frac{1}{1254}$$

$$-171c_{11} + 2023.5c_{12} = 0.$$

We multiply the first equation by -171, to give both equations the same leading coefficient

$$-171c_{11} + \frac{(171)(171)}{1254}c_{12} = \frac{-171}{1254}$$

$$-171c_{11} + 2023.5c_{12} = 0.$$

Next we subtract the second equation from the first; this results in one equation in one unknown:

$$2000.1818c_{12} = 0.13636 \qquad \text{or} \qquad c_{12} = 0.0000682.$$

We now substitute $c_{12} = 0.0000682$ in the first of the two equations (11.9):

$$1254c_{11} - (171)(0.0000682) = 1$$

or

$$1254c_{11} = 1.011662$$

$$c_{11} = 0.000807.$$

We check our solution by substituting $c_{11} = 0.000807$ and $c_{12} = 0.0000682$ in the second of equations (11.9):

$$(-171)(0.000807) + 2024(0.0000682) = -0.138 + 0.138 = 0.$$

Thus the solutions to (11.9) are

$$c_{11} = .000807 \qquad \text{and} \qquad c_{12} = .0000682.$$

ILLUSTRATIVE EXAMPLE

We consider a study undertaken in order to determine how well diastolic blood pressure in males 50 to 59 years of age can be predicted from a set of measurements made 5 years earlier. A sample of n men is selected; the men are given a physical examination, and various measures are recorded. Five years later, diastolic blood pressures are measured on the same men. Provided the measures used as X variables and the diastolic blood pressures Y are approximately normally distributed, the second multiple regression model is appropriate, since a single sample of n men was used, rather than samples of men at certain specified levels of the X variates. As mentioned earlier, however, we can nevertheless do a multiple regression of diastolic blood pressure on a set of X variables.

First we must decide which measurements taken at the outset of the study should be included as X variates. Variables such as diastolic and systolic blood pressure, weight, height, heart rate, and cholesterol level are considered because they may be correlated with blood pressure at the end of the study. A first step is to use some program such as EPID41 to print out scatter diagrams of Y versus each possible X variable. Using the scatter diagrams, we discard obvious outliers or incorrect observations. In order for Y and $X_1,\ldots,X_k$ to have a joint normal distribution, each of the scatter diagrams of Y versus an X variable should look something like Figure 10.3-*c* (i.e., with points falling roughly in an ellipse). If a scatter diagram looks as in Figure 10.3*d*, a transformation such as a logarithmic one on that X variate can be tried and a new scatter diagram examined.

After deciding on a set of X variates, a stepwise regression program such as BMD02R can be run. With standard options, BMD02R usually includes all the variables. In examining the output, we should first look at R^2 for the final set of variables. If R^2 is small—say, less than .04 ($R<.2$)—interpretation or use of the regression analysis is inadvisable; the variance around the regression is almost as large as the original variance of Y.

If the largest R^2 value is sizable, the investigator should not simply accept the regression equation and use it for his ultimate purpose. In order to interpret his equation, he should try various sets of variates, deleting some and adding others. For example, if diastolic and systolic blood pressure at outset are both used as possible X variates, they may be highly correlated. In this case, the program might pick diastolic blood pressure as a first X variable and perhaps not include systolic blood pressure at all. Or, in the final equation, the regression coefficient on diastolic blood pressure may be sizable and positive, whereas that on systolic blood pressure may be small and negative. If, on the other hand, diastolic blood pressure is omitted in starting the stepwise program, systolic blood pressure may be the first variate selected, and it may have a sizable positive regression coefficient. In such a situation, the regression equation obtained from the first run, where both blood pressures were entered, *may* be the best one for predicting Y. Certainly, however, it is unsafe to make any attempt at interpreting the signs of the various regression coefficients without looking at many different sets of variables.

Even for prediction purposes, we do well to examine regressions obtained by starting with several subsets of X variables, in order to seek a small set of variates giving a multiple correlation almost as high with Y as a larger set. It is well known that a regression equation based on many variables is more disappointing when used for prediction than one based on fewer variables, other things being equal. The "other things" here are the size of the sample used to form the regression and the multiple correlation.

If a desk calculator must be used in the regression analysis, it is necessary to choose a small number of X variates carefully. First the array of correlations in (11.23) is calculated. The array is examined and several X variables are chosen that are highly correlated with Y and whose correlations among themselves are small in magnitude.

Frequently in such studies the set of possible X variables includes several that are clearly not normally distributed, nor can they be transformed to normality. For example, each individual in the physical examination may have been rated on amount of physical activity on a 4-point scale from very light to very heavy. It is reasonable to fit a regression plane to a set of data containing such nonnormal variables, provided we believe that the population means for diastolic blood pressure at the end of the study lie on a plane. This, of course, is a difficult assumption to check, but a partial verification can be made by means of the same scatter diagrams used when the X variables were continuous. The scatters now may look like Figure 10.3*a* or 10.3*b*. If they indicate linearity, as in Figure 10.3*a*, or if a transformation can be found to linearity, the variable can be included.

When such nonnormal X's are included in the study, prediction intervals and confidence intervals for the β's and for population means still apply, for they apply to the fixed X model. Since the usual confidence intervals for correlation coefficients depend on joint normality, they must be used with caution if at all—they are at best merely approximations.

It is clear that no variable such as religion or occupation should be included unless it can be placed along a single-dimensional scale.

SUMMARY

Techniques have been presented for predicting values of Y from k different $X_1, X_2, \ldots, X_k$ by means of a plane in a $(k+1)$ dimensional space.

As in the previous chapter, the X_i can be considered fixed, or n individuals can be sampled and $X_1, X_2, \ldots, X_k$ and Y measured on each. An alternative method of performing F tests has been given. The multiple correlation coefficient and the partial correlation coefficients under the joint normal model were introduced.

REFERENCES

APPLIED STATISTICS

11.1. Bennett, C. A., and N. L. Franklin, *Statistical Analysis in Chemistry and the Chemical Industry*, Wiley, New York, 1954.

11.2. Box, G. E. P., "Use and Abuse of Regression," *Technometrics*, Vol. 8, No. 4 (1966), pp. 625–630.

11.3. Daniel, C., and F. S. Wood, *Fitting Equations to Data*, Wiley, New York, 1971.

11.4. Dixon, W. J., and F. J. Massey, Jr., *Introduction to Statistical Analysis*, 3rd ed., McGraw-Hill, New York, 1969.

11.5. Draper, N. R., and H. Smith, *Applied Regression Analysis*, Wiley, New York, 1966.

11.6. Dwyer, P., *Linear Computations*, Wiley, New York, 1951.

11.7. Efroymson, M. A., *Multiple Regression Analysis, Mathematical Methods for Digital Computers*, Part V, (17), A. Ralston and H. S. Wilf (Ed.), Wiley, New York, 1960.

11.8. Morrison, D. F., *Multivariate Statistical Methods*, McGraw-Hill, New York, 1967.

11.9. Ostle, B., *Statistics in Research*, Iowa State University Press, Ames, Iowa, 1963.

11.10. Snedecor, G. W., and W. G. Cochran, *Statistical Methods*, 6th ed., Iowa State University Press, Ames, Iowa, 1967.

11.11. Sprent, P., *Models in Regression and Related Topics*, Methuen, London, 1969.

11.12. Winer, B. J., *Statistical Principles in Experimental Design*, McGraw-Hill, New York, 1962.

MATHEMATICAL STATISTICS

11.13. Rao, C. R., *Linear Statistical Inference and Its Applications*, Wiley, New York, 1965.

11.14. Williams, E. J., *Regression Analysis*, Wiley, New York, 1959.

PROBLEMS

11.1. On the basis of eight patients who underwent a thyroid operation, an attempt was made to determine factors related to blood loss. Measurements of weight, duration of operation, and blood loss for the eight cases were as follows:

X_1, Weight (kg)	X_2, Time (min)	Y, Blood Loss (ml)
44.3	105	503
40.6	80	490
69.0	86	471
43.7	112	505
50.3	109	482
50.2	100	490
35.4	96	513
52.2	120	464

(*a*) Find the least squares plane, fitting blood loss to weight and time.
(*b*) Make an analysis of variance table.
(*c*) Test $H_0: \beta_1 = \beta_2 = 0\ (\alpha = .05)$.
(*d*) Find the least squares line, fitting blood loss to weight.
(*e*) Test $H_0: \beta_2 = 0\ (\alpha = .05)$.
(*f*) Test $H_0: \rho_{y2.1} = 0\ (\alpha = .05)$, where $\rho_{y2.1}$ denotes the partial correlation coefficient of Y and X_2 for X_1 fixed.
(*g*) Give a 95% confidence interval for $\rho_{y2.1}$.

11.2. Autopsies of 1 to 3 day-old infants who died of crythroblastosis kernicterus were performed. Their body weight, heart weight, and finger breadth were measured.

Infant	X_1, Body Weight (g)	X_2, Finger Breadth (cm)	Y, Heart Weight (g)
1	3250	2	32
2	2400	1	20
3	3360	2	19
4	3100	2	21
5	2600	1	23
6	2620	1	18
7	2775	1	15
8	3710	2	29
9	3900	2	24
10	2200	2	17

(*a*) Write the equation predicting heart weight Y from body weight X_1 and finger breadth X_2.
(*b*) Complete the following analysis of variance table.

	SS	d.f.	MS
Due X_1			
Due X_2 after X_1			
Residual			
Total			

(*c*) Assuming $E(Y|X_1,X_2)=\alpha+\beta_1X_1+\beta_2X_2$, test $H_0:\beta_2=0\,(\alpha=.05)$.
(*d*) Write down the least squares line predicting Y from X_1.
(*e*) Assuming $E(Y|X_1,X_2)=\alpha+\beta_1X_1$, test $\beta_1=0\,(\alpha=.05)$.
(*f*) From (*d*) give the estimate of Y for $X_1=3000$ and obtain a 95% confidence interval for $E(Y|X_1=3000)$.

11.3. It was of interest to predict the change of hemoglobin (g) resulting from thyroid operations. Blood loss, duration of operation, and hemoglobin percentile change recorded on eight patients were as follows:

Patient	X_1, Blood Loss (ml)	X_2, Duration (min)	Y, Hemoglobin Percentage Change
1	405	140	−7.0
2	401	90	−6.2
3	710	99	− .5
4	280	112	−1.1
5	502	97	−4.3
6	691	108	−1.0
7	370	120	−5.9
8	486	65	−10.2

(*a*) Find the least squares equation predicting hemoglobin percentage change, using blood loss and duration of the operation as independent variables.
(*b*) Make two F tests, stating the assumptions, null hypothesis, and conclusion for each test.
(*c*) Find the correlation coefficients: $r_{1y}, r_{2y}, r_{12}, r_{2y.1}$ and give a 95% confidence interval for $\rho_{2y.1}$.
(*d*) Find $R_{Y\,12}$.

11.4. To predict the yield of rye Y from yearly rainfall X_1 and temperature X_2, data were collected as follows:

	Yield of Rye (bushels/acre)	Rainfall (in.)	Temperature (°F)
1	21.0	45.	54.1
2	20.0	47.	61.6
3	21.0	33.	50.8
4	24.0	39.	52.1
5	20.0	30.	50.2
6	12.5	28.	57.1
7	19.0	41.	55.7
8	23.0	44.	57.6
9	23.0	31.	50.1
10	19.0	29.	38.0
11	21.0	34.	56.2
12	12.0	27.	51.5
13	21.0	42.	54.1
14	27.0	35.	46.7
15	17.5	43.	60.8
16	26.0	39.	56.9
17	11.0	31.	60.3
18	24.0	42.	54.6
19	26.0	43.	53.5
20	18.5	47.	64.0
21	15.5	25.	45.7
22	16.5	50.	61.5
23	18.0	45.	59.7
24	20.5	34.	53.2
25	22.0	29.	45.1

(*a*) Compute the least squares equation fitting rye yields to rainfall and temperature.
(*b*) State the model.
(*c*) Test the hypothesis $H_0: \beta_2 = 0$ $(\alpha = .05)$.
(*d*) Estimate the partial correlation coefficient $\rho_{2y.1}$ and test $H_0: \rho_{2y.1} = 0$.
(*e*) Test $H_0: R_{Y.12} = 0$. What other test is the equivalent of this one?

11.5. Data on altitude, longitude, latitude, and mean temperature were recorded for 16 cities:

	X_1, Altitude	X_2, Longitude	X_3, Latitude	Y, Temperature
1	1083	112	33	55.7
2	457	86	38	37.8
3	312	118	34	56.4
4	305	90	32	51.0
5	5221	105	40	34.5
6	2842	116	44	34.0
7	807	94	41	36.7
8	4260	112	41	33.4
9	815	83	40	32.6
10	3920	106	32	49.1
11	1054	84	34	46.6
12	4397	120	39	36.3
13	830	93	45	18.2
14	465	90	39	36.7
15	1162	92	47	13.3
16	787	82	41	30.1

(*a*) Find the least squares plane fitting temperature to altitude, longitude, and latitude.
(*b*) Find the least squares line fitting temperature to latitude.
(*c*) Fill in a complete analysis of variance table.
(*d*) Make the several F tests which the table given enables you to make. For each test, state the assumptions and the null hypothesis in terms of $E(Y|X_1,X_2,X_3)$.
(*e*) Assuming that $E(Y|X_1,X_2,X_3)=\alpha+\beta_1x_1+\beta_2x_2+\beta_3x_3$, give a 95% confidence interval for the temperature of a city whose altitude is 1700 feet, whose longitude is 100°F, and whose latitude is 40°.
(*f*) Find r_{2y}, $R_{Y.23}$, $R_{Y.123}$. Give a 95% confidence interval for ρ_{2y}.
(*g*) Calculate $r_{2y.3}$ using the formula

$$r_{2y.3}=\frac{r_{2y}-r_{3y}r_{32}}{\sqrt{(1-r_{3y}^2)(1-r_{32}^2)}}.$$

Give a 95% confidence interval for $\rho_{y2.3}$.

11.6. The following data are measurements of vital capacity (Y), age (X_1), weight (X_2) and height (X_3).
(*a*) Find the equation of the regression plane of Y on X_1, X_2, and X_3.
(*b*) Compute and interpret r_{y2} and $r_{y2.3}$.
(*c*) Calculate the standardized partial regression coefficients. From these, decide which variable might be best excluded from the regression.
(*d*) Test the hypothesis that the population regression coefficients are all zero.

Case Number	Y	X_1	X_2	X_3	Case Number	Y	X_1	X_2	X_3
1	4.74	36	153	65	25	2.00	43	153	61
2	3.00	40	132	63	26	3.38	67	187	67
3	3.64	32	159	64	27	5.33	32	198	72
4	3.80	46	118	64	28	3.64	51	205	72
5	2.15	65	128	63	29	3.84	34	151	66
6	2.47	60	197	64	30	2.99	58	167	65
7	3.44	38	124	65	31	3.35	60	128	63
8	3.38	33	121	64	32	4.81	39	163	72
9	1.56	71	126	62	33	4.36	72	159	71
10	2.90	38	217	61	34	5.33	67	137	69
11	3.38	32	138	64	35	4.62	38	190	72
12	2.86	40	118	60	36	4.62	51	141	69
13	3.18	33	128	58	37	3.38	76	176	65
14	3.10	41	125	63	38	5.66	30	178	70
15	2.66	46	145	63	39	4.88	34	160	67
16	2.99	36	134	62	40	4.70	32	208	72
17	2.54	45	123	60	41	4.03	30	149	65
18	4.42	32	169	64	42	4.03	48	185	67
19	3.12	52	214	62	43	2.60	60	155	66
20	4.94	36	141	66	44	2.80	46	132	64
21	3.18	38	158	59	45	3.25	61	191	68
22	4.00	34	178	63	46	3.20	71	176	65
23	3.18	38	163	62	47	4.74	40	165	68
24	2.90	33	240	63	48	2.47	60	197	64

CHAPTER 12

POLYNOMIAL REGRESSION

We return now to the regression of Y on a single X variable, with a discussion of fitting a polynomial. Polynomial regression might logically be considered in Chapter 10, but we prefer to introduce it after multiple regression.

Sometimes a polynomial is fitted because there are theoretical reasons for believing that the population regression curve is a polynomial of a certain degree. On the other hand, often it is fitted without any theoretical reason but simply because a straight line appears not to give a good fit to the data. It is possible to approximate any regression curve as closely as is desired by using a polynomial of sufficiently high degree. However, in fitting a polynomial to a set of data, we are not just interested in obtaining a curve that lies very close to the sample data points. Rather, we are attempting to obtain a curve that approximates the population regression curve (line of population means). If our primary interest *were* in fitting the curve to the data points, we could actually fit a polynomial of such high degree that it would pass *exactly* through every data point. A straight line (polynomial of degree 1) can be passed through any two points; a quadratic (degree 2) can be passed through any three points; a cubic (degree 3) can be passed through any four points; and so on. Usually k, the degree of the fitted polynomial, is chosen to be 2 or 3; certainly k should be quite small compared with n, the sample size.

THE LEAST SQUARES SOLUTION

The polynomial of degree k can be written

$$Y^{(k)}=a_0+b_1(X-\bar{X})+b_2(X-\bar{X})^2+\cdots+b_k(X-\bar{X})^k. \quad (12.1)$$

We now seek the curve of this form which minimizes the sum of squared vertical deviations

$$\sum_{i=1}^{n} (Y_i - Y_i^{(k)})^2.$$

Instead of deriving new equations for obtaining the b_i's, we use the equations from the multiple regression chapter. If we define $X_1,\ldots,X_k$ by $X_1 = X - \bar{X}, X_2 = (X - \bar{X})^2, \ldots, X_k = (X - \bar{X})^k$, (12.1) can be written

$$Y^{(k)} = a_0 + b_1X_1 + b_2X_2 + \cdots + b_kX_k, \qquad (12.2)$$

which is now in the form of a multiple regression equation. To find the best values for $a_0, b_1, \ldots, b_k$, we put the equation in terms of $x_1 = X_1 - \bar{X}_1, \ldots, x_k = X_k - \bar{X}_k$.

From $X_1 = X - \bar{X}$, we know that $\bar{X}_1 = 0$; therefore, $x_1 = X_1 - \bar{X}_1$ implies that $x_1 = X_1$. The means of the n observations on $X_2 = (X - \bar{X})^2, \ldots, X_k = (X - \bar{X})^k$ are not zero, however; thus we substitute $x_1 = X_1$, $x_2 = X_2 - \bar{X}_2, \ldots, x_k = X_k - \bar{X}_k$. Equation 12.2 becomes

$$Y^{(k)} = a + b_1x_1 + b_2x_2 + \cdots + b_kx_k, \qquad (12.3)$$

where

$$a = a_0 + b_2\bar{X}_2 + \cdots + b_k\bar{X}_k. \qquad (12.4)$$

From (11.6) we can find the numbers $b_1, b_2, \ldots, b_k$ using the methods for multiple regression. The least squares solution for a is $\bar{Y}$. The least squares plane must then be

$$Y^{(k)} = \bar{Y} + b_1x_1 + b_2x_2 + \cdots + b_kx_k. \qquad (12.5)$$

From (12.4) we have

$$a_0 = \bar{Y} - b_2\bar{X}_2 - \cdots - b_k\bar{X}_k, \qquad (12.6)$$

and the polynomial can be written in the form of (12.1).

Note first that it is permissible to change the name of $(X - \bar{X})^2$ to X_2. The X_2 variable is merely a characteristic measured on each individual; it can perfectly well be $(X - \bar{X})^2$ measured on each individual. Note also that, (12.1) was given in terms of $(X - \bar{X})^2, \ldots, (X - \bar{X})^k$ rather than in terms of $X, X^2, \ldots, X^k$. In the form we have chosen, a_0 is the ordinate of the polynomial curve at the point $\bar{X}$. If we wish, the polynomial can also be put in the form

$$Y^{(k)}=c+d_1X+d_2X^2+\cdots+d_kX_k^k, \tag{12.7}$$

where c is the height of the polynomial at $X=0$. The quantities $c,d_1,\ldots,d_k$ can be found in terms of $a,b_1,\ldots,b_k$ by equating the coefficients of powers of X in the two forms of the equation. Thus for $k=2$, after a little manipulation, we obtain

$$c=a_0-b_1\bar{X}+b_2\bar{X}^2$$

$$d_1=b_1-2b_2\bar{X}$$

$$d_2=b_2. \tag{12.8}$$

AN EXAMPLE

As an illustration of polynomial curve fitting, we use a very small example with whole numbers for X and Y. Such an example, although highly uninteresting, may be easier to follow in that the numbers can more readily be matched to the formulae. We use $n=4$, with $Y=3,4,6,7$, $X=2,2,3,9$, and fit $Y''=a_0+b_1(X-\bar{X})+b_2(X-\bar{X})^2$. The data, transformed data, and preliminary calculations of means and sums of squares and cross products are given in Table 12.1.

From the table we have $\sum x_1^2=34$, $\sum x_2^2=369$, $\sum x_1x_2=108$, $\sum x_1y=15$, $\sum x_2y=39$, and $\sum y^2=10$. Using these values in (11.6), we have

$$34b_1+108b_2=15$$

$$108b_1+369b_2=39,$$

which could be solved directly for b_1 and b_2. Instead, we follow (11.9) and (11.10) and solve the two sets of equations

$$34c_{11}+108c_{12}=1$$

$$108c_{11}+369c_{12}=0$$

and

$$34c_{21}+108c_{22}=0$$

$$108c_{21}+369c_{22}=1.$$

TABLE 12.1. COMPUTATION OF A POLYNOMIAL: EXAMPLE

	Y	$y=$ $(Y-\bar{Y})$	X	$X_1=$ $(X-\bar{X})$	$X_2=$ $(X-\bar{X})^2$	x_1	x_2	x_1^2	x_2^2	x_1x_2	x_1y	x_2y	y^2
	3	-2	2	-2	4	-2	-4.5	4	20.25	9	4	9.0	4
	4	-1	2	-2	4	-2	-4.5	4	20.25	9	2	4.5	1
	6	1	3	-1	1	-1	-7.5	1	56.25	7.5	-1	-7.5	1
	7	2	9	5	25	5	16.5	25	272.25	82.5	10	33.0	4
Σ	20	0	16	0	34			34	369	108	15	39	10
	$\bar{Y}=5$		$\bar{X}=4$	$\bar{X}_1=0$	$\bar{X}_2=8.5$								

From the first set of equations, $c_{11}=0.418$, $c_{12}=-0.1224$, and $b_1=(0.418)(15)+(-0.1224)(39)=1.50$. From the second set, $c_{21}=-0.1224$ and $c_{22}=0.03855$; $b_2=(-0.1224)(15)+(0.03855)(39)=-0.333$. We then have

$$Y''=5+1.50x_1-0.333x_2$$

or

$$=5+1.50\left(X_1-\bar{X}_1\right)-0.333\left(X_2-\bar{X}_2\right)$$

$$=5+1.50X_1-0.333(X_2-8.5)$$

$$=7.83+1.50X_1-0.333X_2$$

$$=7.83+1.50(X-4)-0.333(X-4)^2$$

and finally

$$Y''=-3.50+4.16X-0.333X^2, \qquad (12.9)$$

which is the curved line plotted on the scatter diagram in Figure 12.1. We note from the graph that with only four data points, the fitted quadratic lies very close to all the points; however, it is quite unconvincing as an estimate of a population regression line. We could not expect additional observations to lie close to the quadratic, simply because these four points fit the quadratic well.

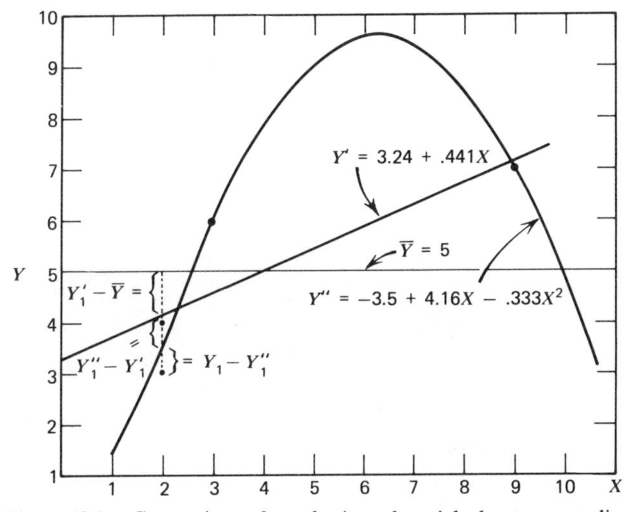

Figure 12.1. Comparison of quadratic and straight least squares lines.

THE MODEL

The model underlying polynomial regression is that for each fixed value of X, there is a normally distributed population of Y's, and

1. the Y populations are normally distributed;
2. their means are on a polynomial in X of known degree k; and
3. their variances are equal.

Our data consist of n independent observations from these populations. We can write this by saying that our data are of the form

$$Y_i = \alpha_0 + \beta_1\left(X_i - \bar{X}\right) + \beta_2\left(X_i - \bar{X}\right)^2 + \cdots + \beta_k\left(X_i - \bar{X}\right)^k + \varepsilon_i, \quad (12.10)$$

where ε_i IND$(0, \sigma_e^2)$, $i = 1, \ldots, n$.

This (with $X - \bar{X} = X_1, \ldots, (X - \bar{X})^k = X_k$) is the same model as the multiple regression model. Thus we can make an analysis of variance table, F tests, and parameter estimates exactly as in multiple regression.

We return to our example as illustration. We have already found the quadratic that, under the model (12.10), fits the data best in the least squares sense. If on the other hand, regression is linear, we fit the least squares line of Chapter 10 and obtain

$$Y' = 5 + 0.441(X - 4)$$
$$= 3.236 + 0.441X.$$

If the expected value of Y does not depend on X, the best fitting horizontal line is $\bar{Y} = 5$. These two straight lines are also plotted in Figure 12.1.

THE ANALYSIS OF VARIANCE AND F TESTS

The basic sum of squared deviations can be expressed in parts as follows:

$$\sum_{i=1}^{n}\left(Y_i - \bar{Y}\right)^2 = \sum_{i=1}^{n}(Y_i - Y_i'')^2 + \sum_{i=1}^{n}(Y_i'' - Y_i')^2 + \sum_{i=1}^{n}\left(Y_i' - \bar{Y}\right)^2. \quad (12.11)$$

We can now make a complete analysis of variance table for the example, as in Table 12.2. In the table, we have differentiated among the regression coefficients by using primes. Thus the best quadratic is $Y'' = a_0'' + b_1''(X - \bar{X}) + b_2''(X - \bar{X})^2$. For the first point $(Y_1 = 3, X_1 = 2)$, Figure 12.1 shows the

TABLE 12.2. ANALYSIS OF VARIANCE TABLE FOR QUADRATIC REGRESSION ILLUSTRATION

Source of Variation (1)	Sum of Squares (2)	d.f. (3)	MS (4)
Due Regression on $X(X_1)$	$b_1' \Sigma x_1 y = 6.615$	1	$MS_X = 6.615$
Due Regression on $X^2(X_2)$ after X	$(b_1'' \Sigma x_1 y + b_2'' \Sigma x_2 y) - b_1' \Sigma x_1 y = 2.885$	1	$MS_{X^2\|X} = 2.885$
Residual	$\Sigma y^2 - b_1'' \Sigma x_1 y - b_2'' \Sigma x_2 y = .5$	$n-3 = 1$	$s_e^2 = .5$
Total	$\Sigma y^2 = 10$	$n-1 = 3$	

division of $Y_1 - \bar{Y}$ into the three parts which enter the sums of squares in (12.11).

The "due regression on X" sum of squares [the third term on the right-hand side of (12.11)], is calculated exactly as in linear regression. The sum of the second and third terms on the right-hand side of (12.11) is "due regression on X and X^2" and is most easily calculated using $b_1''\sum x_1 y + b_2''\sum x_2 y = 1.5(15) + (-0.33)(39) = 9.5$. Subtracting the "due regression on X" from the "due regression on X and X^2," we obtain $9.5 - 6.615 = 2.885 = \sum_{i=1}^{n}(Y_i'' - Y_i')^2$.

Either Table 12.2 or Table 12.3 enables us to make several F tests from the analysis of variance table. For example, if we wish to test $H_0: \beta_2 = 0$, assuming that the regression is of second degree $(E(Y|X) = \alpha_0 + \beta_1 X +$

TABLE 12.3. TABLE FOR OBTAINING F TESTS IN QUADRATIC REGRESSION

$E(Y\|X)$	Best Estimate	Minimum Sum of Squared Deviations	Number of Parameters Estimated	df
$H_1: \alpha_0 + \beta_1 X + \beta_2 X^2$	Y''	$\Sigma y^2 - b_1''\Sigma x_1 y - b_2''\Sigma x_2 y$	3	$n-3=1$
$H_2: \alpha_0 + \beta_1 X$	Y'	$\Sigma y^2 - b_1'\Sigma x_1 y$	2	$n-2=2$
$H_3: \alpha_0$	$\bar{Y}$	Σy^2	1	$n-1=3$

β_2X^2), we are assuming H_1 of Table 12.3 and testing H_2, and our F statistic is

$$F[1-\alpha;1,n-3]$$

$$=\frac{\left[\left(\sum y^2-b_1'\sum x_1y\right)-\left(\sum y^2-b_1''\sum x_1y-b_2''\sum x_2y\right)\right]/[(n-2)-(n-3)]}{\left(\sum y^2-b_1''\sum x_1y-b_2''\sum x_2y\right)/(n-3)}$$

$$=\frac{(3.385-0.5)/1}{0.5/1}=\frac{2.885}{0.5}=5.77$$

to be compared with $F[.95;1,1]=161$.

We see from Table 12.2 that this test is actually a comparison of the "due regression on X^2 after X" mean square with the residual mean square.

The test statistic for testing H_3 ($H_0:\beta_1=0$) given H_2 ($E(Y|X)=\alpha_0+\beta_1X$) is

$$\frac{\sum y^2-\left(\sum y^2-b_1'\sum x_1y\right)}{\left(\sum y^2-b_1'\sum x_1y\right)/(n-2)}=\frac{6.615}{(10-6.615)/2}=3.91,$$

which is compared with $F[1-\alpha;1,n-2]=18.5$, for $\alpha=.05$.

Neither null hypothesis is rejected; with this small sample size, it is difficult to reject any hypothesis. The other F test is formed in an analogous fashion.

CONFIDENCE INTERVALS FOR PARAMETERS

Confidence intervals for the regression coefficients, for points on the population regression curve, and for the Y value of a new individual are exactly the same as in multiple regression (see Table 11.5). We make intervals assuming that the regression curve is of second degree. Under this assumption, a 95% confidence interval for β_2 is

$$b_2\pm t[1-\alpha/2m;n-3]\sqrt{c_{22}s_e^2}\ .$$

For the example, the 95% confidence interval with $m=1$ is

$$-0.333\pm t[.975;1]\sqrt{(0.03855)(0.5)}\qquad\text{or}\qquad -2.10<\beta_2<1.43.$$

We next obtain a 95% confidence interval for α_0, the height of the regression at $X=\overline{X}$:

$$a_0=\overline{Y}-b_2\overline{X}_2;$$

therefore, the variance of a_0 is

$$\operatorname{Var} a_0=\sigma_e^2\left(\frac{1}{n}+\overline{X}_2^2c_{22}\right).$$

The confidence interval for α_0 is thus

$$a_0\pm t[1-\alpha/2m;n-3]\sqrt{s_e^2\left(1/n+\overline{X}_2^2c_{22}\right)}\ .$$

In the example, $a_0=5-(-.333)(8.5)=7.83$, and a 95% confidence interval with $m=1$ is

$$7.83\pm(12.706)\sqrt{(0.5)[.25+(72.25)(0.03855)]}\quad\text{or}\quad -7.82<\alpha_0<23.48.$$

To construct a confidence interval for $E(Y|X=X_*)$ we use $Y_*''=a_0+b_1x_*+b_2x_*^2=a+b_1x_*+b_2(x_3^2-\overline{X}_2)$, where $x_*=X_*-\overline{X}$. The expression for $\operatorname{Var} Y_*''$ is the same, therefore, as that given in Table 11.5, with x_* and $x_*^2-\overline{X}_2$ substituted for x_1^* and x_2^*; the confidence interval follows.

To construct a prediction interval for an individual's Y value if its X value is X_*, we again proceed just as in multiple regression, with x_* and $x_*^2-\overline{X}_2$ replacing x_1^* and x_2^*.

We may also wish a confidence interval for a regression coefficient when the equation is put in terms of powers of X rather than powers of $X-\overline{X}$. If we use as the model $E(Y|X)=\gamma+\delta_1X+\delta_2X^2+\cdots+\delta_kX^k$, we need a confidence interval for the δ_i's.

In the quadratic, we have from (12.8), $d_1=b_1-2b_2\overline{X}=4.16$. The variance of d_1 is therefore equal to $\operatorname{Var} b_1-4\overline{X}\operatorname{Cov}(b_1,b_2)+4\overline{X}^2\operatorname{Var} b_2$, which is estimated by

$$\left(c_{11}-4\overline{X}c_{12}+4\overline{X}^2c_{22}\right)s_e^2$$

$$=\left[0.418-4(4)(-0.1224)+4(4)^2(0.03855)\right](0.5)$$

$$=2.422,$$

and the 95% confidence interval for δ_1 with $m=1$ is

$$4.16 \pm (12.706)\sqrt{2.422} \qquad \text{or} \qquad -15.61 < \delta_1 < 23.93.$$

ORTHOGONAL POLYNOMIALS

When the values of the X variable are equally spaced, a method of fitting a polynomial curve by means of orthogonal polynomials can be used. The use of orthogonal polynomials simplifies calculations, therefore this approach is important for fitting polynomials of degree larger than 3 when work is being done on a desk calculator. Because such problems now are usually worked on a large computer, we do not give the method. For curve fitting using orthogonal polynomials, see Bennett and Franklin [12.1] and Anderson and Bancroft [12.2].

Programs are available (e.g., BMD05R) that perform polynomial regressions. Any standard multiple regression program that allows the user to make transformations on the data can be used. Such programs can also be used for fitting more complicated surfaces such as $\hat{Y}=a_0+b_1(X_1-\bar{X}_1)+b_2(X_1-\bar{X}_1)^2+b_3X_2{}^2+b_4\log X_3$.

ILLUSTRATIVE EXAMPLE

As an illustration of a situation in which polynomial regression might be useful, we consider a study of the yield of a particular crop in relation to water level. Over a very limited range of water level, it might be reasonable to expect yield to increase in a nearly linear fashion with water level. The investigator would expect, however, that either too little or too much water would result in low yields. Thus if he wishes to gather yield data over a wide range of water level, he would expect the regression curve to be of the form pictured in Figure 12.2. He therefore begins the analysis of his data by plotting a scatter diagram and fitting a quadratic of yield to water level. Besides fitting the quadratic to his data, he probably also fits the best straight line, for the water levels he has chosen may not be sufficiently high to reach the sharply curving part of the regression curve. The two regression lines can be plotted on the scatter diagram. He decides, perhaps by eye from the scatter diagram or possibly by a test ($H_0:\beta_2=0$), which of these two underlying models is more realistic within the water level range being considered. If it appears that a quadratic curve is really an improvement over a straight line, he can use the fitted quadratic to locate the

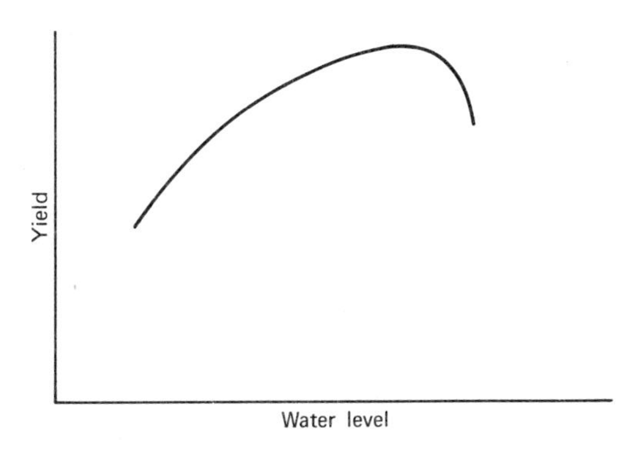

Figure 12.2. Hypothetical relationship between water level and yield.

water level that gives the maximum yield. This he can do either graphically or using the following formula (obtained by calculus methods):

$$X_{\max} = \bar{X} - \frac{b_1}{2b_2}. \tag{12.12}$$

He can proceed to use his quadratic equation for prediction purposes.

SUMMARY

In this chapter we returned to a single fixed X variable, and used multiple regression methods to fit a least squares polynomial of any degree. Under the assumptions of normality and equality of variances, we gave F tests and found confidence intervals.

REFERENCES

APPLIED STATISTICS

12.1. Bennett, C. A., and N. L. Franklin, *Statistical Analysis in Chemistry and the Chemical Industry*, Wiley, New York, 1954.

MATHEMATICAL STATISTICS

12.2. Anderson, R. L., and T. A. Bancroft, *Statistical Theory in Research*, McGraw-Hill, New York, 1952.

PROBLEMS

12.1. The following data represent X, yearly rainfall (in.), and Y, yield of rye (bushels/acre), for 24 states.

State No.	X	Y	State No.	X	Y	State No.	X	Y
1	45	21	9	30	20	17	31	23
2	47	20	10	25	18	18	29	19
3	33	21	11	41	19	19	34	21
4	39	24	12	44	23	20	42	21
5	35	27	13	42	24	21	50	16
6	43	17	14	43	26	22	45	18
7	39	26	15	47	18	23	34	20
8	27	20	16	25	16	24	29	22

(*a*) Plot a scatter diagram.

(*b*) Find the least squares regression line $Y' = a_0 + b_1(X - \bar{X})$ and draw it on the scatter diagram.

(*c*) Find the least squares quadratic $Y'' = a_0 + b_1(X - \bar{X}) + b_2(X - \bar{X})^2$ and draw it on the scatter diagram.

(*d*) Discuss the fit of the two regression lines of (*b*) and (*c*).

(*e*) Assume that $E(Y|X)$ is a quadratic and test $H_0: \beta_2 = 0$ ($\alpha = .05$). Does the test agree with your conclusions in (*d*)?

(*f*) What assumptions are made in this regression problem? Which of these may be unrealistic in this example?

12.2. Assuming that the population regression in Problem 12.1 is of degree 2, make confidence intervals with an overall confidence level of .05 for:

(*a*) The height of the regression at $X = \bar{X}$.

(*b*) The coefficient of X in the population regression $E(Y|X) = \gamma + \delta_1 X + \delta_2 X^2$.

(*c*) The rye yield in a state whose rainfall is 30 in.

CHAPTER 13

ANALYSIS OF COVARIANCE

In the analysis of variance designs in earlier chapters, we compared various linear combinations of sample means. In Chapters 10 through 12, we used regression techniques to study the relation of a dependent variable Y to one or more independent variables $X_1,\ldots,X_k$. A combination of these two techniques is known as *covariance analysis*; it is the simultaneous study of several regressions.

Because analysis of covariance techniques are complicated, we confine ourselves mainly to the simplest type of covariance problem—namely, a combination of a one-way analysis of variance with linear regression on a single X variable.

The purpose of an analysis of covariance is to remove the effect of one or more unwanted factors in an analysis of variance. For example, in studying the heights of three populations of children (cyanotic heart disease children, sibs of heart disease children, and "well" children), we may wish to eliminate the effect of age.

A second example could be an experiment in which two diets are given to two groups of rats. A one-way analysis of variance might be done on the final average weights of the rats under the two diets. If, however, measurements are available on the weight of each rat at the outset of the experiment, a covariance analysis can be performed. For further discussion see Cochran [13.4].

In Chapter 6 we studied the randomized complete block design which is another method of eliminating the effect of an unwanted variable. In the diet experiment, the rats could have been divided on the basis of initial weight into blocks each consisting of two rats. The diets could then have been assigned at random within each block. At the outset of the study, the experimenter chooses whether to eliminate the effect of initial weight by using a block design or by using a covariance analysis.

In the example of the heights of three samples of children, on the other hand, a block design cannot be used to remove the effect of age. The "treatment" (if the illness of a child may be termed a treatment) cannot be assigned by the researcher; he cannot form blocks, and thus he attempts to eliminate age by a covariance analysis.

A variable whose effect one wishes to eliminate by means of a covariance analysis is called a *covariate*; sometimes it is called a *concomitant variable*.

AN EXAMPLE

As an example of a covariance analysis we use some data (Table 13.1) on language scores (Y) for students taught by three different methods. Measurements on IQ (X) are also available. Since the students are not assigned at random to the three teaching methods, there may easily be differences in IQ among the three groups. (If the students *had* been assigned at random to the three teaching methods, differences among groups in IQ should be small in a large experiment, but a covariance analysis might still be appropriate.)

Our object is to examine differences in language scores among the three methods after the effect of IQ has been eliminated. Otherwise, if we claim that method I is superior to method II, we may not be able to refute the statement that the observed difference between the methods occurs because the IQs of the students using method I were higher than those using method II.

TABLE 13.1. DATA ON LANGUAGE SCORES (Y) USING THREE TEACHING METHODS AND IQ SCORES (X)

Method		Student 1	2	3	4	5	6	7	8	9	10
1	Y	72	75	85	70	73	86	92	68	91	75
	X	87	119	121	112	100	133	135	109	139	105
2	Y	90	98	73	88	83	90	98	81	84	79
	X	110	128	117	94	107	125	111	80	123	95
3	Y	59	65	67	71	59	61	58	70	59	48
	X	95	120	125	107	85	98	100	138	112	90

AN UNDERLYING MODEL

Underlying the analysis of covariance, we make certain assumptions. We need to know that for any fixed IQ, the population of language scores for a given teaching method are normally distributed, have means lying on a specified type of regression curve (here a straight line), and have equal variances.

The underlying model is expressed by saying that each observation is of the form

$$Y_{ij} = \alpha_{i0} + \beta_i X_{ij} + \varepsilon_{ij}, \qquad i = 1,\ldots,a; \qquad j = 1,\ldots,n_i, \tag{13.1}$$

where ε_{ij} IND$(0, \sigma_e^2)$. In (13.1), Y_{ij} and X_{ij} denote the Y and X values for the jth individual on the ith treatment. The number of treatments is a and the number of individuals in the ith sample is n_i. In the example, $a = 3$, $n_1 = 10$, $n_2 = 10$, and $n_3 = 10$. The model expressed by (13.1) is clearly that of a separate linear regressions. Note that α_{i0} is the height of the ith population regression line at $X = 0$. We have chosen to write the regression lines in this form in preference to a form involving deviations from the mean, as was done in the regression chapters. In discussing the ordinate of the line at the mean, we must always specify whether we wish the height at the mean $\bar{X}_{i.} = \sum_{j=1}^{n_i} X_{ij}/n_i$ or at the overall mean $\bar{X}_{..} = \sum_{i=1}^{a}\sum_{j=1}^{n_i} X_{ij}/N$, where $N = \sum_{i=1}^{a} n_i$.

As stated earlier, covariance analysis is the simultaneous study of several regressions. The model expressed in (13.1) implies exactly the same assumptions as in a linear regression: independent observations, normal distributions of Y at particular values of X, equal variances, and, for each treatment, population means on a straight line.

PRELIMINARY PROCEDURES IN A COVARIANCE ANALYSIS

Before proceeding with the analysis, it is important to decide whether the underlying assumptions are tenable. For discussion of methods of checking these assumptions, see Chapter 14. Methods of Chapter 12 are also helpful in deciding whether the population regression curves are straight lines. As in other regression problems, transformations can be considered if straight line regression is clearly inappropriate or if variances are grossly unequal. In preliminary consideration of the data, the scatter diagram is particularly recommended.

If the model expressed by (13.1) holds, the three population regression lines may be pictured as in Figure 13.1a. We have three regression lines, each with a different slope. Three such lines, or their sample estimates, are quite inconvenient for studying differences in Y among the three populations. The vertical distance separating any two lines is different for different values of X. In the language score example, we would like to be able to make statements such as "the mean language score is higher for method III than method I." Instead, with regression lines as in Figure 13.1*a*, we can make only such statements as "for high IQ, mean language score is higher for method III than for method I." Under the model expressed in (13.1), the population regression lines are in general not parallel; therefore, we do not succeed in making simple statements on differences in population mean language scores without regard to IQ.

We prefer to think that the three lines are parallel. If $\beta_1 = \beta_2 = \beta_3$, the expected value of the Y_i population at X is $\alpha_{i0} + \beta X$, where β designates

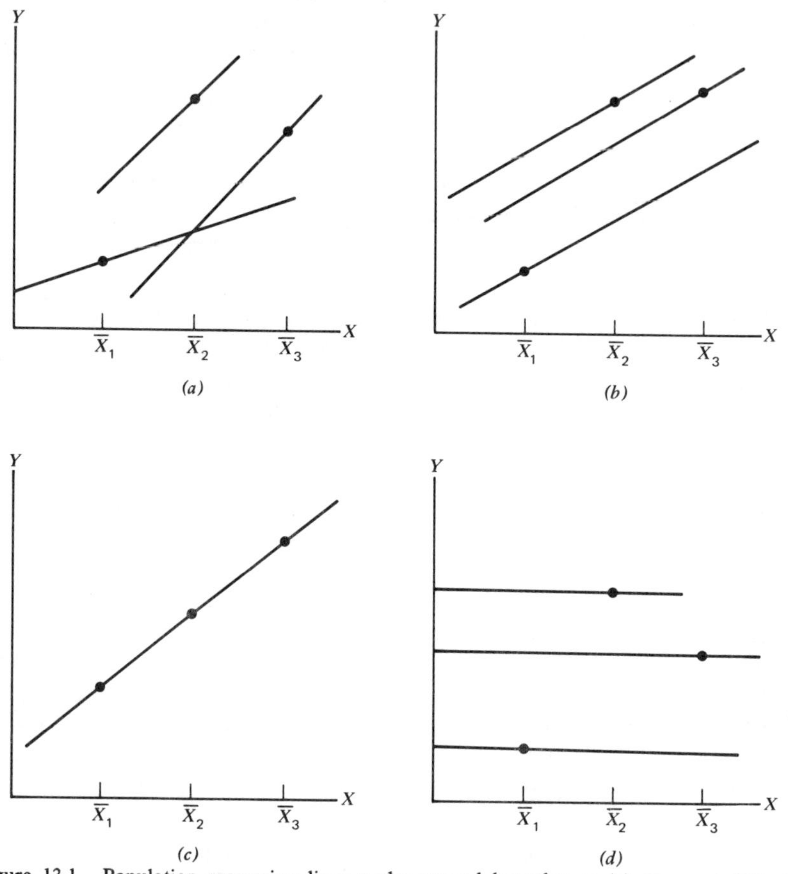

Figure 13.1. Population regression lines under several hypotheses: (*a*) $H_1 : \alpha_{0i} + \beta_i X$; (*b*) $H_2 : \alpha_{0i} + \beta X$; (c) $H_3 : \alpha_0 + \beta X$; (*d*) $H_4 : \alpha_{0i}$.

the common slope of the three lines. With parallel regression lines, the difference in mean Y values between populations 1 and 3 is simply $\alpha_{01}-\alpha_{03}$ at every value of X. As an underlying model then, we prefer to assume [instead of (13.1)]:

$$Y_{ij}=\alpha_{i0}+\beta X+\varepsilon_{ij}, \qquad \varepsilon_{ij}\ \mathrm{IND}(0,\sigma_e^2). \tag{13.2}$$

Figure 13.1*b* shows three parallel population regression lines.

As a next step in his analysis, the investigator should decide (perhaps theoretically or perhaps on the basis of a test) whether the regression lines are parallel. If they are *not* parallel, he estimates the parameters for the a regression lines separately and must content himself with making statements concerning differences between means at particular values of X. If it seems likely that they are close to being parallel, he obtains a parallel regression lines having the least squares property. These lines are used to estimate differences among the Y's for *any* value of X and to estimate the mean value of the Y's for a given treatment at a given value of X. He makes confidence intervals for all parameters of interest (all based on the assumption of parallel regression lines). He frequently tests whether the three population regression lines coincide (as in Figure 13.1*c*). He may also test whether the three parallel population regression lines are horizontal (as in Figure 13.1*d*).

We must now obtain sample regression lines under the four hypotheses pictured in Figure 13.1. After obtaining these lines, we give point estimates of parameters and then we present tests which the experimenter may wish to make. Finally, with the assumptions of the parallel regression line model (13.2), we estimate parameters with confidence intervals.

THE ANALYSIS OF COVARIANCE TABLE

The computations necessary to obtain equations of the sample regression lines are facilitated by forming an analysis of covariance table. Notation for the analysis table appears in Table 13.2; Table 13.4 is the analysis of covariance for the data on language score and IQ score. The preliminary calculations of various sums of squares and cross products on the language score data are found in Table 13.3.

The first three columns of an analysis of covariance table are essentially the first three columns of an analysis of variance table for Y alone—the source of variation, degrees of freedom, and sums of squares columns. Above the double horizontal line, the residual sum of squares appears, broken into a parts (one for each of the a treatments). We have used an

TABLE 13.2. ANALYSIS OF COVARIANCE TABLE: NOTATION

Source of Variation (1)	Total					Due Regression		Devs from Regression	
	df (2)	Σy^2 (3)	Σxy (4)	Σx^2 (5)	b (6)	d.f. (7)	SS (8)	df (9)	SS (10)
Residual (1)	n_1-1	$\Sigma y^2_{(1)}$	$\Sigma xy_{(1)}$	$\Sigma x^2_{(1)}$	$b_{(1)}$	1	$b_{(1)}\Sigma xy_{(1)}$	n_1-2	$\Sigma y^2_{(1)} - b_{(1)}\Sigma xy_{(1)}$
	$\vdots$	$\vdots$	$\vdots$	$\vdots$	$\vdots$	$\vdots$	$\vdots$	$\vdots$	$\vdots$
Residual (a)	n_a-1	$\Sigma y^2_{(a)}$	$\Sigma xy_{(a)}$	$\Sigma x^2_{(a)}$	$b_{(a)}$	1	$b_{(a)}\Sigma xy_{(a)}$	n_a-2	$\Sigma y^2_{(a)} - b_{(a)}\Sigma xy_{(a)}$
Σ	N-a	Σy^2_r	Σxy_r	Σx^2_r		a	$\sum_{i=1}^{a} b_{(i)}xy_{(i)}$	N-2a	D_1
Due A	a-1	Σy^2_a	Σxy_a	Σx^2_a	b_a	1	$b_a\Sigma xy_a$	a-2	$\Sigma y^2_a - b_a\Sigma xy_a$
Residual	N-a	$\Sigma y^2_r = D_4$	Σxy_r	Σx^2_r	b	1	$b\Sigma xy_r$	N-a-1	$\Sigma y^2_r - b\Sigma xy_r = D_2$
Total	N-1	Σy^2_t	Σxy_t	Σx^2_t	b_t	1	$b_t\Sigma xy_t$	N-2	$\Sigma y^2_t - b_t\Sigma xy_t = D_3$

TABLE 13.3. PRELIMINARY CALCULATIONS FOR COVARIANCE ANALYSIS ON LANGUAGE SCORES (Y) AND IQ SCORES (X)

Method	$\bar{Y}_{i.}$	$\bar{X}_{i.}$	$\bar{Y}^2_{i.}$	$\bar{X}_{i.}\bar{Y}_{i.}$	$\bar{X}^2_{i.}$	$\sum_{j=1}^{10} Y^2_{ij}$	$\sum_{j=1}^{10} X_{ij}Y_{ij}$	$\sum_{j=1}^{10} X^2_{ij}$	$\Sigma y^2_{(j)}$	$\Sigma xy_{(i)}$	$\Sigma x^2_{(i)}$
1	78.70	116.00	6,193.69	9,129.20	13,456.00	62,653.00	92,408.0	137,056.00	716.10	1,116.00	2,496.00
2	86.40	109.0	7,464.96	9,417.60	11,881.00	75,228.0	94,583.0	120,958.00	578.40	407.00	2,148.0
3	61.70	107	3,806.89	6,601.90	11,449.00	38,487.00	66,758.0	117,016.00	418.10	739.00	2,526.0
Total	226.80	332.00	17,465.54	25,148.70	36,786.00	176,368.00	253,749.0	375,030.00	1,712.60	2,262.00	7,170.00

$$\Sigma y_t^2 = \sum_{ij}\sum(Y-\bar{Y}_{..})^2 = \sum_{ij}\sum Y^2 - (\sum_{ij}\sum Y)^2/N = 176{,}368.00 - \frac{(2268)^2}{30} = 4{,}907.20$$

$$\Sigma xy_t = \sum_{ij}\sum(X-\bar{X}_{..})(Y-\bar{Y}_{..}) = \sum_{ij}\sum XY - \frac{\sum_{ij}\sum X \sum_{ij}\sum Y}{N} = 253{,}749 - \frac{(3320)(2268)}{30} = 2{,}757.00$$

$$\Sigma x_t^2 = \sum_{ij}\sum(X-\bar{X}_{..})^2 = \sum_{ij}\sum X^2 - (\sum_{ij}\sum X)^2/N = 375{,}030.00 - \frac{(3320)^2}{30} = 7{,}616.67$$

$$\bar{Y}_{..} = 75.60$$

$$\bar{X}_{..} = 110.67$$

$$\Sigma y_a^2 = \sum_{i=1}^{a} n_i\bar{Y}^2_{i.} - \frac{(\sum_{ij}\sum Y)^2}{N} = 10(17{,}465.54) - \frac{(2268)^2}{30} = 3{,}194.60$$

$$\Sigma xy_a = \sum_{i=1}^{a} n_i\bar{X}_{i.}\bar{Y}_{i.} - \frac{\sum_{ij}\sum X \sum_{ij}\sum Y}{N} = 10(25{,}148.70) - \frac{(3320)(2268)}{30} = 495.00$$

$$\Sigma x_a^2 = \sum_{i=1}^{a} n_i\bar{X}^2_{i.} - \frac{(\sum_{ij}\sum X)^2}{N} = 10(36{,}786.00) - \frac{(3320)^2}{30} = 446.67$$

TABLE 13.4. ANALYSIS OF COVARIANCE FOR LANGUAGE SCORES (Y) AND IQ SCORES (X)

Source of Variation	df	y^2	xy	x^2	b	Due Regression df	Due Regression SS	Devs from Regression df	Devs from Regression SS
(1)	(2)	(3)	(4)	(5)	(6)	(7)	(8)	(9)	(10)
Residual (1)	9	716.10	1116.00	2496.00	.44712	1	499.0	8	217.1
Residual (2)	9	578.40	407.00	2148.00	.18948	1	77.1	8	501.3
Residual (3)	9	418.10	739.00	2526.00	.29256	1	216.2	8	201.9
Σ	27	1712.60	2262.00	7170.00	——	3	792.3	24	$920.3 = D_1$
Due A	2	3194.60	495.00	446.67	1.10820	1	548.2	1	2646.4
Residual	27	$1712.60 = D_4$	2262.00	7170.00	.31548	1	713.6	26	$999.0 = D_2$
Total	29	4907.20	2757.00	7616.67	.36197	1	998.0	28	$3909.2 = D_3$

abbreviated notation in Table 13.2. We let

$$\sum_{i=1}^{a} n_i = N$$

$$\sum y_r^2 = \sum_{i=1}^{a} \sum_{j=1}^{n_i} \left(Y_{ij} - \bar{Y}_{i.}\right)^2$$

$$\sum y_a^2 = \sum_{i=1}^{a} n_i \left(\bar{Y}_{i.} - \bar{Y}_{..}\right)^2$$

$$\sum y_t^2 = \sum_{i=1}^{a} \sum_{j=1}^{n_i} \left(Y_{ij} - \bar{Y}_{..}\right)^2,$$

where t denotes total, a denotes the due A, and r the residual sums of squares. With this notation, the usual analysis of variance division of the basic sum of squares for Y becomes simply

$$\sum y_t^2 = \sum y_a^2 + \sum y_r^2.$$

Considering just one treatment at a time, we let

$$\sum y_{(i)}^2 = \sum_{j=1}^{n_i} \left(Y_{ij} - \bar{Y}_{i.}\right)^2,$$

and can write

$$\sum y_r^2 = \sum_{i=1}^{a} \left(\sum y^2{}_{(i)}\right).$$

The deviations $X_{ij} - \bar{X}_{..}$ can be broken up into two parts in the same way. We can then express the sums of squares as follows:

$$\sum x_t^2 = \sum x_a^2 + \sum x_r^2,$$

where

$$\sum x_r^2 = \sum_{i=1}^{a} \sum_{j=1}^{n_i} \left(X_{ij} - \bar{X}_{i.}\right)^2$$

$$\sum x_a^2 = \sum_{i=1}^{a} n_i \left(\bar{X}_{i.} - \bar{X}_{..}\right)^2$$

$$\sum x_t^2 = \sum_{i=1}^{a} \sum_{j=1}^{n_i} \left(X_{ij} - \bar{X}_{..}\right)^2$$

We let

$$\sum x_{(i)}^2 = \sum_{j=1}^{n_i} \left(X_{ij} - \overline{X}_{i.} \right)^2$$

and have

$$\sum x_r^2 = \sum_{i=1}^{a} \left(\sum x_{(i)}^2 \right).$$

These quantities are shown in columns 3 and 5 of Table 13.2.

To obtain column 4, we form the cross products of the X and Y deviations $(X_{ij} - \overline{X}_{..})(Y_{ij} - \overline{Y}_{..})$. When these cross products are summed over all the observations it can be shown that

$$\sum_{i=1}^{a} \sum_{j=1}^{n_i} \left(X_{ij} - \overline{X}_{..} \right)\left(Y_{ij} - \overline{Y}_{..} \right) = \sum_{i=1}^{a} n_i \left(\overline{X}_{i.} - \overline{X}_{..} \right)\left(\overline{Y}_{i.} - \overline{Y}_{..} \right)$$

$$+ \sum_{i=1}^{a} \sum_{j=1}^{n_i} \left(X_{ij} - \overline{X}_{i.} \right)\left(Y_{ij} - \overline{Y}_{i.} \right)$$

or

$$\sum xy_t = \sum xy_a + \sum xy_r,$$

where

$$\sum xy_r = \sum_{i=1}^{a} \sum_{j=1}^{n_i} \left(X_{ij} - \overline{X}_{i.} \right)\left(Y_{ij} - \overline{Y}_{i.} \right)$$

$$\sum xy_a = \sum_{i=1}^{a} n_i \left(\overline{X}_{i.} - \overline{X}_{..} \right)\left(\overline{Y}_{i.} - \overline{Y}_{..} \right)$$

$$\sum xy_t = \sum_{i=1}^{a} \sum_{j=1}^{n_i} \left(X_{ij} - \overline{X}_{..} \right)\left(Y_{ij} - \overline{Y}_{..} \right).$$

The sum $\sum xy_a$ is broken into a sums, with

$$\sum xy_{(i)} = \sum_{j=1}^{n_i} \left(X_{ij} - \overline{X}_{i.} \right)\left(Y_{ij} - \overline{Y}_{i.} \right)$$

$$\sum xy_a = \sum_{i=1}^{a} \left(\sum xy_{(i)} \right).$$

This completes the discussion of columns 1 to 5. The remaining columns of the analysis of covariance table are concerned with fitting straight lines in various ways; they are really analysis of variance tables for regression written sideways. Columns 7 to 10 of the row marked $\sum$ are obtained by adding the numbers in the columns above them. The other rows are formed horizontally, and from each of these other rows we have an analysis of variance table for regression. Each regression coefficient b is formed by dividing the corresponding $\sum xy$ by $\sum x^2$. For example, we write

$$b_{(1)} = \frac{\sum xy_{(1)}}{\sum x^2{}_{(1)}}.$$

LEAST SQUARES LINES UNDER SEVERAL HYPOTHESES

Having completed the analysis of covariance table, we are ready to use it to obtain sample regression lines under each of the four hypotheses concerning $E(Y_i|X)$ depicted in Figure 13.1. Under each hypothesis, we discuss how to estimate the slope of the least squares lines, how to find a point through which the sample regression line must pass, and how to write down the equation of the least squares line. We also obtain the minimum sum of squared deviations from regression under each hypothesis.

$$H_1 : E(Y_i|X) = \alpha_{i0} + \beta_i X$$

If the population regression lines have different slopes, the a lines are fitted individually using methods of Chapter 10. The appropriate regression slope is $b_{(i)} = \sum xy_{(i)} / \sum x^2_{(i)}$, which can be read from column 6 in the analysis of covariance table. The ith line must pass through the sample mean point $(\overline{X}_{i.}, \overline{Y}_{i.})$. Therefore, its equation is

$$Y_i' = \overline{Y}_{i.} + b_{(i)}\left(X - \overline{X}_{i.}\right). \tag{13.3}$$

For the example, the lines obtained by fitting three separate lines (H_1) are

$$Y_1' = 78.7 + 0.4471(X - 116.0) = 26.8 + 0.447X$$

$$Y_2' = 86.4 + 0.1895(X - 109.0) = 65.7 + 0.189X$$

$$Y_3' = 61.7 + 0.2926(X - 107.0) = 30.4 + 0.293X$$

To obtain the minimum sum of squared deviations under this hypothesis, we again use the analysis of covariance table. The due regression degrees of freedom for each of the a lines is 1, and the due regression sum of squares for each of the a individual lines is $b_{(i)}\sum xy_{(i)}$. The sum of squared deviations from regression for each line, which is obtained from columns 3 and 8 by subtraction, is $\sum y_{(i)}^2 - b_{(i)}\sum xy_{(i)}$. Because each of the a lines fits its own set of data the best, the a lines together form the least squares solution to the problem. Adding the a sums of squared deviations from regression, we obtain the minimum sum of squared deviations from regression under H_1:

$$D_1 = \sum_{i=1}^{a} \left(\sum y_{(i)}^2 - b_{(i)} \sum xy_{(i)} \right).$$

In the example, the minimum sum of squares is $D_1 = 217.1 + 501.3 + 201.9 = 920.3$.

$$H_2 : E(Y_i|X) = \alpha_i + \beta X$$

To obtain slopes for the least squares lines under the second hypothesis, it would be possible to minimize the sum of squared deviations

$$\sum_{i=1}^{a} \sum_{j=1}^{n_i} (Y_{ij} - \alpha_{i0} - \beta X_{ij})^2$$

by calculus methods. Instead, we reason as follows.

The a samples must somehow be combined in order to obtain a common slope. To do this, we subtract $\overline{Y}_{i.}$ from each Y_{ij} and $\overline{X}_{i.}$ from each X_{ij}. The new data table so formed has zero means for each of the a samples. Because all the samples have the same mean, we can find the slope of the line best fitting the combined data on $Y_{ij} - \overline{Y}_{i.}$ and $X_{ij} - \overline{X}_{i.}$. To obtain the common slope, we use $b = \sum xy_r / \sum x_r^2$ from the residual row of Table 13.2; this is the slope of the line best fitting the $Y_{ij} - \overline{Y}_{i.}$ and $X_{ij} - \overline{X}_{i.}$.

The parallel lines best fitting the original data are the lines with slope $b = \sum xy_r / \sum x_r^2$ which pass through the sample mean points $(\overline{X}_{i.}, \overline{Y}_{i.})$, because the fit of any line with slope b can be improved by raising or lowering its level until it passes through the mean point. The equation of the ith regression line is therefore

$$Y_i'' = \overline{Y}_{i.} + b\left(X - \overline{X}_{i.}\right). \tag{13.4}$$

As mentioned earlier, these lines are the most useful regression lines; they enable us to estimate population differences in Y without regard to X. In the analysis of covariance table, the "analysis of variance table for regression" for these lines is found in the row marked "Residual." The deviations from regression denoted by D_2 are $\sum y_r^2 - b\sum xy_r$. In insisting that the sample lines be parallel, we cannot possibly improve the fit to the data; thus D_2 is always at least as large as D_1.

In the example, the three sample regression lines are

$$Y_1'' = 78.7 + 0.315(X - 116.0) = 42.2 + 0.315X$$

$$Y_2'' = 86.4 + 0.315(X - 109.0) = 52.1 + 0.315X$$

$$Y_3'' = 61.7 + 0.315(X - 107.0) = 28.0 + 0.315X.$$

The minimum sum of squared deviations from regression is $D_2 = 999.0$.

$$H_3: E(Y_i|X) = \alpha_0 + \beta X$$

If the a population regression lines coincide, all the data from the a samples should be combined to fit a single line. To obtain this line, we use the "Total" row of the analysis of covariance table; the slope is $b_t = \sum xy_t / \sum x_t^2$. The line must pass through the overall mean point $(\bar{X}_{..}, \bar{Y}_{..})$, thus its equation is

$$Y''' = \bar{Y}_{..} + b_t\left(X - \bar{X}_{..}\right). \tag{13.5}$$

The minimum sum of squared deviations under this hypothesis is denoted by D_3; it is $D_3 = \sum y_t^2 - b_t \sum xy_t$. A single line certainly cannot fit the data better than three parallel lines, and this implies that, D_3 is at least as large as D_2.

In the example, the best-fitting single line is

$$Y''' = 75.60 + 0.3620(X - 110.67) = 35.54 + 0.362X \qquad \text{and} \qquad D_3 = 3909.2.$$

$$H_4: E(Y_i|X) = \alpha_{i0}$$

If the population regression lines do not depend on X, we have a simple one-way analysis of variance problem. The best-fitting straight lines are horizontal and are simply

$$\mathrm{Y}_{\mathrm{i}}'''' = \bar{Y}_{\mathrm{i}.}. \tag{13.6}$$

The sum of squared deviations from the best-fitting lines (D_4), found in column 3 of the analysis of covariance table, is $D_4 = \sum y_r^2$. We can fit the data at least as well using three parallel lines with arbitrary slope as with three horizontal lines; therefore, D_4 cannot be smaller than D_2. The relative size of D_3 and D_4 is uncertain; either one may be the larger of the two. In the example, $Y_1'''' = 78.7$, $Y_2'''' = 86.4$, $Y_3'''' = 61.7$, and $D_4 = 1712.60$.

Point estimates of the various parameters are presented in Table 13.5 under the four hypotheses. The slopes in column 3 of the table have already been discussed. As an estimate of the height of the ith population regression line at any point X^*, we use the height of the proper sample regression line at X^*, as given in column 4. The expressions in column 4 have been manipulated algebraically in order to display the point estimate of the height of the population regression line at $X^* = 0$. For example, $\overline{Y}_{i.} - b_{(i)}\overline{X}_{i.}$ estimates the parameter α_{i0}. Point estimates for σ_e^2 are found under each hypothesis by dividing the minimum sum of squared deviations by the proper degrees of freedom.

A DIVISION OF THE BASIC SUM OF SQUARES

Before proceeding to F tests, we set down a possible division of the basic sum of squares. Each deviation from the overall mean can be written as follows:

$$\left(Y_{ij} - \overline{Y}_{..}\right) = \left(Y_{ij} - Y_i'\right) + \left(Y_i' - Y_i''\right) + \left(Y_i'' - Y_i'''\right) + \left(Y_i''' - \overline{Y}_{..}\right). \quad (13.7)$$

Squaring and adding over all the data, we have

$$\sum\left(Y_{ij} - \overline{Y}_{..}\right)^2 = \overbrace{\sum\left(Y_{ij} - Y_i'\right)^2}^{D_1} + \overbrace{\sum\left(Y_i' - Y_i''\right)^2}^{D_2 - D_1} + \overbrace{\sum\left(Y_i'' - Y_i'''\right)^2}^{D_3 - D_2} + \overbrace{\sum\left(Y_i''' - \overline{Y}_{..}\right)^2}^{b_t \sum xy_t}$$

H_3: D_3 (first three terms); due regression $= b_t \sum xy_t$

H_2: D_2 (first two terms); due regression $= (D_3 - D_2) + b_t \sum xy_t$

H_1: D_1 (first term); due regression $= (D_2 - D_1) + (D_3 - D_2) + b_t \sum xy_t$.

(13.8)

Beneath (13.8) are displayed the terms entering into the due regression and deviations from regression sum of squares under each of these three hypotheses.

TABLE 13.5. ESTIMATES OF PARAMETERS UNDER VARIOUS HYPOTHESES

$E(Y_i \mid X^*)$	Point on Line	Estimate of Slope	Estimate of $E(Y_i \mid X^*)$	Min. Sum of Squared Dev.	Number of Parameters	d.f.	Estimate of σ_e^2
(1)	(2)	(3)	(4)	(5)	(6)	(7)	(8)
H_1: $\alpha_{io} + \beta_i X^*$	$\overline{X}_{i.}, \overline{Y}_{i.}$	$b_{(i)}$	$Y_i^{*'} = (\overline{Y}_{i.} - b_{(i)}\overline{X}_{i.}) + b_{(i)}X^*$	D_1	2a	N-2a	$D_1/(N-2a)$
H_2: $\alpha_{io} + \beta X^*$	$\overline{X}_{i.}, \overline{Y}_{i.}$	b	$Y_i^{*''} = (\overline{Y}_{i.} - b\overline{X}_{i.}) + bX^*$	D_2	a+1	N-a-1	$D_2/(N-a-1)$
H_3: $\alpha_o + \beta X^*$	$\overline{X}_{..}, \overline{Y}_{..}$	b_t	$Y_i^{*'''} = (\overline{Y}_{..} - b_t\overline{X}_{..}) + b_t X^*$	D_3	2	N-2	$D_3/(N-2)$
H_4: α_{io}	$\overline{X}_{i.}, \overline{Y}_{i.}$	0	$Y_i^{*''''} = \overline{Y}_{i.}$	D_4	a	N-a	$D_4/(N-a)$

The geometric picture of the division into four terms of a single deviation $Y_{ij} - \overline{Y}_{..}$ appears in Figure 13.2. For simplicity, the point X_{ij}, Y_{ij}; has been selected in such a way that each term is positive; however, terms can be negative.

F TESTS

With the help of Table 13.5 we now write down some F tests that can be made before confidence intervals are constructed.

To decide whether the population regression lines have a common slope, we assume H_1 and test H_2. The F statistic is thus

$$F = \frac{(D_2 - D_1)/[(N-a-1)-(N-2a)]}{D_1/(N-2a)}$$

or

$$F = \frac{(D_2 - D_1)/(a-1)}{D_1/(N-2a)},$$

which is to be compared with $F[1-\alpha; a-1, N-2a]$.

In the example, we have

$$F = \frac{(999.0 - 920.3)/(26-24)}{920.3/24} = 1.026,$$

$F[.99; 2, 24] = 3.40$. Thus we accept the null hypothesis; the lines may be equal in slope.

To test whether a single regression line can be used for all the a populations, assuming that the lines have a common slope, we assume H_2 and test H_3. This gives

$$F = \frac{(D_3 - D_2)/[(N-2)-(N-a-1)]}{D_2/(N-a-1)},$$

which is to be compared with $F[1-\alpha; a-1, N-a-1)]$.

For the language score data, we have

$$F = \frac{(3909.2 - 999.0)/(28-26)}{999.0/26} = 37.87$$

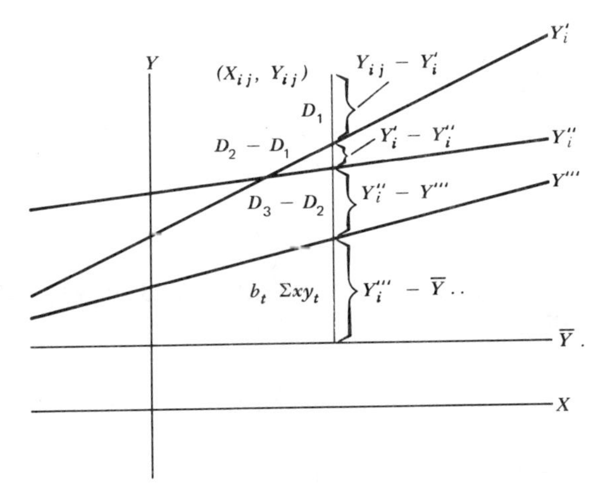

Figure 13.2. Division of a deviation from overall mean into four parts.

and $F[.95;2,26]=3.37$. We reject the null hypothesis of no difference among the three teaching methods and conclude that a single regression line does not suffice. This test is usually the most important test to be made.

We may also wish a test of whether X makes any difference at all. This could be made assuming either H_1 or H_2.

In the example, we may prefer to assume H_2 and test H_4, since we have accepted the hypothesis that the slopes of the three lines are equal. The test statistic then becomes

$$F=\frac{(D_4-D_2)/[(N-a)-(N-a-1)]}{D_2/(N-a-1)},$$

which is to be compared with $F[1-\alpha;1,N-a-1]$.

In the example, we have

$$F=\frac{(1712.60-999.0)/(27-26)}{999.0/26}=18.57$$

and $F[.95;1,26]=4.23$; we conclude that the population regression lines are not horizontal.

Other possible tests can be designed from Table 13.5.

CONFIDENCE INTERVAL ESTIMATES OF PARAMETERS

We return now to the summary in Table 13.5 of estimates of parameters in order to obtain confidence intervals. We construct them under the preferred hypothesis $H_2: E(Y_i|X) = \alpha_{i0} + \beta X$.

CONFIDENCE INTERVAL FOR σ_e^2

The variance σ_e^2 is estimated by $D_2/(N-a-1)$, and a $1-\alpha$ level interval is formed as usual:

$$\frac{D_2}{\chi^2[1-\alpha/2; N-a-1]} < \sigma_e^2 < \frac{D_2}{\chi^2[\alpha/2; N-a-1]}. \qquad (13.9)$$

In the example, the 95% confidence interval is

$$\frac{999.0}{41.92} \quad \text{to} \quad \frac{999.0}{13.84} \quad \text{or} \quad 23.83 < \sigma_e^2 < 72.18. \qquad (13.10)$$

To derive confidence intervals for other parameters, we need the variances of their estimates. As in ordinary linear regression, the variance of $\bar{Y}_{i.}$ is σ_e^2/n_i, the variance of b is $\sigma_e^2/\sum x_r^2$, and the two estimates are statistically independent.

CONFIDENCE INTERVAL FOR β

The confidence interval for β is

$$b \pm t[1-\alpha/2m; N-a-1]\sqrt{[D_2/(N-a-1)](1/\sum x_r^2)}\ ; \qquad (13.11)$$

here m intervals are being made with an overall level of $1-\alpha$.

With $m=5$ and $1-\alpha=.95$, the confidence interval for β in the example is

$$.315 \pm t[.995; 26]\sqrt{(999.0/26)(1/7170.00)} \quad \text{or} \quad 0.112 < \beta < 0.518.$$

CONFIDENCE INTERVAL FOR $E(Y_i|\bar{X}_{i.})$

Often the experimenter wishes a confidence interval for the ordinate of the population regression at $\bar{X}_{i.}$. At $X = \bar{X}_{i.}$, $Y'' = \bar{Y}_{i.}$ and, because Var $\bar{Y}_{i.}$

$=\sigma_e^2/n_i$, the interval is

$$\bar{Y}_{i.} \pm t[1-\alpha/2m; N-a-1]\sqrt{[D_2/(N-a-1)](1/n_i)}\ ; \quad (13.12)$$

again, m intervals are being formed with an overall $1-\alpha$ confidence level.

For $m=5$ and $1-\alpha=.95$, the confidence interval for the height of the population regression line for teaching method I at $X=\bar{X}_{1.}=116.0$ is

$$78.7 \pm t[.995; 26]\sqrt{(999.0/26)(1/10)}$$

or

$$73.3 < E(Y_1|X=116.0) < 84.1.$$

CONFIDENCE INTERVAL FOR $E(Y_i|X^*)$

If we wish an estimate of a population mean at any other value of X,—say, $X=X^*$, we write

$$\operatorname{Var} Y_i^{*\prime\prime} = \sigma_e^2\left[\frac{1}{n_i} + \frac{(X^* - \bar{X}_{i.})^2}{\sum x_r^2}\right], \qquad (13.13)$$

and the confidence interval is

$$Y_i^{*\prime\prime}\, t[1-\alpha/2m; N-a-1]\left[\left(\frac{1}{n_i} + \frac{(X^* - \bar{X}_{i.})^2}{\sum x_r^2}\right)\frac{D_2}{N-a-1}\right]^{1/2} \qquad (13.14)$$

where $Y_i^{*\prime\prime} = \bar{Y}_{i.} + b(X^* - \bar{X}_{i.})$.

In the example, at $X^*=100$, the confidence interval for the mean for teaching method I ($m=5$ and $1-\alpha=.95$) is

$$78.7 + .315(100-116.0) \pm 2.779\sqrt{\left[\frac{1}{10} + \frac{(100-116)^2}{7170}\right]\frac{(999.0)}{26}},$$

or

$$67.31 < E(Y_1|X^* = 100) < 80.00.$$

PREDICTION INTERVAL FOR Y^*

A confidence interval or prediction interval for the Y value of a new individual in the ith population with $X = X^*$ is similar to the interval given in Chapter 10. It is

$$\bar{Y}_{i.} + b(X^* - \bar{X}_{i.}) \pm t[1 - \alpha/2m, N - a - 1]$$

$$\times \left[\left(1 + \frac{1}{n_i} + \frac{(X^* - \bar{X}_{i.})^2}{\sum x_r^2} \right) \frac{D_2}{N - a - 1} \right]^{1/2}. \qquad (13.15)$$

In the example, again for the first teaching method and for an individual whose IQ is 100, we have

$$78.7 + .315(100 - 116.) \pm 2.779 \sqrt{\left[1 + \frac{1}{10} + \frac{(100 - 116)^2}{7170} \right] \frac{999.0}{26}}$$

or

$$55.30 < Y^* < 92.02.$$

CONFIDENCE INTERVAL FOR DIFFERENCE BETWEEN TWO TREATMENTS

Possibly the most important of all the quantities to be estimated is the difference between two population means at any particular value of X. As mentioned earlier, if H_2 is correct and $E(Y_i|X^*) = \alpha_{i0} + \beta X^*$, the difference between the two population means at $X = X^*$ does not involve X^*; it is simply $\alpha_{i0} - \alpha_{i'0}$. We need a confidence interval for $\alpha_{i0} - \alpha_{i'0}$.

Our estimate of $\alpha_{i0} - \alpha_{i'0}$ is $Y_i^{*\prime\prime} - Y_{i'}^{*\prime\prime} = (\bar{Y}_{i.} - b\bar{X}_{i.} + bX^*) - (\bar{Y}_{i'.} - b\bar{X}_{i'.} + bX^*) = \bar{Y}_{i.} - \bar{Y}_{i'.} - b(\bar{X}_{i.} - \bar{X}_{i'.})$. The statistics $\bar{Y}_{i.}$, $\bar{Y}_{i'.}$, and b are statistically independent; to obtain the variance of $Y_i^{*\prime\prime} - Y_{i'}^{*\prime\prime}$, therefore, we merely add the variances of these three terms. The variance thus becomes

$$\text{Var}(Y_i^{*\prime\prime} - Y_{i'}^{*\prime\prime}) = \left[\frac{1}{n_i} + \frac{1}{n_{i'}} + \frac{(\bar{X}_{i.} - \bar{X}_{i'.})^2}{\sum x_r^2} \right] \sigma_e^2. \qquad (13.16)$$

A confidence interval for $\alpha_{i0}-\alpha_{i'0}$ is

$$\left[\bar{Y}_{i.}-\bar{Y}_{i'.}-b\left(\bar{X}_{i.}-\bar{X}_{i'.}\right)\right] \pm t[1-\alpha/2m; N-a-1]$$

$$\times\left[\left(\frac{1}{n_i}+\frac{1}{n_{i'}}+\frac{\left(\bar{X}_{i.}-\bar{X}_{i'.}\right)^2}{\sum x_r^2}\right)\frac{D_2}{N-a-1}\right]^{1/2}, \qquad (13.17)$$

where m intervals are being made with overall level $1-\alpha$.

In the example, a confidence interval $(1-\alpha=.95, m=5)$ for the difference in population mean language scores for any particular IQ between methods I and II is

$$(78.7-86.4)-.315(116.0-109.0)\pm 2.779$$

$$\times\sqrt{\left[\frac{2}{10}+\frac{(116.0-109.0)^2}{7170}\right]\frac{999.0}{26}}$$

or

$$-17.74<\alpha_{10}-\alpha_{20}<2.07.$$

ADJUSTED MEANS

Frequently the researcher calculates the heights of the sample regression lines at the overall mean $\bar{X}_{..}$; these are called *adjusted means*. His purpose is simply to have a means that can be compared. Calling the adjusted mean for ith population adj $\bar{Y}_{i.}$, we use the foregoing definition to obtain

$$\text{adj } \bar{Y}_{i.}=\bar{Y}_{i.}+b\left(\bar{X}_{..}-\bar{X}_{i.}\right). \qquad (13.18)$$

The variance of the adjusted mean is obtained by setting $X^*=\bar{X}_{..}$ in (13.13). The confidence interval, for m intervals with an overall confidence level of $1-\alpha$, is

$$\bar{Y}_{i.}+b\left(\bar{X}_{..}-\bar{X}_{i.}\right) \pm t[1-\alpha/2m; N-a-1]$$

$$\times\left[\left(\frac{1}{n_i}+\frac{\left(\bar{X}_{..}-\bar{X}_{i.}\right)^2}{\sum x_r^2}\right)\frac{D_2}{N-a-1}\right]^{1/2}. \qquad (13.19)$$

CONFIDENCE INTERVALS WHEN POPULATION REGRESSION LINES ARE NOT PARALLEL

If the experimenter decides that the population regression lines are not parallel, he can still make confidence intervals. For the variance σ_e^2 he now uses as a point estimate $D_1/(N-2a)$ and has a $1-\alpha$ level interval:

$$\frac{D_1}{\chi^2[1-\alpha/2;N-2a]}<\sigma_e^2<\frac{D_1}{\chi^2[\alpha/2;N-2a]}.$$

The intervals for the other parameters are as might be expected. For example, the confidence interval for the population mean at a point X^* is

$$Y_i^{*\prime} \pm t[1-\alpha/2m;N-2a]\left[\left(\frac{1}{n_i}+\frac{(X^*-\bar{X}_{i.})^2}{\sum x_r^2}\right)\frac{D_1}{N-2a}\right]^{1/2} \tag{13.20}$$

where $Y_i^{*\prime}=\bar{Y}_{i.}+b_{(i)}(X^*-\bar{X}_{i.})$.

This confidence interval is exactly that of (13.14), with $Y_i^{*\prime}$ for $Y_i^{*\prime\prime}$, D_1 substituted for D_2, and $N-2a$ substituted for $N-a-1$.

We note that the intervals obtained under H_1 differ from those which we might write down for each single regression line using Chapter 10 methods. Here we take advantage of the assumption that the variances around the a regression lines are equal, and we combine the information from a samples to estimate σ_e^2.

FITTING A LINE TO WEIGHTED MEANS

Occasionally the experimenter wishes to fit a straight line to the means $(\bar{Y}_{i.},\bar{X}_{i.})$, each weighted by the sample size n_i. Weighted least squares was covered in Chapter 10. We should mention here, however, that after an analysis of covariance table has been constructed, the line can be written down very easily using the due A row of Table 13.2. The proper slope is

$$b_a=\frac{\sum xy_a}{\sum x_a^2},$$

and the line must go thru the point $(\bar{X}_{..},\bar{Y}_{..})$. The line is thus

$$\tilde{Y}=\bar{Y}_{..}+b_a(X-\bar{X}_{..}),$$

where $\tilde{Y}$ is the height of the line at a point X.

DISCUSSION

In the example, three different teaching methods were used; language score was the response variate, and IQ the covariate whose effect we wished to remove. In a one-way analysis of covariance, we tested whether the three population regression lines were equal. If we had not known the students' IQs, we could have tested simply whether the three population means on language scores were equal. Thus covariance analysis can be thought of as a refinement of ordinary analysis of variance in which we adjust in a linear fashion for a covariate. As in regression, the covariate must be measurable on an interval scale.

Covariance techniques can be extended to include any of the analysis of variance designs discussed so far. For example, we could have a two-way analysis of covariance in which three teaching methods were crossed with two instructors. In these more complicated designs, the analysis is usually done on a large-scale computer using a packaged program. For example, the BMD03V program allows the investigator to perform an analysis of covariance with as many as six factors. The program also allows the user more than one covariate. The investigator may have available both IQ scores and scores on a test designed to measure language aptitude. He then fits planes using Y=language score, X_1=IQ, and X_2=language aptitude. In a covariance analysis, he tests whether these planes coincide and estimates the differences among teaching methods when the linear effects of IQ and language aptitude are removed.

Covariance programs such as BMD03V usually assume that the regression coefficients for the different planes (or lines) are equal (i.e., that the planes are parallel). On the basis of this assumption the programs test whether the levels of the planes coincide. Before using these programs, the investigator should perform a regression analysis on each treatment combination and decide informally whether his sample regression planes have slopes sufficiently similar to warrant proceeding. BMDP1V, a new program for a one-way analysis of covariance, includes a test of equality of slopes.

Often investigators decide whether to perform a covariance analysis or a simpler analysis of variance by first examining r_{yx}, the correlation coefficient between the response variate and the covariate, or $R_{Y.x_1,\ldots,x_k}$, the multiple correlation coefficient between the response variate and the k covariates. Here r_{yx} (or $R_{Y.x_1,\ldots,x_k}$) can be calculated from the separate regressions; most regression programs print these correlations. Because $r_{yx}^2(R_{Y.x_1,\ldots,x_k}^2)$ measures the proportion of the sum of squares $\sum y^2$ which is removed fitting a line (plane), it is not worthwhile to perform a covariance analysis if the correlation coefficient is small. A rough rule of thumb [13.4] is that unless the correlation coefficient is at least .3 (a reduction of about 10% in the residual variance), it is sensible to perform a simple analysis of variance.

OBJECTIVES IN PERFORMING A COVARIANCE ANALYSIS

There are two reasons for performing the more complicated covariance analysis instead of the simpler analysis of variance: (1) to reduce the variance of the sample means obtained, and (2) to remove the effect of unwanted variables. Here we can distinguish between surveys of two or more populations and experiments in which assignment to treatment group is made at random. The main reason for using covariates in the experiment is to make it more sensitive by reducing the variance; in the survey, the major reason is to eliminate the effect of unwanted variables.

The reduction in variance by using a covariate can be considerable. We recall from Chapter 10 that $\sigma_e^2 = \sigma_y^2(1-\rho^2)$. In an agricultural station, for example, the same plots of land can be used year after year for experiments on corn. Good land tends to yield large crops rather consistently, and correlations as high as .8 have been found between yields of consecutive years. Thus in a study of this year's yield, the use of last year's yield may result in a reduction of as much as 64% in the variance. Confidence intervals for differences among treatments are shorter, and F tests are more sensitive.

In the survey, however, the primary purpose of the covariate analysis is to remove the effect of unwanted variables. In the example of teaching method, IQ was chosen as a covariate because of its probable correlation with language score. With just two teaching methods, and with one class using method I and another class using method II, the scatter diagram might appear as in Figure 13.3*a*. A simple analysis of variance shows method II superior to method I, but the covariance analysis indicates that the difference reflects the difference in IQ between the two groups.

We try in a survey to eliminate variables such as age, IQ, and initial condition that may differ markedly from one study group to another. We want to remove them and then be able to conclude that the difference observed is due to the teaching method. In a survey, unfortunately, there is always the possibility of unrecognized differences in composition between the groups which may cause the observed difference in response.

The removal of the effects of unwanted variables is less important in an experiment than in a survey (especially when sample sizes are large), since the random assignment of experimental units to the treatment groups tends to equalize such variables. In the study of the two teaching methods, if

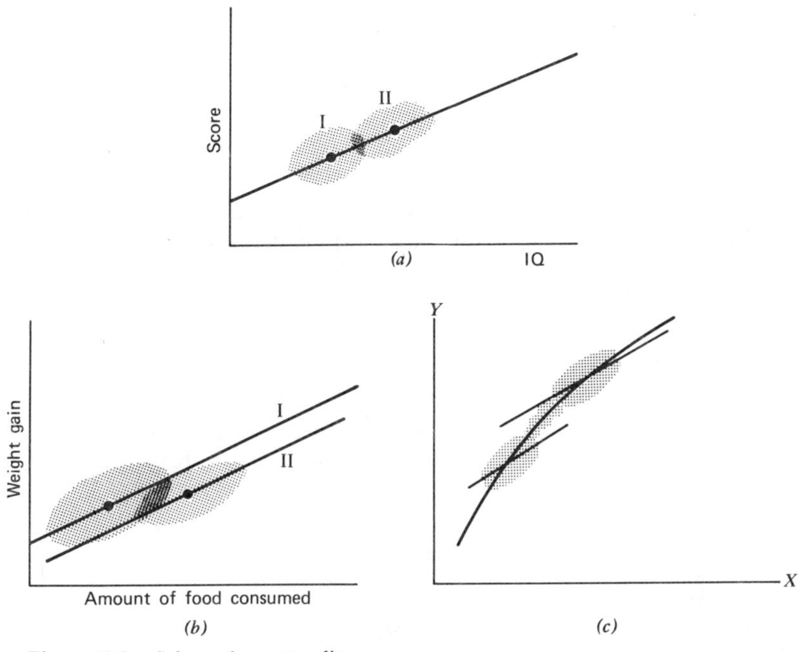

Figure 13.3. Schematic scatter diagrams.

students had been randomly assigned to the two methods, the study would be called an experiment. Although with small sample sizes there may be a sizable difference in IQ between the two samples, with larger sample sizes, the average IQs of the two groups are probably nearly equal.

SELECTION OF COVARIATES

The statistician should be alert at the outset of the study for possible covariates. For example, if the response variable Y is the weight of a rat after an experiment, the statistician makes sure that the animal is weighed before treatment. Whenever possible, the variable to be measured after treatment should be measured before treatment. Sometimes this is impossible; for example, it may be necessary to sacrifice the animal in order to make the observation. In such cases, we must measure another variable which is correlated with the response variable and is measurable before treatment. As mentioned earlier, it is inadvisable to bother using covariates in the analysis if the correlation with the response variate is low.

It is highly desirable that covariates be measured before treatment because of the possibility that any variate measured after treatment has been affected by the treatment. In such cases, the interpretation of the covariance analysis becomes more difficult. In an experiment involving two diets for babies, diet I might be highly nutritious but unpalatable, diet II less nutritious but palatable. If gain in weight is the response variate and average amount of food consumed the covariate, the scatter diagram might resemble Figure 13.3*b*. From this covariance analysis, we might carelessly conclude from the difference in adjusted means that diet I results in larger weight gains than diet II. For a given food intake this conclusion is true; diet II is nevertheless the diet that results in higher weight gain.

The values of a covariate should overlap among the groups. If the groups are widely separated, an apparent treatment difference may be merely an indication of lack of linearity (Figure 13.3*c*; see Cochran [13.4]). Because of the difficulties in interpreting the more complicated analysis of covariance, examination of scatter diagrams is particularly important.

SUMMARY

This chapter gave a detailed description of a one-way analysis of covariance with a single covariate. Covariance analysis is a useful technique in experiments or in surveys when the investigator wishes to reduce the variance and to remove the effect of a factor, such as age, which may have an appreciable effect on his response variable. The technique can be used on any type of analysis of variance. Since it requires extensive calculations, the use of a large-scale computer should be considered.

REFERENCES

APPLIED STATISTICS

13.1. Bennett, C. A., and N. L. Franklin, *Statistical Analysis in Chemistry and the Chemical Industry*, Wiley, New York, 1954.

13.2. Brownlee, K. A., *Statistical Theory and Methodology in Science and Engineering*, 2nd ed., Wiley, New York, 1965.

13.3. Cochran, W. G., and G. M. Cox, *Experimental Designs*, Wiley, New York, 1957.

13.4. Cochran, W. G., "Analysis of Covariance: Its Nature and Uses," *Biometrics*, Vol. 13 (1957).

13.5. Dixon, W. J., and F. J. Massey, Jr., *Introduction to Statistical Analysis*, 3rd ed., McGraw-Hill, New York, 1969.

13.6. Ostle, B., *Statistics in Research*, Iowa State University Press, Ames, Iowa, 1963.

13.7. Snedecor, G. W., and W. G. Cochran, *Statistical Methods*, 6th ed., Iowa State

University Press, Ames, Iowa, 1967.

13.8. Winer, B. J., *Statistical Principles in Experimental Design*, McGraw-Hill, New York, 1962.

MATHEMATICAL STATISTICS

13.9. Rao, C. R., *Linear Statistical Inference and Its Applications*, Wiley, New York, 1965.

13.10. Scheffé, H., *The Analysis of Variance*, Wiley, New York, 1967.

PROBLEMS

13.1. In an agricultural experiment comparing yields Y of three varieties of corn (bushels/acre), rainfall X was measured (in.).

Variety I		Variety II		Variety III	
X	Y	X	Y	X	Y
14	68.5	30	81.5	15	70.0
36	83.0	33	85.2	11	84.0
31	83.0	39	87.1	42	75.2
27	66.5	26	69.3	14	67.5
15	58.1	43	73.5	15	66.0
17	82.4	31	65.5	42	79.1
		18	73.4	26	75.5
		14	56.1		

(*a*) Make an analysis of covariance table.

(*b*) Under the assumption that the population regression lines are parallel, test whether there are differences in mean yield among the three varieties of corn.

13.2. Glucose measurements Y were made at four different times X after glucose load on three types of individual (A, B, C). Four individuals of each type were used, and they were randomly assigned to the four time levels.

A		B		C	
X	Y	X	Y	X	Y
0.5	118	0.5	120	0.5	114
1.	111	1.	118	1.	113
2.	97	2.	101	2.	94
3.	85	3.	90	3.	82

(*a*) Complete an analysis of covariance table.
(*b*) Test the hypothesis that the three population regression lines have a common slope, and compare the results with scatter diagrams of the points.
(*c*) Assuming that the three population regression lines have a common slope β, test to see whether the population means change with time.
(*d*) Find a 95% confidence interval for β.
(*e*) Find a 95% confidence interval for σ_e^2.

13.3. Two methods of estimating glucose level were used on a group of patients. Patients were randomly assigned to method and to different time levels after glucose load.

Method	Results							
Folin-Wu								
Time, X	0.5	0.5	0.5	1.	1.	2.	2.	2.
Glucose level, Y	140	148	133	122	133	107	109	118
True glucose								
Time, X	0.5	0.5	0.5	1.	1.	2.	2.	2.
Glucose level, Y	91	109	97	95	92	77	82	95

(*a*) Complete an analysis of covariance table.
(*b*) Assuming that $E(Y_i|X)=\alpha_{0i}+\beta_i X$, test $H_0: E(Y_i|X)=\alpha_{0i}+\beta X$.
(*c*) Assuming that $E(Y_i|X)=\alpha_{0i}+\beta X$, test $H_0: \beta=0$.
(*d*) Find the best-fitting straight lines under the assumption that $E(Y_i|X)=\alpha_{0i}+\beta X$.
(*e*) Estimate the difference between glucose level as estimated by the two methods, and give its 95% confidence interval.

13.4. In studying intelligence of children with heart disease, two groups of patients were considered: those who underwent surgery and those who did not. Investigations measured IQ at the beginning of study (X) and IQ at the end of the study (Y).

(*a*) Complete the analysis of covariance table.
(*b*) Write down the two regression lines under H_1 (two straight lines not necessarily parallel).
(*c*) Write down the two regression lines under H_2 (two parallel lines).
(*d*) Plot the regression lines from (*b*) and (*c*) on a scatter diagram.
(*e*) Assuming H_2, estimate the population mean IQ for children undergoing surgery and having an initial IQ of 100. Find a 95% confidence interval.
(*f*) Give a 95% confidence interval for the final IQ of a child whose initial IQ was 100 and who underwent surgery.
(*g*) Find the adjusted means for final IQs for the two populations.

Nonsurgery				Surgery			
Y	X	Y	X	Y	X	Y	X
127	124	82	95	86	78	99	90
124	127	111	125	99	87	106	114
100	90	89	100	119	100	90	81
104	93	96	92	102	103	114	108
130	115	100	96	113	114	107	107
94	100	94	85	99	97	88	98
121	111	100	112	105	113	83	95
100	97	90	98	111	111	88	88
101	109	105	109	119	104	100	105
110	95	105	99	102	92	87	85
104	110	112	115	113	107	107	72
88	117	100	88	103	118	109	100

CHAPTER 14

DATA SCREENING

An unfortunate by-product of changing from the desk calculator to the large-scale computer is that now analyses of variance and regression analyses can be performed without proper scrutiny of the data. When calculations were made on a desk calculator by an individual thoroughly familiar with the procedures used in obtaining the basic data, errors and peculiarities of the data could be seen during the course of the computations. With the large-scale computer, on the other hand, intermediate steps are seldom seen by knowledgeable individuals; the end results often do not display the errors, and the comment "garbage in, garbage out" applies all too often.

In this chapter we discuss some techniques for screening data. Screening is done with two objectives in mind: to detect errors in the data and to detect departures from the assumptions the investigator wishes to make in his analysis. In a sense, errors in the data can be viewed as departures from the assumptions; a set of data on heights measured in inches containing as errors several heights measured in centimeters cannot be considered to be a random sample from a single normally distributed population of heights measured in inches. Although particularly important for work with large-scale computers, some screening methods can be used to advantage with desk calculators.

POST HOC REVIEW OF DATA

Ideally, data screening should take place before analysis, but often in practice it is done because the investigator obtains unexpected results from his analysis. In Chapter 7, the case of the unexpected interaction was discussed. Another common problem is an exceptionally large s_e (larger than experience indicates), which may be an indication of outliers. Occasionally an extremely small computed F ratio is obtained. The computed F ratio tends to be greater than or equal to 1 because the expected value of the numerator contains a term in addition to those in the denominator. In the simplest case, a one-way analysis of variance, the F ratio is an estimate of the ratio $(\sigma_e^2 + n\sigma_a^2)/\sigma_e^2$ which is greater than or equal to 1. When the F value is smaller than $F[\alpha; a-1, a(n-1)]$, the sample means of the a populations are closer together than would be expected considering the basic variability of the data, and it is advisable to seek an explanation. The small F value may be an indication of some error, although more likely it points to lack of independence of the observations. For example, if the observations on the a treatments were taken over a period of time, and if the jth observation on all treatments was made at the jth time, $j = 1, \ldots, n$, the observations on the various treatments may be highly correlated. When an unusually low value of F occurs, the data should be examined and plotted.

DATA SCREENING IN ANALYSIS OF VARIANCE PROBLEMS

We deal first with data screening in relation to the analysis of variance. Data screening for regression analysis is discussed separately later in the chapter. The screening process in regression analysis is complicated because there is often only one Y value for a given value of X; in analysis of variance, we usually have several observations on each treatment combination.

To understand the data screening process, we review the assumptions underlying the analysis of variance. In Chapter 5, the three assumptions for a one-way analysis of variance problem were:

1. Normally distributed data.
2. Equal population variances.
3. Independent random samples.

More complicated analysis of variance problems sometimes require additional assumptions, but these three are always present and form a basis for data screening.

Screening can be done either by informal techniques or by formally testing that the assumptions just listed hold. Informal techniques are often sufficient; we give them first, and later we describe or reference more formal methods. One problem with the formal tests is that testing one assumption usually necessitates assuming that the other two hold; thus if the data are really "dirty," it is difficult to decide where to begin.

INFORMAL PRELIMINARY SCREENING TECHNIQUES

When there is a large amount of data on a computer, a sensible first step is to make sure that no letters of the alphabet were keypunched in place of numerical data. Verification of keypunching can avoid much of this type of error. Its occurrence can also be detected by running program BMD04D, which makes counts of all the different entries, either alphabetical or numerical, in a specified column of the data cards.

INFORMAL TECHNIQUES FOR FINDING OUTLIERS

After a preliminary check has been performed, the next step in looking for errors may well be a search for outliers. Observations that are far removed in some sense from the rest of the data are usually called outliers. As already mentioned, errors in the data can be viewed as departures from one or another of the underlying assumptions. It is possible to use the same methods for detecting outliers as are used in checking the assumptions; for if errors exist in the data, we do not have independent random samples from normally distributed populations.

Gross outliers can be detected visually by simply scanning the data. If the data set is so large that careful scanning is impractical, gross outliers can be spotted from printouts of high and low values of the data. Often comparison of these printouts of high and low values with known or reasonable ranges of the data (e.g., a weight of 13 lbs. for an adult male is unreasonable), detects outliers caused by a misplaced decimal point, keypunch error, and so on.

When a gross outlier is detected, we must decide what to do with the observation. Sometimes it is possible to replace the value with a corrected value or to retake the observation. If no reason for the unusual value can be found, the problem is more difficult. Investigators vary in their opinions; there are some investigators of the "if in doubt, throw it out" school of thought and some who believe that it is unethical to throw observations away ([14.1]–[14.3], [14.5].) It seems foolish to use observations that are clearly errors in the statistical analysis, yet discarding observations without a convincing explanation can reduce the confidence felt in the results of the analysis. We suggest that gross outliers be eliminated in the statistical analysis but that they be reported along with the statistical analysis. This allows the reader of the report to judge for himself; if he wishes, he can replace the outliers and rerun the analysis.

There are many explanations for outliers besides the suggestion that they are simply errors. One possibility is that a single sample is actually taken from a mixture of two or more populations. For example, several technicians may have obtained the observations on a single treatment, and the low readings from one of the technicians may appear as outliers. Clearly, we do not have a simple random sample from a normally distributed population. The presence of outliers may indicate that the sample is from a nonnormal distribution, perhaps with a long upper tail.

FORMAL TESTS FOR OUTLIERS

If the investigator is fairly sure of the normality of his data, he can make tests for outliers based on this assumption. His justification for the assumption of normality may be theoretical; or it may come from an examination of his own data or from a study of large sets of similar data found in the literature.

The easiest test for detecting outliers is that of Dixon ([14.5], [14.6]); a more complete set of tables is given in Ref. 14.7. Dixon's test is suitable for small samples. The assumption underlying the test is that the sample is a random one from a normal distribution. Such a sample might be the observations on a single treatment combination in an analysis of variance. In performing the test, the sample data are first ordered from smallest to largest. Let $Y_{(1)}$ be the smallest observation, $Y_{(2)}$ the next smallest, and $Y_{(n)}$ the largest. The test statistic for testing whether $Y_{(1)}$ is an outlier is

$$r_{10}=\frac{Y_{(2)}-Y_{(1)}}{Y_{(n)}-Y_{(1)}}. \tag{14.1}$$

To test whether $Y_{(n)}$ is an outlier, the statistic

$$r_{01}=\frac{Y_{(n)}-Y_{(n-1)}}{Y_{(n)}-Y_{(1)}} \tag{14.2}$$

is used. The calculated r_{10} (or r_{01}) is compared with the entries in Table A.7, and the observation is rejected if the ratio is too large. The usual significance level is $\alpha=.10$, but a smaller value of α is used if it is reasonably certain that no outliers exist.

As an example of Dixon's outlier test, we order the data from tree 1 and spray 1 in Table 6.1. We have $Y_{(1)}=4.50$, $Y_{(2)}=4.98$, $Y_{(3)}=5.48$, $Y_{(4)}=6.54$, $Y_{(5)}=7.04$, and $Y_{(6)}=7.20$, and

$$r_{10}=\frac{4.98-4.50}{7.20-4.50}=\frac{.48}{2.70}=.18.$$

Using $\alpha=.10$, we compare .18 with .482 from Table A.7; we cannot conclude that the $Y_{(1)}$ is an outlier. When an outlier is found, the ratio test can be used on the remaining observations to detect additional outliers. Note that a large value of r_{10} will be found whenever $Y_{(1)}$ is very different from $Y_{(2)}$ compared with the difference $Y_{(n)} - Y_{(1)}$.

Another test for outliers based on the same assumption of normality is Grubbs's test [14.10], which is useful when a single outlier is suspected. If the largest observation is suspect, we calculate the test statistic

$$T_n = \frac{Y_{(n)} - \overline{Y}}{s}, \tag{14.3}$$

where s is the standard deviation of the sample and $Y_{(n)}$ is the largest observation. To test whether the smallest observation $Y_{(1)}$ is an outlier, we use the statistic

$$T_1 = \frac{\overline{Y} - Y_{(1)}}{s}. \tag{14.4}$$

Several percentage points of the distribution of T_n and T_1 are given in Table A.8. If the calculated statistic for either test is larger than the tabled value, $Y_{(n)}$ (or $Y_{(1)}$) is an outlier.

Additional discussion and other outlier techniques can be found in a summary article by Grubbs [14.10]. Other tests for labeling outliers are given by Ferguson [14.9] or Mickey [14.11]; the use of these tests involves a somewhat greater amount of work.

INDEPENDENCE

When the investigator has detected and in some fashion taken care of gross outliers in the data, he often considers next whether his observations are actually independent. Violation of this assumption can have an appreciable effect on the outcome of F tests (see Refs. 14.14 and 14.20).

In verifying the assumption of independence, there are two types we may wish to check—independence of the observations within a single sample, and independence of the observations of two or more different samples.

We define a simple random sample as a subset of the population chosen in such a way that any subset of equal size is equally likely to be chosen. When a random sample is chosen from a small population, it seems clear that two individuals in the sample are *not* statistically independent of each other. Knowledge of the first individual chosen gives us information concerning the second individual (e.g., it cannot be the first individual).

For an infinite population, however, any two observations in a random sample are statistically independent of each other.

In checking the assumption of independence in random samples, we usually assume that each sample is from an infinite population, and we attempt to ascertain whether the observations within a sample are independent of each other. In other words, we are mainly checking to determine whether each sample from an infinite population is actually a random sample. We also sometimes check whether two random samples are "independent random samples"; that is, whether each observation of one sample is statistically independent of the observations in the other sample. A one-way analysis of variance in which experimental animals are blocked according to litter is an example of a situation lacking among sample independence.

The observations within a sample are statistically independent of each other, and the observations in one sample are statistically independent of those in another sample because of the fashion in which the samples were chosen. It is difficult to decide whether observations are independent by looking at the data without investigating the actual experimental situation. Sometimes, however, because of lack of information, the investigator can only look at the data themselves. In any case, investigation of independence serves as a check of one's knowledge of the experimental situation. For an extensive discussion of tests of independence, see Chapter 11, Bennett and Franklin [14.12].

Lack of independence among observations within a sample occasionally arises because measurements have not been taken in a random order and a dependence exists from observation to observation. Such dependence can be caused by factors such as equipment drifting out of calibration, physical effects of the location of experimental animal cages, or the learning curve on the part of the investigator performing the experiment.

The simplest method for detecting gross lack of independence is a graph of the observations in each sample. If we suspect, for example, that a certain piece of measuring equipment is drifting out of calibration, we plot the observations in the order in which they were taken, as in Figure 14.1; this plot leads us to suspect that Y_i for $i=7,\ldots,20$ are not independent.

Another technique for detecting lack of independence among observations within a sample consists of computing the serial correlation and testing to decide whether it is significantly different from zero. Serial correlation is defined as the simple correlation between adjacent observations Y_i and Y_{i+1}, for $i=1,\ldots,n-1$. It is thus a measure of the association between adjacent observations, and it should be large if observations tend, say, to increase as the measuring instrument gradually reads higher and higher. It is calculated from

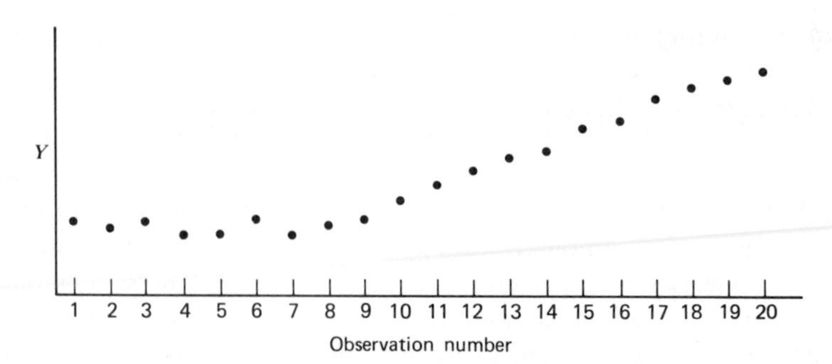

Figure 14.1. Data indicating lack of independence among successive observations.

$$r_s = \frac{(n-1)\sum_{i=1}^{n-1} Y_{i+1}Y_i - \sum_{i=1}^{n-1} Y_{i+1} \sum_{i=1}^{n-1} Y_i}{\sqrt{\left[(n-1)\sum_{i=1}^{n-1} Y_{i+1}^2 - \left(\sum_{i=1}^{n-1} Y_{i+1}\right)^2\right]\left[(n-1)\sum_{i=1}^{n-1} Y_i^2 - \left(\sum_{i=1}^{n-1} Y_i\right)^2\right]}} \tag{14.5}$$

For example, for $Y_1 = 1, Y_2 = 2, Y_3 = 2, Y_4 = 4$, we have

	Y_{i+1}	Y_i	Y_{i+1}^2	Y_i^2	$Y_{i+1}Y_i$	
	$Y_2 = 2$	$Y_1 = 1$	4	1	2	
	$Y_3 = 2$	$Y_2 = 2$	4	4	4	
	$Y_4 = 4$	$Y_3 = 2$	16	4	8	
Total	8	5	24	9	14	$n-1=3$

$$r_s = \frac{(3)(14) - (8)(5)}{\sqrt{[(3)(24) - (8)^2][(3)(9) - (5)^2]}}$$

$$= \frac{2}{\sqrt{(8)(2)}} = .5.$$

For large n, an approximate test of the null hypothesis that the population serial correlation is zero can be made using univariate normal tables. The test statistic is

$$z = \frac{r_s + [1/(n-1)]}{(n-2)^2/(n-1)^3}, \tag{14.6}$$

which is compared with $z[1-\alpha/2]$ for a test with significance level that is approximately α. For small n, this test may be poor. For mathematical treatment of serial correlation, see Ref. 14.13.

NORMALITY

Finally, there is the question of the normality of the data. The simplest technique consists in drawing a histogram of the data from a sample and noting whether it looks like a normal distribution (symmetric and bell-shaped). For small n, this approach is unsatisfactory; the histogram simply looks irregular.

To increase the number of observations used in drawing a histogram, the investigator sometimes examines simultaneously residuals from the entire study, rather than working with the observations themselves. By residuals, we mean the difference between an observation and the estimate for the mean of the population from which it was drawn. In the one-way analysis of variance, the estimate of the $\mu+\alpha_i$ is $\overline{Y}_{i.}$; thus the residual for Y_{ij} is

$$e_{ij}=Y_{ij}-\overline{Y}_{i.}.$$

For a two-way classification with n larger than 1, the residual for Y_{ijk} is

$$e_{ijk}=Y_{ijk}-\overline{Y}_{ij.}.$$

Tests for Normality. The various methods for testing for normality can be performed separately for the observations in each sample. On the other hand, they are frequently done using all the residuals. Unfortunately, the residuals do not form a single random sample, since they are not all statistically independent of one another. For example, using the methods of Chapter 2 to find the covariance of $e_{ij}=Y_{ij}-\overline{Y}_{i.}$ and $e_{ij'}=Y_{ij'}-\overline{Y}_{i.}$ in the one-way analysis of variance, we find

$$\operatorname{Cov} e_{ij}, e_{ij'}=\frac{-\sigma_e^2}{n};$$

the correlation between them is

$$\operatorname{Corr} e_{ij}, e_{ij'}=\frac{-1}{n-1}.$$

Strictly speaking, then, the entire set of residuals does not satisfy the requirements for any of the following tests; in using residuals for any test, we have some uncertainty concerning our significance level.

A χ^2 test of goodness of fit can be used to test the null hypothesis of normality (14.6). Even when using residuals, however, this test is not well suited to the smaller sample sizes frequently found in analysis of variance problems; therefore, it is not discussed here.

A test useful with small samples ($n < 50$) is the W test given by Shapiro and Wilk [14.17], [14.18]. The null hypothesis for this test is that the observations (or residuals) form a random sample from a normally distributed population. To make the W test, we order the observations from smallest to largest. The test statistic, based on n observations, is

$$W = \frac{b^2}{S^2}, \tag{14.7}$$

where

$$S^2 = \sum_{i=1}^{n} \left(Y_{(i)} - \bar{Y} \right)^2$$

and

$$b = \sum_{i=1}^{k} a_{n-i+1} \left(Y_{(n-i+1)} - Y_{(i)} \right). \tag{14.8}$$

The quantity k in (14.8) is defined by $n = 2k$ for n even and by $n = 2k + 1$ for n odd. The coefficients a_{n-i+1} in (14.8) are found in Table A.9. (For n odd, the middle observation, although used in the calculation of S^2, is not used in calculating b; thus $a_{(n+1)/2} = a_{k+1}$ does not appear in the table.) Table A.10 contains several percentage points of the W statistic. The hypothesis of normality is rejected when the calculated W is smaller than the value given in Table A.10.

As an example of the W statistic, we use the following ordered observations of weights of adult males:

$$Y_{(1)} = 148, \quad Y_{(2)} = 154, \quad Y_{(3)} = 158, \quad Y_{(4)} = 160,$$

$$Y_{(5)} = 161, \quad Y_{(6)} = 162, \quad Y_{(7)} = 165, \quad Y_{(8)} = 170,$$

$$Y_{(9)} = 182, \quad Y_{(10)} = 195, \quad Y_{(11)} = 236$$

$$S^2 = 6239$$

$$b = .56(236 - 148) + .33(195 - 154) + .23(182 - 158)$$

$$+ .14(170 - 160) + .07(165 - 161)$$

or

$$b = 70$$

$$W = 4900/6239 = 0.79.$$

From Table A.10 we find that the fifth percentile in the distribution of W is 0.850 and we reject the null hypothesis that the data are from a normal distribution.

Transformations. In analysis of variance problems, it is often better to look for suitable transformations of the observations than to test for normality. Sometimes a researcher knows the theoretical distribution of his data. As a rule, the square root transformation given in (14.9) is used if the data have a Poisson distribution ([14.6], [14.12]). If the data are distributed as a log normal distribution, the logarithmic transformation (14.10) is appropriate. Or, the investigator may know that traditionally a particular transformation has been used with a particular type of data. If no obvious transformation is apparent to the investigator, there are two approaches to finding suitable transformations.

The older of the two methods of seeking a transformation is based on the fact that for the normal distribution, the mean and variance (or standard deviation) of a sample are statistically independent. A transformation is sought that equalizes the variance regardless of the size of the mean [14.21]. The first step in this procedure is to plot the sample means versus the sample variances and also versus the sample standard deviations of the separate treatment groups. Several schematic plots of this type appear in Figure 14.2.

Figure 14.2*a* is a case of samples that are likely to be from normal populations, since the means and variances appear to be independent. In Figure 14.2*b* there seems to be a linear relation between the sample variances and the sample means; in Figure 14.2*c* the linear relation is between the standard deviation and the mean.

For the case of Figure 14.2*a* of the data graphed, we decide that no transformation is necessary. For Figure 14.2*b*, the following square root transformation is useful:

$$Y_i' = \sqrt{Y_i} + \sqrt{Y_i + 1}\,. \tag{14.9}$$

This transformation is the appropriate one if the original data follow a Poisson distribution [(14.6), (14.10)]. Count data (14.10) or occurrence of rare events often follow the Poisson distribution (14.6).

For Figure 14.2*c*, the following logarithmic transformation is used:

$$Y_i' = \log_{10}(Y_i + A). \tag{14.10}$$

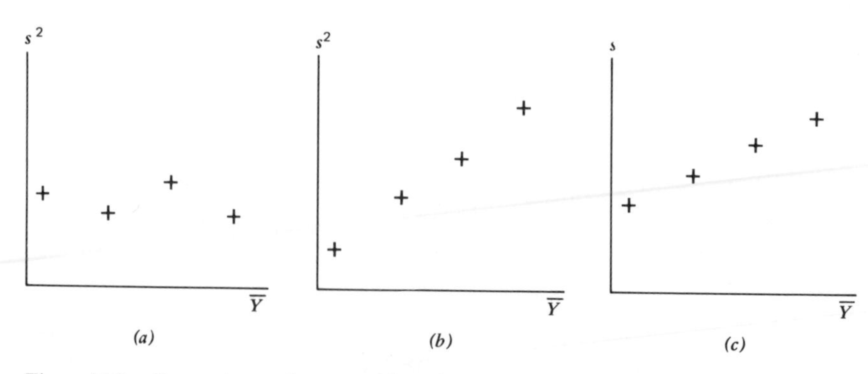

Figure 14.2. Comparisons of means with variances and standard deviations.

The constant A is always chosen so that $Y_i + A$ is greater than zero for every observation. Often A is chosen to be zero. Sometimes A is selected by trying several values and transforming the data with each value. The final value of A is selected by examining histograms of residuals or by making tests with the W statistic on the transformed data.

Since the means and variances are always computed in an analysis of variance, a check on their independence takes little additional effort. If we do not wish to compute the standard deviations, the sample ranges can be plotted instead of the variances; if the ranges show a linear relation with the means, the logarithm transformation is appropriate. Indeed, a quick look at the data can often reveal whether large ranges occur with large means and small ranges with small means; thus we have a crude indication of whether it is useful to consider making transformations.

Bartlett [14.21] gives other transformations that may be helpful. If transformations of the observations are made prior to the analysis of variance, and if the analysis is performed on the transformed data, the statements about the outcome of the analysis of variance apply directly to the *transformed* populations.

More recently, for cross classifications, emphasis has been placed on finding a transformation that reduces the interaction mean square. If there is little interaction, the results of an analysis of variance are simple to interpret. The transformations (14.9) and (14.10) sometimes have this effect, although they are not specifically designed for this purpose. Draper and Hunter [14.24] suggest considering transformations such as $(Y_i^p - 1)/p$ for $p \neq 0$ or $\log_{10}(Y_i - A)$. Tukey [14.27] has proposed $(Y_i + A)^p$, where $(Y_i + A) > 0$. If the analysis is done on a large-scale computer, it is possible to try empirically various values of p and A, to perform the analysis of variance, and to use the particular transformation that results in a small value for the mean square of the interactions relative to the main effects.

Further discussions on transformations can be found in Refs. 14.22 to 14.26.

Finally, it is sometimes possible simply to avoid the problem by using means of observations in the analysis rather than the observations themselves. As is well known, means tend to be normally distributed for moderate sample size. With the use of means, fewer observations are available.

We have discussed techniques for screening data to assess the presence of outliers or to determine whether the data are normally independently distributed. Of the three basic assumptions made in analysis of variance, we have not tested the assumption of equal variances. Conventional tests for equality of variances [14.6] are grossly affected by any lack of normality that may exist in the data. It is therefore more usual to search for a transformation that stabilizes the variance, rather than to attempt formal tests of the hypothesis of equal variances.

ROBUSTNESS

The investigator should always consider the relative seriousness of departures from the assumptions. Considerable discussion of this matter has been published [14.12], [14.14], [14.19], [14.20], but the following statements may serve as a rough guide. Lack of independence can be the most serious, since it is usually impossible to correct and invalidates the levels of significance. Nonnormality of the data is less serious than lack of independence; such nonnormality may be reduced by a transformation. Furthermore, nonnormality of the data is not very troublesome in testing equality of means. Inequality of variances is seldom serious if there are equal sample sizes in the treatment groups [14.19]; it is also amenable to transformations. The effect of outliers depends on the size and number of the outliers. Hence outliers may be of only minor consequence or they may completely change the results. Because outliers have gross effects on variances, their presence sometimes results in an unexplainable inequality of variances. They can also produce spurious interactions and main effects.

DATA SCREENING IN REGRESSION ANALYSIS

Two models for regression analysis were discussed in Chapters 10 and 11. In the first model, it is assumed that for each fixed value of X (or of $X_1,\ldots,X_k$), there exists a population of Y values and that the means of the various populations of Y values lie on a straight line (or on a plane). A sample regression line or plane is fitted which minimizes the sum of squared vertical deviations $\sum_{i=1}^{n}(Y_i - Y_i')^2$, where Y_i is the Y value of the ith data point and Y_i' is the height of the regression line at the point X_i (or $X_{1i},\ldots,X_{ki}$). With this model, if distributional assumptions are met, we can make tests and confidence intervals on regression coefficients and on the height of the population regression line (plane) at any X value. In the second model, a bivariate or multivariate normal distribution is assumed. A regression line or plane is fitted in the same manner, and it is also possible to test hypotheses about the various correlation coefficients, means, and variances.

THE FIXED X MODEL

Several Y Values at Each X Value. We consider first the simplest case of data screening in regression analysis (i.e., the case of a single X variate, with N_i values of Y_{ij} chosen at each fixed value X_i, $i=1,\ldots,n$). Here we wish to assume that the n_i values of Y_{ij} corresponding to X_i form a random sample from a normal distribution with mean $\alpha+\beta(X_i-\bar{X})$, that the random samples are independent, and that the populations all have equal variances.

In this situation, the Y_{ij} can be checked for outliers using the methods of Dixon or Grubbs. After removal of outliers, a common indication of nonnormality or inequality of variances is an increasing range of Y_{ij} or residuals with increasing X_i. Figure 14.3*b* reveals such a pattern. A transformation such as the log transformation on the Y_i can be considered in this case, although we must remember that the straight line fits the

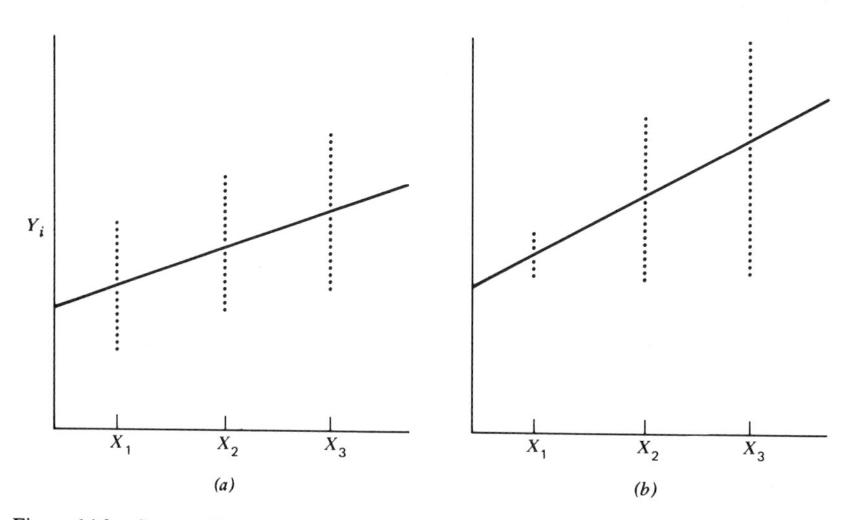

Figure 14.3. Scatter diagrams: (*a*) data with equal variances for fixed values of X; (*b*) data with unequal variances for fixed values of X.

logarithms of the Y_{ij} to the X_i and not the basic data themselves. Excellent graphs representing the before and after appearance of various transformations in regression analysis can be found in Ref. 14.29. The investigator should devote a great deal of thought to the type of relationship he expects.

The residuals can be graphed as histograms, and any of the tests of normality can be performed. Scatter diagrams of the residuals against such factors as the time when the observations were taken can be made to check gross lack of independence. Examples of these time sequence plots are given in Ref. 14.28. A test of independence can be performed using the serial correlation.

The adequacy of the fit can be checked by using residuals. For example, if a curvilinear relation holds, as in Figure 14.4, a graph of the residuals would show mostly positive residuals for X_2 and mixtures of positive and negative residuals for X_1 and X_3. For such a case it is appropriate to fit a quadratic if the variances of the three sets of residuals appear to be equal; if the variances increase with X_i, a transformation such as $\log X_i$ may be appropriate.

A Single Y Value at each X Value. If only one Y_i exists at each fixed X_i, we have simply n samples of size 1. The residuals $Y_i - Y_i'$, however, can be considered to be a single sample of size n. It is important to examine them, for an outlier can have an overwhelming effect on the fitted line, as demonstrated in Figure 14.5. Residuals from the least squares line are best examined on the scatter diagram. It is often possible to spot gross outliers, such as the × marked on Figure 14.5*a*, by their failure to fall within the range of most of the points.

After gross outliers have been removed, the least squares line should be plotted and the new residuals reexamined. If it seems that a straight line may fit the data, the normality and independence of the residuals can be checked using the techniques given earlier in this chapter. If a transformation is indicated or if a polynomial is fitted, the residuals should be studied after the appropriate function has been fitted.

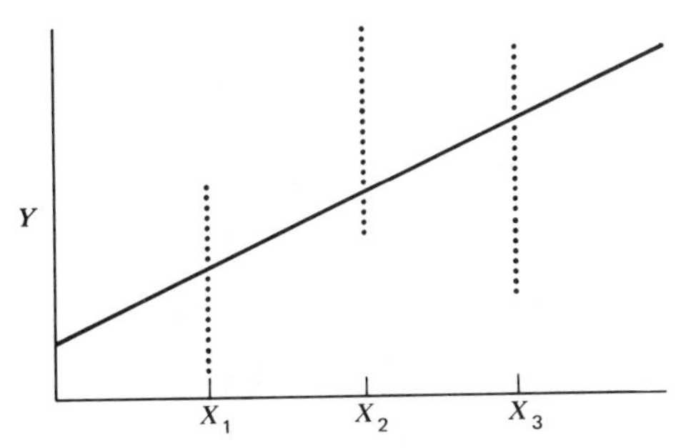

Figure 14.4. Scatter diagram representing nonlinear regression of Y on X.

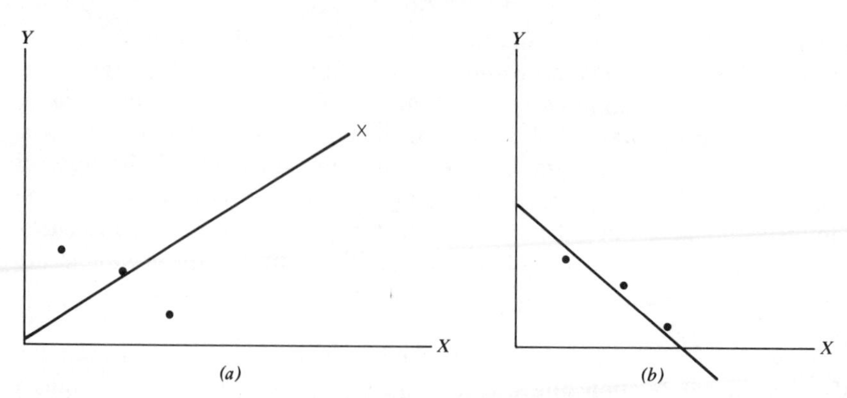

Figure 14.5. Schematic scatter diagram and least squares line: (*a*) with outlier; (*b*) outlier removed.

THE JOINT DISTRIBUTION MODEL

The problem of determining whether the assumptions are met in the second model of regression analysis (bivariate or multivariate regression analysis) is also rather difficult. Sampling from a bivariate normal distribution, we expect to see data somewhat as in Figure 14.6*a*. The elliptical contours are meant to show frequency of observations in much the same manner as the contours of a topological map represent the altitude of land. After we have examined the scatter diagram, X and Y can be considered separately. If the observations are actually from a bivariate normal distribution, the observations of X form a random sample from a normal distribution; similarly, the observations on Y form a random sample from a normally distributed population. Thus separate histograms for the X values and the Y values may reveal gross departures from normality. We

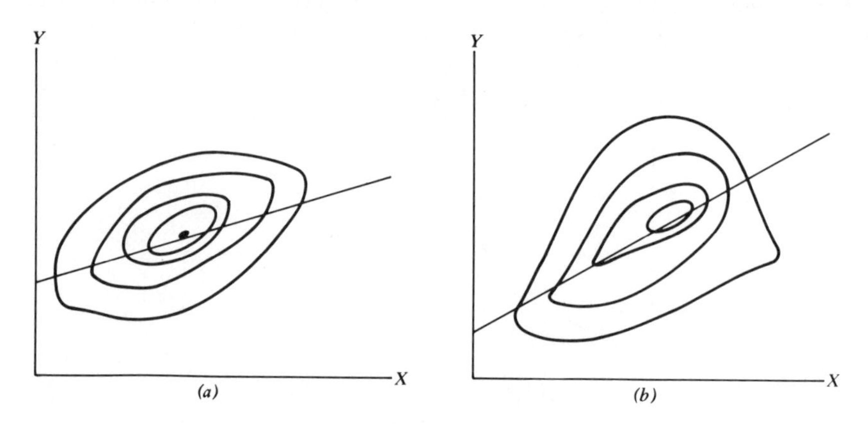

Figure 14.6. Samples from bivariate distributions: (*a*) normal; (*b*) nonnormal.

can also apply any tests for normality separately to the X values and to the Y values. It is theoretically possible that X and Y are both normally distributed but that jointly they do not have a bivariate normal distribution; in actual practice, we do not expect this to occur.

Figure 14.6*b* shows a common departure from normality that is found in practice. The fan-shaped distribution with increasing variation in Y_i values with X is a clear indication of nonnormality.

After deciding that normality is reasonable and that a straight line gives a reasonable fit, we can examine residuals and make tests in a search for outliers [14.11].

THE MULTIVARIATE CASE

For the multivariate situation, where several X variates are being used to fit a regression plane, it is simplest to look at the variables two at a time. Because paper has two dimensions, most investigators visualize their data best in just two dimensions, and more complex presentations are seldom helpful. Large-scale computers make it possible to obtain in only a few seconds scatter diagrams for Y versus each X variate; for all combinations of the X variates, two at a time; and for residuals versus the value on the regression plane (Y_i'). For further discussion, see Refs. 14.1, 14.15, and 14.30. These visual plots are frequently helpful in identifying outliers or in deciding on transformations (when combined with knowledge of the actual data). They are also important in giving the investigator a feel for the fit of his plane or line that cannot be achieved by simply looking at the equations or at the results of statistical tests. It is possible to test for multivariate normality formally, but the practice is of questionable worth in messy real-life situations, which are likely to contain many different types and degrees of nonnormality [14.28]. Visual examination of the various X variates versus Y in a scatter diagram usually reveals whether a transformation or a quadratic term should be considered for any of the X variables.

SUMMARY

In this chapter, we have discussed methods of determining whether data meet the assumptions made in analysis of variance or regression analysis. It is always possible to fit regression lines or to find means of treatment groups, regardless of whether any assumptions are met. It is only when the investigator wishes to make a statistical test or to find a confidence interval that he needs to be sure that his data meet the appropriate assumptions.

An increase in the sample size does not diminish the problem. Usually, after an investigator has discarded or corrected any data that are clearly in error (or that are not from the population from which he intended to sample), the most serious assumption to check is that of independence. A careful review of the method of obtaining the data is the most important procedure; plots are second in importance, and statistical analysis third. If the assumption of independent random samples is violated, all statistical analysis should be viewed with extreme caution. Finally, the data can be tested to determine whether they are normally distributed. If the data are not normally distributed, transformations can be considered.

REFERENCES

DETECTION OF OUTLIERS

14.1. Anscombe, F. J., and J. W. Tukey, "The Examination and Analysis of Residuals," *Technometrics*, Vol. 5 (1963), pp. 141–160.

14.2. Anscombe, F. J., "Rejection of Outliers," *Technometrics*, Vol. 2 (1963), pp. 123–148.

14.3. Bross, I. D. J., "Outliers in Patterned Experiments: A Strategic Appraisal," *Technometrics*, Vol. 3 (1961), pp. 91–102.

14.4. Daniel, C., "Locating Outliers in Factorial Experiments," *Technometrics*, Vol. 2 (1960), pp. 149–156.

14.5. Dixon, W. J., "Processing Data for Outliers," *Biometrics*, Vol. 9 (1953), pp. 74–89.

14.6. Dixon, W. J., and F. J. Massey, Jr., *Introduction to Statistical Analysis*, 3rd ed., McGraw-Hill, New York, 1969, pp. 328–330.

14.7. Dixon, W. J., "Ratios Involving Extreme Values," *Annals of Mathematical Statistics*, Vol. 22 (1951), pp. 68–78.

14.8. Dixon, W. J. (Ed.), *Biomedical Computer Programs,* No. 2, University of California Press, Los Angeles, 1967.

14.9. Ferguson, T., "On the Rejection of Outliers," *Fourth Berkeley Symposium*, University of California Press, Los Angeles, 1961.

14.10. Grubbs, F. E., "Procedures for Detecting Outlying Observations in Samples" *Technometrics*, Vol. 11 (1969), pp. 1–21.

14.11. Mickey, M. R., O. J. Dunn, and V. A. Clark, "Note on the Use of Stepwise Regression in Detecting Outliers," *Computer and Biomedical Research*, Vol. 1 (1967), pp. 105–111.

TESTS OF INDEPENDENCE

14.12. Bennett, C. A., and N. L. Franklin, *Statistical Analysis in Chemistry and the Chemical Industry*, Wiley, New York, 1954.

14.13. Kendall, M. G., and A. Stuart, *The Advanced Theory of Statistics*, Vol. III, Hafner, New York, 1966.

14.14. Scheffé, H., *The Analysis of Variance*, Wiley, New York, 1959.

REFERENCES

TESTS OF NORMALITY

14.15. Anscombe, F. J., "Examination of Residuals," *Fourth Berkeley Symposium*, University of California Press, Los Angeles, 1961, pp. 1–36.

14.16. Pearson, E. S., "Tables of Percentage Points of $\sqrt{b_1}$ and $\sqrt{b_2}$ in Normal Samples; A Rounding Off," *Biometrika*, Vol. 52 (1965), pp. 282–284.

14.17. Shapiro, S. S., and M. B. Wilk, "An Analysis of Variance Test for Normality," *Biometrika*, Vol. 52 (1965), pp. 591–612.

14.18. Shapiro, S. S., M. B. Wilk, and H. J. Chen, "A Comparative Study of Various Tests for Normality," *Journal of American Statistical Association*, Vol. 63 (1968), pp. 1343–1372.

ROBUSTNESS

14.19. Box, G. E. P., "Some Theorems on Quadratic Forms Applied in the Study of Analysis of Variance Problems: I. Effect of Inequality of Variance in the One-Way Classification," *Annals of Mathematical Statistics*, Vol. 25 (1954), pp. 290–302.

14.20. Box, G. E. P., "Some Theorems on Quadratic Forms Applied in the Study of Analysis of Variance Problems: II. Effect of Inequality of Variance and of Correlation of Errors in the Two-Way Classification," *Annals of Mathematical Statistics*, Vol. 25 (1954), pp. 484–498.

USE OF TRANSFORMATIONS

14.21. Bartlett, M. S., "The Use of Transformations," *Biometrics*, Vol. 3 (1947), pp. 39–52.

14.22. Box, G. E. P., and D. R. Cox, "An Analysis of Transformations," *Royal Statistical Society, Series B*, Vol. 26 (1964), pp. 211–243.

14.23. Dolby, J. L., "A Quick Method for Choosing a Transformation," *Technometrics*, Vol. 5 (1963), pp. 317–326.

14.24. Draper, N. R., and W. G. Hunter, "Transformations: Some Examples Revisited," *Technometrics*, Vol. 11 (1969), pp. 23–40.

14.25. Kruskal, J. B., "Analysis of Factorial Experiments by Estimating Monotone Transformations of the Data," *Royal Statistical Society, Series B*, Vol. 27 (1965), pp. 251–263.

14.26. Mueller, C. G., "Numerical Transformations in the Analysis of Experimental Data," *Psych. Bulletin*, Vol. 46 (1949), pp. 198–223.

14.27. Tukey, J. W., "On the Comparative Anatomy of Transformations," *Annals of Mathematical Statistics*, Vol. 28 (1957), pp. 602–632.

FAILURE OF ASSUMPTIONS IN REGRESSION ANALYSIS

14.28. Draper, N. R., and H. Smith, *Applied Regression Analysis*, Wiley, New York, 1966.

14.29. Hald, A., *Statistical Theory with Engineering Applications*, Wiley, New York, 1952.

14.30. Kowalski, C. J., "The Performance of Some Rough Tests for Bivariate Normality Before and After Coordinate Transformations to Normality," *Technometrics*, Vol. 12 (1970), pp. 517–544.

APPENDIX

TABLE A.1. THE STANDARD NORMAL DISTRIBUTION*

z[λ]	λ	z[λ]	λ	z[λ]	λ	z[λ]	λ
0.00	0.5000						
0.01	0.5040	0.51	0.6950	1.01	0.8438	1.51	0.9345
0.02	0.5080	0.52	0.6985	1.02	0.8461	1.52	0.9357
0.03	0.5120	0.53	0.7019	1.03	0.8485	1.53	0.9370
0.04	0.5160	0.54	0.7054	1.04	0.8508	1.54	0.9382
0.05	0.5199	0.55	0.7088	1.05	0.8531	1.55	0.9394
0.06	0.5239	0.56	0.7123	1.06	0.8554	1.56	0.9406
0.07	0.5279	0.57	0.7157	1.07	0.8577	1.57	0.9418
0.08	0.5319	0.58	0.7190	1.08	0.8599	1.58	0.9429
0.09	0.5359	0.59	0.7224	1.09	0.8621	1.59	0.9441
0.10	0.5398	0.60	0.7257	1.10	0.8643	1.60	0.9452
0.11	0.5438	0.61	0.7291	1.11	0.8665	1.61	0.9463
0.12	0.5478	0.62	0.7324	1.12	0.8686	1.62	0.9474
0.13	0.5517	0.63	0.7357	1.13	0.8708	1.63	0.9484
0.14	0.5557	0.64	0.7389	1.14	0.8729	1.64	0.9495
0.15	0.5596	0.65	0.7422	1.15	0.8749	1.65	0.9505
0.16	0.5636	0.66	0.7454	1.16	0.8770	1.66	0.9515
0.17	0.5675	0.67	0.7486	1.17	0.8790	1.67	0.9525
0.18	0.5714	0.68	0.7517	1.18	0.8810	1.68	0.9535
0.19	0.5753	0.69	0.7549	1.19	0.8830	1.69	0.9545
0.20	0.5793	0.70	0.7580	1.20	0.8849	1.70	0.9554
0.21	0.5832	0.71	0.7611	1.21	0.8869	1.71	0.9564
0.22	0.5871	0.72	0.7642	1.22	0.8888	1.72	0.9573
0.23	0.5910	0.73	0.7673	1.23	0.8907	1.73	0.9582
0.24	0.5948	0.74	0.7704	1.24	0.8925	1.74	0.9591
0.25	0.5987	0.75	0.7734	1.25	0.8944	1.75	0.9599
0.26	0.6026	0.76	0.7764	1.26	0.8962	1.76	0.9608
0.27	0.6064	0.77	0.7794	1.27	0.8980	1.77	0.9616
0.28	0.6103	0.78	0.7823	1.28	0.8997	1.78	0.9625
0.29	0.6141	0.79	0.7852	1.29	0.9015	1.79	0.9633
0.30	0.6179	0.80	0.7881	1.30	0.9032	1.80	0.9641
0.31	0.6217	0.81	0.7910	1.31	0.9049	1.81	0.9649
0.32	0.6255	0.82	0.7939	1.32	0.9066	1.82	0.9656
0.33	0.6293	0.83	0.7967	1.33	0.9082	1.83	0.9664
0.34	0.6331	0.84	0.7995	1.34	0.9099	1.84	0.9671
0.35	0.6368	0.85	0.8023	1.35	0.9115	1.85	0.9678
0.36	0.6406	0.86	0.8051	1.36	0.9131	1.86	0.9686
0.37	0.6443	0.87	0.8078	1.37	0.9147	1.87	0.9693
0.38	0.6480	0.88	0.8106	1.38	0.9162	1.88	0.9699
0.39	0.6517	0.89	0.8133	1.39	0.9177	1.89	0.9706
0.40	0.6554	0.90	0.8159	1.40	0.9192	1.90	0.9713
0.41	0.6591	0.91	0.8186	1.41	0.9207	1.91	0.9719
0.42	0.6628	0.92	0.8212	1.42	0.9222	1.92	0.9726
0.43	0.6664	0.93	0.8238	1.43	0.9236	1.93	0.9732
0.44	0.6700	0.94	0.8264	1.44	0.9251	1.94	0.9738
0.45	0.6736	0.95	0.8289	1.45	0.9265	1.95	0.9744
0.46	0.6772	0.96	0.8315	1.46	0.9279	1.96	0.9750
0.47	0.6808	0.97	0.8340	1.47	0.9292	1.97	0.9756
0.48	0.6844	0.98	0.8365	1.48	0.9306	1.98	0.9761
0.49	0.6879	0.99	0.8389	1.49	0.9319	1.99	0.9767
0.50	0.6915	1.00	0.8413	1.50	0.9332	2.00	0.9772

TABLE A.1. (CONTINUED)

z[λ]	λ	z[λ]	λ	z[λ]	λ	z[λ]	λ
2.01	0.9778	2.51	0.9940	3.01	0.9987	3.51	0.9998
2.02	0.9783	2.52	0.9941	3.02	0.9987	3.52	0.9998
2.03	0.9788	2.53	0.9943	3.03	0.9988	3.53	0.9998
2.04	0.9793	2.54	0.9945	3.04	0.9988	3.54	0.9998
2.05	0.9798	2.55	0.9946	3.05	0.9989	3.55	0.9998
2.06	0.9803	2.56	0.9948	3.06	0.9989	3.56	0.9998
2.07	0.9808	2.57	0.9949	3.07	0.9989	3.57	0.9998
2.08	0.9812	2.58	0.9951	3.08	0.9990	3.58	0.9998
2.09	0.9817	2.59	0.9952	3.09	0.9990	3.59	0.9998
2.10	0.9821	2.60	0.9953	3.10	0.9990	3.60	0.9998
2.11	0.9826	2.61	0.9955	3.11	0.9991	3.61	0.9998
2.12	0.9830	2.62	0.9956	3.12	0.9991	3.62	0.9999
2.13	0.9834	2.63	0.9957	3.13	0.9991	3.63	0.9999
2.14	0.9838	2.64	0.9959	3.14	0.9992	3.64	0.9999
2.15	0.9842	2.65	0.9960	3.15	0.9992	3.65	0.9999
2.16	0.9846	2.66	0.9961	3.16	0.9992	3.66	0.9999
2.17	0.9850	2.67	0.9962	3.17	0.9992	3.67	0.9999
2.18	0.9854	2.68	0.9963	3.18	0.9993	3.68	0.9999
2.19	0.9857	2.69	0.9964	3.19	0.9993	3.69	0.9999
2.20	0.9861	2.70	0.9965	3.20	0.9993	3.70	0.9999
2.21	0.9864	2.71	0.9966	3.21	0.9993	3.71	0.9999
2.22	0.9868	2.72	0.9967	3.22	0.9994	3.72	0.9999
2.23	0.9871	2.73	0.9968	3.23	0.9994	3.73	0.9999
2.24	0.9875	2.74	0.9969	3.24	0.9994	3.74	0.9999
2.25	0.9878	2.75	0.9970	3.25	0.9994	3.75	0.9999
2.26	0.9881	2.76	0.9971	3.26	0.9994	3.76	0.9999
2.27	0.9884	2.77	0.9972	3.27	0.9995	3.77	0.9999
2.28	0.9887	2.78	0.9973	3.28	0.9995	3.78	0.9999
2.29	0.9890	2.79	0.9974	3.29	0.9995	3.79	0.9999
2.30	0.9893	2.80	0.9974	3.30	0.9995	3.80	0.9999
2.31	0.9896	2.81	0.9975	3.31	0.9995	3.81	0.9999
2.32	0.9898	2.82	0.9976	3.32	0.9996	3.82	0.9999
2.33	0.9901	2.83	0.9977	3.33	0.9996	3.83	0.9999
2.34	0.9904	2.84	0.9977	3.34	0.9996	3.84	0.9999
2.35	0.9906	2.85	0.9978	3.35	0.9996	3.85	0.9999
2.36	0.9909	2.86	0.9979	3.36	0.9996	3.86	0.9999
2.37	0.9911	2.87	0.9979	3.37	0.9996	3.87	0.9999
2.38	0.9913	2.88	0.9980	3.38	0.9996	3.88	0.9999
2.39	0.9916	2.89	0.9981	3.39	0.9997	3.89	1.0000
2.40	0.9918	2.90	0.9981	3.40	0.9997	3.90	1.0000
2.41	0.9920	2.91	0.9982	3.41	0.9997	3.91	1.0000
2.42	0.9922	2.92	0.9982	3.42	0.9997	3.92	1.0000
2.43	0.9925	2.93	0.9983	3.43	0.9997	3.93	1.0000
2.44	0.9927	2.94	0.9984	3.44	0.9997	3.94	1.0000
2.45	0.9929	2.95	0.9984	3.45	0.9997	3.95	1.0000
2.46	0.9931	2.96	0.9985	3.46	0.9997	3.96	1.0000
2.47	0.9932	2.97	0.9985	3.47	0.9997	3.97	1.0000
2.48	0.9934	2.98	0.9986	3.48	0.9997	3.98	1.0000
2.49	0.9936	2.99	0.9986	3.49	0.9998	3.99	1.0000
2.50	0.9938	3.00	0.9986	3.50	0.9998		

TABLE A.1. (CONTINUED)

$^{*}\lambda$ = Area under curve from $-\infty$ to $z[\lambda]$.

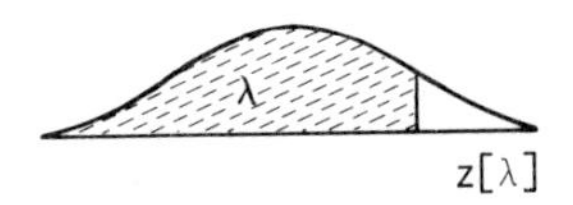

$z(.50) = 0$, and for values of λ less than .50, $z(\lambda)$ is found by symmetry, i.e., $z(\lambda) = -z(1-\lambda)$.

Data in the table are extracted from Owen, D.B., HANDBOOK OF STATISTICAL TABLES, 1962, Addison-Wesley, Reading, Mass. Courtesy of the U.S. Atomic Energy Commission.

TABLE A.2. PERCENTAGE POINTS $t[\lambda; \nu]$ OF STUDENT'S t DISTRIBUTION*

ν \ λ	.75	.90	.95	.975	.99	.995	.9975	.999
1	1.000	3.078	6.314	12.706	31.821	63.657	127.321	318.309
2	.816	1.886	2.920	4.303	6.965	9.925	14.089	22.327
3	.765	1.638	2.353	3.182	4.541	5.841	7.453	10.214
4	.741	1.533	2.132	2.776	3.747	4.604	5.598	7.173
5	.727	1.476	2.015	2.571	3.365	4.032	4.773	5.893
6	.718	1.440	1.943	2.447	3.143	3.707	4.317	5.208
7	.711	1.415	1.895	2.365	2.998	3.499	4.029	4.785
8	.706	1.397	1.860	2.306	2.896	3.355	3.833	4.501
9	.703	1.383	1.833	2.262	2.821	3.250	3.690	4.297
10	.700	1.372	1.812	2.228	2.764	3.169	3.581	4.144
11	.697	1.363	1.796	2.201	2.718	3.106	3.497	4.025
12	.695	1.356	1.782	2.179	2.681	3.055	3.428	3.930
13	.694	1.350	1.771	2.160	2.650	3.012	3.372	3.852
14	.692	1.345	1.761	2.145	2.624	2.977	3.326	3.787
15	.691	1.341	1.753	2.131	2.602	2.947	3.286	3.733
16	.690	1.337	1.746	2.120	2.583	2.921	3.252	3.686
17	.689	1.333	1.740	2.110	2.567	2.898	3.223	3.646
18	.688	1.330	1.734	2.101	2.552	2.878	3.197	3.610
19	.688	1.328	1.729	2.093	2.539	2.861	3.174	3.579
20	.687	1.325	1.725	2.086	2.528	2.845	3.153	3.552
21	.686	1.323	1.721	2.080	2.518	2.831	3.135	3.527
22	.686	1.321	1.717	2.074	2.508	2.819	3.119	3.505
23	.685	1.319	1.714	2.069	2.500	2.807	3.104	3.485
24	.685	1.318	1.711	2.064	2.492	2.797	3.090	3.467
25	.684	1.316	1.708	2.060	2.485	2.787	3.078	3.450
26	.684	1.315	1.706	2.056	2.479	2.779	3.067	3.435
27	.684	1.314	1.703	2.052	2.473	2.771	3.057	3.421
28	.683	1.313	1.701	2.048	2.467	2.763	3.047	3.408
29	.683	1.311	1.699	2.045	2.462	2.756	3.038	3.396
30	.683	1.310	1.697	2.042	2.457	2.750	3.030	3.385
35	.682	1.306	1.690	2.030	2.438	2.724	2.996	3.340
40	.681	1.303	1.684	2.021	2.423	2.704	2.971	3.307
45	.680	1.301	1.679	2.014	2.412	2.690	2.952	3.281
50	.679	1.299	1.676	2.009	2.403	2.678	2.937	3.261
55	.679	1.297	1.673	2.004	2.396	2.668	2.925	3.245
60	.679	1.296	1.671	2.000	2.390	2.660	2.915	3.232
70	.678	1.294	1.667	1.994	2.381	2.648	2.899	3.211
80	.678	1.292	1.664	1.990	2.374	2.639	2.887	3.195
90	.677	1.291	1.662	1.987	2.368	2.632	2.878	3.183
100	.677	1.290	1.660	1.984	2.364	2.626	2.871	3.174
120	.677	1.289	1.657	1.980	2.351	2.618	2.860	3.153
200	.676	1.286	1.652	1.972	2.345	2.601	2.838	3.131
500	.675	1.283	1.648	1.965	2.334	2.586	2.820	3.107
∞	.674	1.282	1.645	1.960	2.326	2.576	2.807	3.090

TABLE A.2. (CONTINUED)

ν \ λ	.9995	.99975	.9999	.99995	.999975	.99999
1	636.619	1,273.239	3,183.099	6,366.198	12,732.395	31,830.989
2	31.598	44.705	70.700	99.992	141.416	223.603
3	12.924	16.326	22.204	28.000	35.298	47.928
4	8.610	10.306	13.034	15.544	18.522	23.332
5	6.869	7.976	9.678	11.178	12.893	15.547
6	5.959	6.788	8.025	9.082	10.261	12.032
7	5.408	6.082	7.063	7.885	8.782	10.103
8	5.041	5.618	6.442	7.120	7.851	8.907
9	4.781	5.291	6.010	6.594	7.215	8.102
10	4.587	5.049	5.694	6.211	6.757	7.527
11	4.437	4.863	5.453	5.921	6.412	7.098
12	4.318	4.716	5.263	5.694	6.143	6.756
13	4.221	4.597	5.111	5.513	5.928	6.501
14	4.140	4.499	4.985	5.363	5.753	6.287
15	4.073	4.417	4.880	5.239	5.607	6.109
16	4.015	4.346	4.791	5.134	5.484	5.960
17	3.965	4.286	4.714	5.044	5.379	5.832
18	3.922	4.233	4.648	4.966	5.288	5.722
19	3.883	4.187	4.590	4.897	5.209	5.627
20	3.850	4.146	4.539	4.837	5.139	5.543
21	3.819	4.110	4.493	4.784	5.077	5.469
22	3.792	4.077	4.452	4.736	5.022	5.402
23	3.768	4.048	4.415	4.693	4.972	5.343
24	3.745	4.021	4.382	4.654	4.927	5.290
25	3.725	3.997	4.352	4.619	4.887	5.241
26	3.707	3.974	4.324	4.587	4.850	5.197
27	3.690	3.954	4.299	4.558	4.816	5.157
28	3.674	3.935	4.275	4.530	4.784	5.120
29	3.659	3.918	4.254	4.506	4.756	5.086
30	3.646	3.902	4.234	4.482	4.729	5.054
35	3.591	3.836	4.153	4.389	4.622	4.927
40	3.551	3.788	4.094	4.321	4.544	4.835
45	3.520	3.752	4.049	4.269	4.485	4.766
50	3.496	3.723	4.014	4.228	4.438	4.711
55	3.476	3.700	3.986	4.196	4.401	4.667
60	3.460	3.681	3.962	4.169	4.370	4.631
70	3.435	3.651	3.926	4.127	4.323	4.576
80	3.416	3.629	3.899	4.096	4.288	4.535
90	3.402	3.612	3.878	4.072	4.261	4.503
100	3.390	3.598	3.862	4.053	4.240	4.478
120	3.373	3.579	3.838	4.025	4.209	4.442
200	3.340	3.539	3.789	3.970	4.146	4.369
500	3.310	3.504	3.747	3.922	4.091	4.306
∞	3.291	3.481	3.719	3.891	4.056	4.265

TABLE A.2. (CONTINUED)

*λ = Area under curve from $-\infty$ to $t[\lambda;\nu]$.

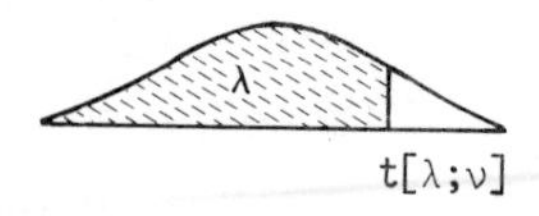

$t[.50;\nu] = 0$, and for values of λ less than .50, $t[\lambda;\nu]$ is found by symmetry, i.e., $t[\lambda;\nu] = -t[1-\lambda;\nu]$. Note also that $t(\lambda;\infty) = z(\lambda)$.

The data in this table are reprinted from E. T. Federighi, Extended Tables of the Percentage Points of Student's t-Distribution, JASA, Sept. 1959, with the kind permission of the author.

TABLE A.3. PERCENTAGE POINTS $\chi^2[\lambda;\nu]$ OF THE χ^2 DISTRIBUTION

ν \ λ	.005	.01	.025	.05	.10	.25
1	-	-	0.001	0.004	0.016	0.102
2	0.010	0.020	0.051	0.103	0.211	0.575
3	0.072	0.115	0.216	0.352	0.584	1.213
4	0.207	0.297	0.484	0.711	1.064	1.923
5	0.412	0.554	0.831	1.145	1.610	2.675
6	0.676	0.872	1.237	1.635	2.204	3.455
7	0.989	1.239	1.690	2.167	2.833	4.255
8	1.344	1.646	2.180	2.733	3.490	5.071
9	1.735	2.088	2.700	3.325	4.168	5.899
10	2.156	2.558	3.247	3.940	4.865	6.737
11	2.603	3.053	3.816	4.575	5.578	7.584
12	3.074	3.571	4.404	5.226	6.304	8.438
13	3.565	4.107	5.009	5.892	7.042	9.299
14	4.075	4.660	5.629	6.571	7.790	10.165
15	4.601	5.229	6.262	7.261	8.547	11.037
16	5.142	5.812	6.908	7.962	9.312	11.912
17	5.697	6.408	7.564	8.672	10.085	12.792
18	6.265	7.015	8.231	9.390	10.865	13.675
19	6.844	7.633	8.907	10.117	11.651	14.562
20	7.434	8.260	9.591	10.851	12.443	15.452
22	8.643	9.542	10.982	12.338	14.042	17.240
24	9.886	10.856	12.401	13.848	15.659	19.037
26	11.160	12.198	13.844	15.379	17.292	20.843
28	12.461	13.565	15.308	16.928	18.939	22.657
30	13.787	14.954	16.791	18.493	20.599	24.478
40	20.707	22.164	24.433	26.509	29.051	33.660
50	27.991	29.707	32.357	34.764	37.689	42.942
60	35.534	37.485	40.482	43.188	46.459	52.294
70	43.275	45.442	48.758	51.739	55.329	61.698
80	51.172	53.540	57.153	60.391	64.278	71.145
90	59.196	61.754	65.647	69.126	73.291	80.625
100	67.328	70.065	74.222	77.929	82.358	90.133
110	75.550	78.458	82.867	86.792	91.471	99.666
120	83.852	86.923	91.573	95.705	100.624	109.220
130	92.222	95.451	100.331	104.662	109.811	118.792
140	100.655	104.034	109.137	113.659	119.029	128.380
150	109.142	112.668	117.985	122.692	128.275	137.983
200	152.241	156.432	162.728	168.279	174.835	186.172
500	422.303	429.388	439.936	449.147	459.926	478.323

TABLE A.3. (CONTINUED)

ν \ λ	.75	.90	.95	.975	.99	.995
1	1.323	2.706	3.841	5.024	6.635	7.879
2	2.773	4.605	5.991	7.378	9.210	10.597
3	4.108	6.251	7.815	9.348	11.345	12.838
4	5.385	7.779	9.488	11.143	13.277	14.860
5	6.626	9.236	11.071	12.833	15.086	16.750
6	7.841	10.645	12.592	14.449	16.812	18.548
7	9.037	12.017	14.067	16.013	18.475	20.278
8	10.219	13.362	15.507	17.535	20.090	21.955
9	11.389	14.684	16.919	19.023	21.666	23.589
10	12.549	15.987	18.307	20.483	23.209	25.188
11	13.701	17.275	19.675	21.920	24.725	26.757
12	14.845	18.549	21.026	23.337	26.217	28.299
13	15.984	19.812	22.362	24.736	27.688	29.819
14	17.117	21.064	23.685	26.119	29.141	31.319
15	18.245	22.307	24.996	27.488	30.578	32.801
16	19.369	23.542	26.296	28.845	32.000	34.267
17	20.489	24.769	27.587	30.191	33.409	35.718
18	21.605	25.989	28.869	31.526	34.805	37.156
19	22.718	27.204	30.144	32.852	36.191	38.582
20	23.828	28.412	31.410	34.170	37.566	39.997
22	26.039	30.813	33.924	36.781	40.289	42.796
24	28.241	33.196	36.415	39.364	42.980	45.559
26	30.435	35.563	38.885	41.923	45.642	48.290
28	32.620	37.916	41.337	44.461	48.278	50.993
30	34.800	40.256	43.773	46.979	50.892	53.672
40	45.616	51.805	55.758	59.342	63.691	66.766
50	56.334	63.167	67.505	71.420	76.154	79.490
60	66.981	74.397	79.082	83.298	88.379	91.952
70	77.577	85.527	90.531	95.023	100.425	104.215
80	88.130	96.578	101.879	106.629	112.329	116.321
90	98.650	107.565	113.145	118.136	124.116	128.299
100	109.141	118.498	124.342	129.561	135.807	140.169
110	119.608	129.385	135.480	140.917	147.414	151.948
120	130.055	140.233	146.567	152.211	158.950	163.648
130	140.482	151.045	157.610	163.453	170.423	175.278
140	150.894	161.827	168.613	174.648	181.840	186.847
150	161.291	172.581	179.581	185.800	193.208	198.360
200	213.102	226.021	233.994	241.058	249.445	255.264
500	520.950	540.930	553.127	563.852	576.493	585.207

* λ = Area under curve from zero to $\chi^2[\lambda;\nu]$.

Data in the table are extracted from: Owen, D.B., HANDBOOK OF STATISTICAL TABLES, 1962, Addison-Wesley, Reading, Mass. Courtesy of the U.S. Atomic Energy Commission.

TABLE A.4. PERCENTAGE POINTS $F[\lambda;\nu_1,\nu_2]$ OF THE F DISTRIBUTION*

ν_2	Cum. Prop. \ ν_1	1	2	3	4	5	6	7	8	9	10	11	12
1	.025	$.0^{2}15$	.026	.057	.082	.100	.113	.124	.132	.139	.144	.149	.153
	.95	161	200	216	225	230	234	237	239	241	242	243	244
	.975	648	800	864	900	922	937	948	957	963	969	973	977
	.99	405^{1}	500^{1}	540^{1}	562^{1}	576^{1}	586^{1}	593^{1}	598^{1}	602^{1}	606^{1}	608^{1}	611^{1}
	.999	406^{3}	500^{3}	540^{3}	562^{3}	576^{3}	586^{3}	593^{3}	598^{3}	602^{3}	606^{3}	609^{3}	611^{3}
	.9995	162^{4}	200^{4}	216^{4}	225^{4}	231^{4}	234^{4}	237^{4}	239^{4}	241^{4}	242^{4}	243^{4}	244^{4}
2	.025	$.0^{2}13$	.026	.062	.094	.119	.138	.153	.165	.175	.183	.190	.196
	.95	18.5	19.0	19.2	19.2	19.3	19.3	19.4	19.4	19.4	19.4	19.4	19.4
	.975	38.5	39.0	39.2	39.2	39.3	39.3	39.4	39.4	39.4	39.4	39.4	39.4
	.99	98.5	99.0	99.2	99.2	99.3	99.3	99.4	99.4	99.4	99.4	99.4	99.4
	.999	998	999	999	999	999	999	999	999	999	999	999	999
	.9995	200^{1}	200^{1}	200^{1}	200^{1}	200^{1}	200^{1}	200^{1}	200^{1}	200^{1}	200^{1}	200^{1}	200^{1}
3	.025	$.0^{2}12$	.026	.065	.100	.129	.152	.170	.185	.197	.207	.216	.224
	.95	10.1	9.55	9.28	9.12	9.01	8.94	8.89	8.85	8.81	8.79	8.76	8.74
	.975	17.4	16.0	15.4	15.1	14.9	14.7	14.6	14.5	14.5	14.4	14.4	14.3
	.99	34.1	30.8	29.5	28.7	28.2	27.9	27.7	27.5	27.3	27.2	27.1	27.1
	.999	167	149	141	137	135	133	132	131	130	129	129	128
	.9995	266	237	225	218	214	211	209	208	207	206	204	204
4	.025	$.0^{2}11$	.026	.066	.104	.135	.161	.181	.198	.212	.224	.234	.243
	.95	7.71	6.94	6.59	6.39	6.26	6.16	6.09	6.04	6.00	5.96	5.94	5.91
	.975	12.2	10.6	9.98	9.60	9.36	9.20	9.07	8.98	8.90	8.84	8.79	8.75
	.99	21.2	18.0	16.7	16.0	15.5	15.2	15.0	14.8	14.7	14.5	14.4	14.4
	.999	74.1	61.2	56.2	53.4	51.7	50.5	49.7	49.0	48.5	48.0	47.7	47.4
	.9995	106	87.4	80.1	76.1	73.6	71.9	70.6	69.7	68.9	68.3	67.8	67.4
5	.025	$.0^{2}11$	.025	.067	.107	.140	.167	.189	.208	.223	.236	.248	.257
	.95	6.61	5.79	5.41	5.19	5.05	4.95	4.88	4.82	4.77	4.74	4.71	4.68
	.975	10.0	8.43	7.76	7.39	7.15	6.98	6.85	6.76	6.68	6.62	6.57	6.52
	.99	16.3	13.3	12.1	11.4	11.0	10.7	10.5	10.3	10.2	10.1	9.96	9.89
	.999	47.2	37.1	33.2	31.1	29.7	28.8	28.2	27.6	27.2	26.9	26.6	26.4
	.9995	63.6	49.8	44.4	41.5	39.7	38.5	37.6	36.9	36.4	35.9	35.6	35.2
6	.025	$.0^{2}11$	.025	.068	.109	.143	.172	.195	.215	.231	.246	.258	.268
	.95	5.99	5.14	4.76	4.53	4.39	4.28	4.21	4.15	4.10	4.06	4.03	4.00
	.975	8.81	7.26	6.60	6.23	5.99	5.82	5.70	5.60	5.52	5.46	5.41	5.37
	.99	13.7	10.9	9.78	9.15	8.75	8.47	8.26	8.10	7.98	7.87	7.79	7.72
	.999	35.5	27.0	23.7	21.9	20.8	20.0	19.5	19.0	18.7	18.4	18.2	18.0
	.9995	46.1	34.8	30.4	28.1	26.6	25.6	24.9	24.3	23.9	23.5	23.2	23.0
7	.025	$.0^{2}10$	.025	.068	.110	.146	.176	.200	.221	.238	.253	.266	.277
	.95	5.59	4.74	4.35	4.12	3.97	3.87	3.79	3.73	3.68	3.64	3.60	3.57
	.975	8.07	6.54	5.89	5.52	5.29	5.12	4.99	4.90	4.82	4.76	4.71	4.67
	.99	12.2	9.55	8.45	7.85	7.46	7.19	6.99	6.84	6.72	6.62	6.54	6.47
	.999	29.2	21.7	18.8	17.2	16.2	15.5	15.0	14.6	14.3	14.1	13.9	13.7
	.9995	37.0	27.2	23.5	21.4	20.2	19.3	18.7	18.2	17.8	17.5	17.2	17.0

TABLE A.4. (CONTINUED)

ν_2	Cum. Prop. \ ν_1	15	20	24	30	40	50	60	100	120	200	500	∞
1	.025	.161	.170	.175	.180	.184	.187	.189	.193	.194	.196	.198	.199
	.95	246	248	249	250	251	252	252	253	253	254	254	254
	.975	985	993	997	100^{1}	101^{1}	101^{1}	101^{1}	101^{1}	101^{1}	102^{1}	102^{1}	102^{1}
	.99	616^{1}	621^{1}	623^{1}	626^{1}	629^{1}	630^{1}	631^{1}	633^{1}	634^{1}	635^{1}	636^{1}	637^{1}
	.999	616^{3}	621^{3}	623^{3}	626^{3}	629^{3}	630^{3}	631^{3}	633^{3}	634^{3}	635^{3}	636^{3}	637^{3}
	.9995	246^{4}	248^{4}	249^{4}	250^{4}	251^{4}	252^{4}	252^{4}	253^{4}	253^{4}	253^{4}	254^{4}	254^{4}
2	.025	.210	.224	.232	.239	.247	.251	.255	.261	.263	.266	.269	.271
	.95	19.4	19.4	19.5	19.5	19.5	19.5	19.5	19.5	19.5	19.5	19.5	19.5
	.975	39.4	39.4	39.5	39.5	39.5	39.5	39.5	39.5	39.5	39.5	39.5	39.5
	.99	99.4	99.4	99.5	99.5	99.5	99.5	99.5	99.5	99.5	99.5	99.5	99.5
	.999	999	999	999	999	999	999	999	999	999	999	999	999
	.9995	200^{1}	200^{1}	200^{1}	200^{1}	200^{1}	200^{1}	200^{1}	200^{1}	200^{1}	200^{1}	200^{1}	200^{1}
3	.025	.241	.259	.269	.279	.289	.295	.299	.308	.310	.314	.318	.321
	.95	8.70	8.66	8.63	8.62	8.59	8.58	8.57	8.55	8.55	8.54	8.53	8.53
	.975	14.3	14.2	14.1	14.1	14.0	14.0	14.0	14.0	13.9	13.9	13.9	13.9
	.99	26.9	26.7	26.6	26.5	26.4	26.4	26.3	26.2	26.2	26.2	26.1	26.1
	.999	127	126	126	125	125	125	124	124	124	124	124	123
	.9995	203	201	200	199	199	198	198	197	197	197	196	196
4	.025	.263	.284	.296	.308	.320	.327	.332	.342	.346	.351	.356	.359
	.95	5.86	5.80	5.77	5.75	5.72	5.70	5.69	5.66	5.66	5.65	5.64	5.63
	.975	8.66	8.56	8.51	8.46	8.41	8.38	8.36	8.32	8.31	8.29	8.27	8.26
	.99	14.2	14.0	13.9	13.8	13.7	13.7	13.7	13.6	13.6	13.5	13.5	13.5
	.999	46.8	46.1	45.8	45.4	45.1	44.9	44.7	44.5	44.4	44.3	44.1	44.0
	.9995	66.5	65.5	65.1	64.6	64.1	63.8	63.6	63.2	63.1	62.9	62.7	62.6
5	.025	.280	.304	.317	.330	.344	.353	.359	.370	.374	.380	.386	.390
	.95	4.62	4.56	4.53	4.50	4.46	4.44	4.43	4.41	4.40	4.39	4.37	4.36
	.975	6.43	6.33	6.28	6.23	6.18	6.14	6.12	6.08	6.07	6.05	6.03	6.02
	.99	9.72	9.55	9.47	9.38	9.29	9.24	9.20	9.13	9.11	9.08	9.04	9.02
	.999	25.9	25.4	25.1	24.9	24.6	24.4	24.3	24.1	24.1	23.9	23.8	23.8
	.9995	34.6	33.9	33.5	33.1	32.7	32.5	32.3	32.1	32.0	31.8	31.7	31.6
6	.025	.293	.320	.334	.349	.364	.375	.381	.394	.398	.405	.412	.415
	.95	3.94	3.87	3.84	3.81	3.77	3.75	3.74	3.71	3.70	3.69	3.68	3.67
	.975	5.27	5.17	5.12	5.07	5.01	4.98	4.96	4.92	4.90	4.88	4.86	4.85
	.99	7.56	7.40	7.31	7.23	7.14	7.09	7.06	6.99	6.97	6.93	6.90	6.88
	.999	17.6	17.1	16.9	16.7	16.4	16.3	16.2	16.0	16.0	15.9	15.8	15.7
	.9995	22.4	21.9	21.7	21.4	21.1	20.9	20.7	20.5	20.4	20.3	20.2	20.1
7	.025	.304	.333	.348	.364	.381	.392	.399	.413	.418	.426	.433	.437
	.95	3.51	3.44	3.41	3.38	3.34	3.32	3.30	3.27	3.27	3.25	3.24	3.23
	.975	4.57	4.47	4.42	4.36	4.31	4.28	4.25	4.21	4.20	4.18	4.16	4.14
	.99	6.31	6.16	6.07	5.99	5.91	5.86	5.82	5.75	5.74	5.70	5.67	5.65
	.999	13.3	12.9	12.7	12.5	12.3	12.2	12.1	11.9	11.9	11.8	11.7	11.7
	.9995	16.5	16.0	15.7	15.5	15.2	15.1	15.0	14.7	14.7	14.6	14.5	14.4

TABLE A.4. (CONTINUED)

ν_2	ν_1 Cum. Prop.	1	2	3	4	5	6	7	8	9	10	11	12
8	.025	$.0^{2}10$	.025	.069	.111	.148	.179	.204	.226	.244	.259	.273	.285
	.95	5.32	4.46	4.07	3.84	3.69	3.58	3.50	3.44	3.39	3.35	3.31	3.28
	.975	7.57	6.06	5.42	5.05	4.82	4.65	4.53	4.43	4.36	4.30	4.24	4.20
	.99	11.3	8.65	7.59	7.01	6.63	6.37	6.18	6.03	5.91	5.81	5.73	5.67
	.999	25.4	18.5	15.8	14.4	13.5	12.9	12.4	12.0	11.8	11.5	11.4	11.2
	.9995	31.6	22.8	19.4	17.6	16.4	15.7	15.1	14.6	14.3	14.0	13.8	13.6
9	.025	$.0^{2}10$	.025	.069	.112	.150	.181	.207	.230	.248	.265	.279	.291
	.95	5.12	4.26	3.86	3.63	3.48	3.37	3.29	3.23	3.18	3.14	3.10	3.07
	.975	7.21	5.71	5.08	4.72	4.48	4.32	4.20	4.10	4.03	3.96	3.91	3.87
	.99	10.6	8.02	6.99	6.42	6.06	5.80	5.61	5.47	5.35	5.26	5.18	5.11
	.999	22.9	16.4	13.9	12.6	11.7	11.1	10.7	10.4	10.1	9.89	9.71	9.57
	.9995	28.0	19.9	16.8	15.1	14.1	13.3	12.8	12.4	12.1	11.8	11.6	11.4
10	.025	$0^{2}10$	.025	.069	.113	.151	.183	.210	.233	.252	.269	.283	.296
	.95	4.96	4.10	3.71	3.48	3.33	3.22	3.14	3.07	3.02	2.98	2.94	2.91
	.975	6.94	5.46	4.83	4.47	4.24	4.07	3.95	3.85	3.78	3.72	3.66	3.62
	.99	10.0	7.56	6.55	5.99	5.64	5.39	5.20	5.06	4.94	4.85	4.77	4.71
	.999	21.0	14.9	12.6	11.3	10.5	9.92	9.52	9.20	8.96	8.75	8.58	8.44
	.9995	25.5	17.9	15.0	13.4	12.4	11.8	11.3	10.9	10.6	10.3	10.1	9.93
11	.025	$.0^{2}10$	.025	.069	.114	.152	.185	.212	.236	.256	.273	.288	.301
	.95	4.84	3.98	3.59	3.36	3.20	3.09	3.01	2.95	2.90	2.85	2.82	2.79
	.975	6.72	5.26	4.63	4.28	4.04	3.88	3.76	3.66	3.59	3.53	3.47	3.43
	.99	9.65	7.21	6.22	5.67	5.32	5.07	4.89	4.74	4.63	4.54	4.46	4.40
	.999	19.7	13.8	11.6	10.3	9.58	9.05	8.66	8.35	8.12	7.92	7.76	7.62
	.9995	23.6	16.4	13.6	12.2	11.2	10.6	10.1	9.76	9.48	9.24	9.04	8.88
12	.025	$.0^{2}10$	.025	.070	.114	.153	.186	.214	.238	.259	.276	.292	.305
	.95	4.75	3.89	3.49	3.26	3.11	3.00	2.91	2.85	2.80	2.75	2.72	2.69
	.975	6.55	5.10	4.47	4.12	3.89	3.73	3.61	3.51	3.44	3.37	3.32	3.28
	.99	9.33	6.93	5.95	5.41	5.06	4.82	4.64	4.50	4.39	4.30	4.22	4.16
	.999	18.6	13.0	10.8	9.63	8.89	8.38	8.00	7.71	7.48	7.29	7.14	7.01
	.9995	22.2	15.3	12.7	11.2	10.4	9.74	9.28	8.94	8.66	8.43	8.24	8.08
15	.025	$.0^{2}10$	.025	.070	.116	.156	.190	.219	.244	.265	.284	.300	.315
	.95	4.54	3.68	3.29	3.06	2.90	2.79	2.71	2.64	2.59	2.54	2.51	2.48
	.975	6.20	4.76	4.15	3.80	3.58	3.41	3.29	3.20	3.12	3.06	3.01	2.96
	.99	8.68	6.36	5.42	4.89	4.56	4.32	4.14	4.00	3.89	3.80	3.73	3.67
	.999	16.6	11.3	9.34	8.25	7.57	7.09	6.74	6.47	6.26	6.08	5.93	5.81
	.9995	19.5	13.2	10.8	9.48	8.66	8.10	7.68	7.36	7.11	6.91	6.75	6.60
20	.025	$.0^{2}10$	.025	.071	.117	.158	.193	.224	.250	.273	.292	.310	.325
	.95	4.35	3.49	3.10	2.87	2.71	2.60	2.51	2.45	2.39	2.35	2.31	2.28
	.975	5.87	4.46	3.86	3.51	3.29	3.13	3.01	2.91	2.84	2.77	2.72	2.68
	.99	8.10	5.85	4.94	4.43	4.10	3.87	3.70	3.56	3.46	3.37	3.29	3.23
	.999	14.8	9.95	8.10	7.10	6.46	6.02	5.69	5.44	5.24	5.08	4.94	4.82
	.9995	17.2	11.4	9.20	8.02	7.28	6.76	6.38	6.08	5.85	5.66	5.51	5.38

TABLE A.4. (CONTINUED)

ν_2	Cum. Prop. \ ν_1	15	20	24	30	40	50	60	100	120	200	500	∞
8	.025	.313	.343	.360	.377	.395	.407	.415	.431	.435	.442	.450	.456
	.95	3.22	3.15	3.12	3.08	3.04	3.02	3.01	2.97	2.97	2.95	2.94	2.93
	.975	4.10	4.00	3.95	3.89	3.84	3.81	3.78	3.74	3.73	3.70	3.68	3.67
	.99	5.52	5.36	5.28	5.20	5.12	5.07	5.03	4.96	4.95	4.91	4.88	4.86
	.999	10.8	10.5	10.3	10.1	9.92	9.80	9.73	9.57	9.54	9.46	9.39	9.34
	.9995	13.1	12.7	12.5	12.2	12.0	11.8	11.8	11.6	11.5	11.4	11.4	11.3
9	.025	.320	.352	.370	.388	.408	.420	.428	.446	.450	.459	.467	.473
	.95	3.01	2.94	2.90	2.86	2.83	2.80	2.79	2.76	2.75	2.73	2.72	2.71
	.975	3.77	3.67	3.61	3.56	3.51	3.47	3.45	3.40	3.39	3.37	3.35	3.33
	.99	4.96	4.81	4.73	4.65	4.57	4.52	4.48	4.42	4.40	4.36	4.33	4.31
	.999	9.24	8.90	8.72	8.55	8.37	8.26	8.19	8.04	8.00	7.93	7.86	7.81
	.9995	11.0	10.6	10.4	10.2	9.94	9.80	9.71	9.53	9.49	9.40	9.32	9.26
10	.025	.327	.360	.379	.398	.419	.431	.441	.459	.464	.474	.483	.488
	.95	2.85	2.77	2.74	2.70	2.66	2.64	2.62	2.59	2.58	2.56	2.55	2.54
	.975	3.52	3.42	3.37	3.31	3.26	3.22	3.20	3.15	3.14	3.12	3.09	3.08
	.99	4.56	4.41	4.33	4.25	4.17	4.12	4.08	4.01	4.00	3.96	3.93	3.91
	.999	8.13	7.80	7.64	7.47	7.30	7.19	7.12	6.98	6.94	6.87	6.81	6.76
	.9995	9.56	9.16	8.96	8.75	8.54	8.42	8.33	8.16	8.12	8.04	7.96	7.90
11	.025	.332	.368	.386	.407	.429	.442	.450	.472	.476	.485	.495	.503
	.95	2.72	2.65	2.61	2.57	2.53	2.51	2.49	2.46	2.45	2.43	2.42	2.40
	.975	3.33	3.23	3.17	3.12	3.06	3.03	3.00	2.96	2.94	2.92	2.90	2.88
	.99	4.25	4.10	4.02	3.94	3.86	3.81	3.78	3.71	3.69	3.66	3.62	3.60
	.999	7.32	7.01	6.85	6.68	6.52	6.41	6.35	6.21	6.17	6.10	6.04	6.00
	.9995	8.52	8.14	7.94	7.75	7.55	7.43	7.35	7.18	7.14	7.06	6.98	6.93
12	.025	.337	.374	.394	.416	.437	.450	.461	.481	.487	.498	.508	.514
	.95	2.62	2.54	2.51	2.47	2.43	2.40	2.38	2.35	2.34	2.32	2.31	2.30
	.975	3.18	3.07	3.02	2.96	2.91	2.87	2.85	2.80	2.79	2.76	2.74	2.72
	.99	4.01	3.86	3.78	3.70	3.62	3.57	3.54	3.47	3.45	3.41	3.38	3.36
	.999	6.71	6.40	6.25	6.09	5.93	5.83	5.76	5.63	5.59	5.52	5.46	5.42
	.9995	7.74	7.37	7.18	7.00	6.80	6.68	6.61	6.45	6.41	6.33	6.25	6.20
15	.025	.349	.389	.410	.433	.458	.474	.485	.508	.514	.526	.538	.546
	.95	2.40	2.33	2.39	2.25	2.20	2.18	2.16	2.12	2.11	2.10	2.08	2.07
	.975	2.86	2.76	2.70	2.64	2.59	2.55	2.52	2.47	2.46	2.44	2.41	2.40
	.99	3.52	3.37	3.29	3.21	3.13	3.08	3.05	2.98	2.96	2.92	2.89	2.87
	.999	5.54	5.25	5.10	4.95	4.80	4.70	4.64	4.51	4.47	4.41	4.35	4.31
	.9995	6.27	5.93	5.75	5.58	5.40	5.29	5.21	5.06	5.02	4.94	4.87	4.83
20	.025	.363	.406	.430	.456	.484	.503	.514	.541	.548	.562	.575	.585
	.95	2.20	2.12	2.08	2.04	1.99	1.97	1.95	1.91	1.90	1.88	1.86	1.84
	.975	2.57	2.46	2.41	2.35	2.29	2.25	2.22	2.17	2.16	2.13	2.10	2.09
	.99	3.09	2.94	2.86	2.78	2.69	2.64	2.61	2.54	2.52	2.48	2.44	2.42
	.999	4.56	4.29	4.15	4.01	3.86	3.77	3.70	3.58	3.54	3.48	3.42	3.38
	.9995	5.07	4.75	4.58	4.42	4.24	4.15	4.07	3.93	3.90	3.82	3.75	3.70

TABLE A.4. (CONTINUED)

ν_2	Cum. Prop. \ ν_1	1	2	3	4	5	6	7	8	9	10	11	12
24	.025	$.0^{2}10$	.025	.071	.117	.159	.195	.227	.253	.277	.297	.315	.331
	.95	4.26	3.40	3.01	2.78	2.62	2.51	2.42	2.36	2.30	2.25	2.21	2.18
	.975	5.72	4.32	3.72	3.38	3.15	2.99	2.87	2.78	2.70	2.64	2.59	2.54
	.99	7.82	5.61	4.72	4.22	3.90	3.67	3.50	3.36	3.26	3.17	3.09	3.03
	.999	14.0	9.34	7.55	6.59	5.98	5.55	5.23	4.99	4.80	4.64	4.50	4.39
	.9995	16.2	10.6	8.52	7.39	6.68	6.18	5.82	5.54	5.31	5.13	4.98	4.85
30	.025	$.0^{2}10$	.025	.071	.118	.161	.197	.229	.257	.281	.302	.321	.337
	.95	4.17	3.32	2.92	2.69	2.53	2.42	2.33	2.27	2.21	2.16	2.13	2.09
	.975	5.57	4.18	3.59	3.25	3.03	2.87	2.75	2.65	2.57	2.51	2.46	2.41
	.99	7.56	5.39	4.51	4.02	3.70	3.47	3.30	3.17	3.07	2.98	2.91	2.84
	.999	13.3	8.77	7.05	6.12	5.53	5.12	4.82	4.58	4.39	4.24	4.11	4.00
	.9995	15.2	9.90	7.90	6.82	6.14	5.66	5.31	5.04	4.82	4.65	4.51	4.38
40	.025	$.0^{3}99$	.025	.071	.119	.162	.199	.232	.260	.285	.307	.327	.344
	.95	4.08	3.23	2.84	2.61	2.45	2.34	2.25	2.18	2.12	2.08	2.04	2.00
	.975	5.42	4.05	3.46	3.13	2.90	2.74	2.62	2.53	2.45	2.39	2.33	2.29
	.99	7.31	5.18	4.31	3.83	3.51	3.29	3.12	2.99	2.89	2.80	2.73	2.66
	.999	12.6	8.25	6.60	5.70	5.13	4.73	4.44	4.21	4.02	3.87	3.75	3.64
	.9995	14.4	9.25	7.33	6.30	5.64	5.19	4.85	4.59	4.38	4.21	4.07	3.95
60	.025	$.0^{3}99$	.025	.071	.120	.163	.202	.235	.264	.290	.313	.333	.351
	.95	4.00	3.15	2.76	2.53	2.37	2.25	2.17	2.10	2.04	1.99	1.95	1.92
	.975	5.29	3.93	3.34	3.01	2.79	2.63	2.51	2.41	2.33	2.27	2.22	2.17
	.99	7.08	4.98	4.13	3.65	3.34	3.12	2.95	2.82	2.72	2.63	2.56	2.50
	.999	12.0	7.76	6.17	5.31	4.76	4.37	4.09	3.87	3.69	3.54	3.43	3.31
	.9995	13.6	8.65	6.81	5.82	5.20	4.76	4.44	4.18	3.98	3.82	3.69	3.57
120	.025	$.0^{3}99$	.025	.072	.120	.165	.204	.238	.268	.295	.318	.340	.359
	.95	3.92	3.07	2.68	2.45	2.29	2.18	2.09	2.02	1.96	1.91	1.87	1.83
	.975	5.15	3.80	3.23	2.89	2.67	2.52	2.39	2.30	2.22	2.16	2.10	2.05
	.99	6.85	4.79	3.95	3.48	3.17	2.96	2.79	2.66	2.56	2.47	2.40	2.34
	.999	11.4	7.32	5.79	4.95	4.42	4.04	3.77	3.55	3.38	3.24	3.12	3.02
	.9995	12.8	8.10	6.34	5.39	4.79	4.37	4.07	3.82	3.63	3.47	3.34	3.22
∞	.025	$.0^{3}98$	.025	.072	.121	.166	.206	.241	.272	.300	.325	.347	.367
	.95	3.84	3.00	2.60	2.37	2.21	2.10	2.01	1.94	1.88	1.83	1.79	1.75
	.975	5.02	3.69	3.12	2.79	2.57	2.41	2.29	2.19	2.11	2.05	1.99	1.94
	.99	6.63	4.61	3.78	3.32	3.02	2.80	2.64	2.51	2.41	2.32	2.25	2.18
	.999	10.8	6.91	5.42	4.62	4.10	3.74	3.47	3.27	3.10	2.96	2.84	2.74
	.9995	12.1	7.60	5.91	5.00	4.42	4.02	3.72	3.48	3.30	3.14	3.02	2.90

TABLE A.4. (CONTINUED)

ν_2	ν_1 / Cum. Prop.	15	20	24	30	40	50	60	100	120	200	500	∞
24	.025	.370	.415	.441	.468	.498	.518	.531	.562	.568	.585	.599	.610
	.95	2.11	2.03	1.98	1.94	1.89	1.86	1.84	1.80	1.79	1.77	1.75	1.73
	.975	2.44	2.33	2.27	2.21	2.15	2.11	2.08	2.02	2.01	1.98	1.95	1.94
	.99	2.89	2.74	2.66	2.58	2.49	2.44	2.40	2.33	2.31	2.27	2.24	2.21
	.999	4.14	3.87	3.74	3.59	3.45	3.35	3.29	3.16	3.14	3.07	3.01	2.97
	.9995	4.55	4.25	4.09	3.93	3.76	3.66	3.59	3.44	3.41	3.33	3.27	3.22
30	.025	.378	.426	.453	.482	.515	.535	.551	.585	.592	.610	.625	.639
	.95	2.01	1.93	1.89	1.84	1.79	1.76	1.74	1.70	1.68	1.66	1.64	1.62
	.975	2.31	2.20	2.14	2.07	2.01	1.97	1.94	1.88	1.87	1.84	1.81	1.79
	.99	2.70	2.55	2.47	2.39	2.30	2.25	2.21	2.13	2.11	2.07	2.03	2.01
	.999	3.75	3.49	3.36	3.22	3.07	2.98	2.92	2.79	2.76	2.69	2.63	2.59
	.9995	4.10	3.80	3.65	3.48	3.32	3.22	3.15	3.00	2.97	2.89	2.82	2.78
40	.025	.387	.437	.466	.498	.533	.556	.573	.610	.620	.641	.662	.674
	.95	1.92	1.84	1.79	1.74	1.69	1.66	1.64	1.59	1.58	1.55	1.53	1.51
	.975	2.18	2.07	2.01	1.94	1.88	1.83	1.80	1.74	1.72	1.69	1.66	1.64
	.99	2.52	2.37	2.29	2.20	2.11	2.06	2.02	1.94	1.92	1.87	1.83	1.80
	.999	3.40	3.15	3.01	2.87	2.73	2.64	2.57	2.44	2.41	2.34	2.28	2.23
	.9995	3.68	3.39	3.24	3.08	2.92	2.82	2.74	2.60	2.57	2.49	2.41	2.37
60	.025	.396	.450	.481	.515	.555	.581	.600	.641	.654	.680	.704	.720
	.95	1.84	1.75	1.70	1.65	1.59	1.56	1.53	1.48	1.47	1.44	1.41	1.39
	.975	2.06	1.94	1.88	1.82	1.74	1.70	1.67	1.60	1.58	1.54	1.51	1.48
	.99	2.35	2.20	2.12	2.03	1.94	1.88	1.84	1.75	1.73	1.68	1.63	1.60
	.999	3.08	2.83	2.69	2.56	2.41	2.31	2.25	2.11	2.09	2.01	1.93	1.89
	.9995	3.30	3.02	2.87	2.71	2.55	2.45	2.38	2.23	2.19	2.11	2.03	1.98
120	.025	.406	.464	.498	.536	.580	.611	.633	.684	.698	.729	.762	.789
	.95	1.75	1.66	1.61	1.55	1.50	1.46	1.43	1.37	1.35	1.32	1.28	1.25
	.975	1.95	1.82	1.76	1.69	1.61	1.56	1.53	1.45	1.43	1.39	1.34	1.31
	.99	2.19	2.03	1.95	1.86	1.76	1.70	1.66	1.56	1.53	1.48	1.42	1.38
	.999	2.78	2.53	2.40	2.26	2.11	2.02	1.95	1.82	1.76	1.70	1.62	1.54
	.9995	2.96	2.67	2.53	2.38	2.21	2.11	2.01	1.88	1.84	1.75	1.67	1.60
∞	.025	.418	.480	.517	.560	.611	.645	.675	.741	.763	.813	.878	1.00
	.95	1.67	1.57	1.52	1.46	1.39	1.35	1.32	1.24	1.22	1.17	1.11	1.00
	.975	1.83	1.71	1.64	1.57	1.48	1.43	1.39	1.30	1.27	1.21	1.13	1.00
	.99	2.04	1.88	1.79	1.70	1.59	1.52	1.47	1.36	1.32	1.25	1.15	1.00
	.999	2.51	2.27	2.13	1.99	1.84	1.73	1.66	1.49	1.45	1.34	1.21	1.00
	.9995	2.65	2.37	2.22	2.07	1.91	1.79	1.71	1.53	1.48	1.36	1.22	1.00

*λ = Area under curve from zero to $F[\lambda;\nu_1,\nu_2]$.

To obtain values of $F(.05;\nu_1,\nu_2)$ and $F(.01;\nu_1,\nu_2)$ one uses $F(\lambda;\nu_1,\nu_2) = 1/F(1-\lambda;\nu_2,\nu_1)$.

Notation: $593^3 = 593 \times 10^3$

$.0^2 11 = .11 \times 10^{-2}$

The data in this table are extracted from Dixon and Massey, Introduction to Statistical Analysis, 3rd Ed., McGraw-Hill, 1969. Used with permission of McGraw-Hill Book Company.

TABLE A.5. PERCENTILES OF THE STUDENTIZED RANGE

ν	cum. prop. \ k	2	3	4	5	6	7	8	9	10	11	12	13	14	15	16	17	18	19	20
1	.95	18.0	27.0	32.8	37.1	40.4	43.1	45.4	47.4	49.1	50.6	52.0	53.2	54.3	55.4	56.3	57.2	58.0	58.8	59.6
	.99	90.0	135	164	186	202	216	227	237	246	253	260	266	272	277	282	286	290	294	298
2	.95	6.09	8.3	9.8	10.9	11.7	12.4	13.0	13.5	14.0	14.4	14.7	15.1	15.4	15.7	15.9	16.1	16.4	16.6	16.8
	.99	14.0	19.0	22.3	24.7	26.6	28.2	29.5	30.7	31.7	32.6	33.4	34.1	34.8	35.4	36.0	36.5	37.0	37.5	37.9
3	.95	4.50	5.91	6.82	7.50	8.04	8.48	8.85	9.18	9.46	9.72	9.95	10.2	10.4	10.5	10.7	10.8	11.0	11.1	11.2
	.99	8.26	10.6	12.2	13.3	14.2	15.0	15.6	16.2	16.7	17.1	17.5	17.9	18.2	18.5	18.8	19.1	19.3	19.5	19.8
4	.95	3.93	5.04	5.76	6.29	6.71	7.05	7.35	7.60	7.83	8.03	8.21	8.37	8.52	8.66	8.79	8.91	9.03	9.13	9.23
	.99	6.51	8.12	9.17	9.96	10.6	11.1	11.5	11.9	12.3	12.6	12.8	13.1	13.3	13.5	13.7	13.9	14.1	14.2	14.4
5	.95	3.64	4.60	5.22	5.67	6.03	6.33	6.58	6.80	6.99	7.17	7.32	7.47	7.60	7.72	7.83	7.93	8.03	8.12	8.21
	.99	5.70	6.97	7.80	8.42	8.91	9.32	9.67	9.97	10.2	10.5	10.7	10.9	11.1	11.2	11.4	11.6	11.7	11.8	11.9
6	.95	3.46	4.34	4.90	5.31	5.63	5.89	6.12	6.32	6.49	6.65	6.79	6.92	7.03	7.14	7.24	7.34	7.43	7.51	7.59
	.99	5.24	6.33	7.03	7.56	7.97	8.32	8.61	8.87	9.10	9.30	9.49	9.65	9.81	9.95	10.1	10.2	10.3	10.4	10.5
7	.95	3.34	4.16	4.68	5.06	5.36	5.61	5.82	6.00	6.16	6.30	6.43	6.55	6.66	6.76	6.85	6.94	7.02	7.09	7.17
	.99	4.95	5.92	6.54	7.01	7.37	7.68	7.94	8.17	8.37	8.55	8.71	8.86	9.00	9.12	9.24	9.35	9.46	9.55	9.65
8	.95	3.26	4.04	4.53	4.89	5.17	5.40	5.60	5.77	5.92	6.05	6.18	6.29	6.39	6.48	6.57	6.65	6.73	6.80	6.87
	.99	4.74	5.63	6.20	6.63	6.96	7.24	7.47	7.68	7.87	8.03	8.18	8.31	8.44	8.55	8.66	8.76	8.85	8.94	9.03
9	.95	3.20	3.95	4.42	4.76	5.02	5.24	5.43	5.60	5.74	5.87	5.98	6.09	6.19	6.28	6.36	6.44	6.51	6.58	6.64
	.99	4.60	5.43	5.96	6.35	6.66	6.91	7.13	7.32	7.49	7.65	7.78	7.91	8.03	8.13	8.23	8.32	8.41	8.49	8.57
10	.95	3.15	3.88	4.33	4.65	4.91	5.12	5.30	5.46	5.60	5.72	5.83	5.93	6.03	6.11	6.20	6.27	6.34	6.40	6.47
	.99	4.48	5.27	5.77	6.14	6.43	6.67	6.87	7.05	7.21	7.36	7.48	7.60	7.71	7.81	7.91	7.99	8.07	8.15	8.22
11	.95	3.11	3.82	4.26	4.57	4.82	5.03	5.20	5.35	5.49	5.61	5.71	5.81	5.90	5.99	6.06	6.14	6.20	6.26	6.33
	.99	4.39	5.14	5.62	5.97	6.25	6.48	6.67	6.84	6.99	7.13	7.25	7.36	7.46	7.56	7.65	7.73	7.81	7.88	7.95
12	.95	3.08	3.77	4.20	4.51	4.75	4.95	5.12	5.27	5.40	5.51	5.62	5.71	5.80	5.88	5.95	6.03	6.09	6.15	6.21
	.99	4.32	5.04	5.50	5.84	6.10	6.32	6.51	6.67	6.81	6.94	7.06	7.17	7.26	7.36	7.44	7.52	7.59	7.66	7.73
13	.95	3.06	3.73	4.15	4.45	4.69	4.88	5.05	5.19	5.32	5.43	5.53	5.63	5.71	5.79	5.86	5.93	6.00	6.05	6.11
	.99	4.26	4.96	5.40	5.73	5.98	6.19	6.37	6.53	6.67	6.79	6.90	7.01	7.10	7.19	7.27	7.34	7.42	7.48	7.55
14	.95	3.03	3.70	4.11	4.41	4.64	4.83	4.99	5.13	5.25	5.36	5.46	5.55	5.64	5.72	5.79	5.85	5.92	5.97	6.03
	.99	4.21	4.89	5.32	5.63	5.88	6.08	6.26	6.41	6.54	6.66	6.77	6.87	6.96	7.05	7.12	7.20	7.27	7.33	7.39

TABLE A.5. (CONTINUED)

ν	cum. prop. \ k	2	3	4	5	6	7	8	9	10	11	12	13	14	15	16	17	18	19	20
15	.95	3.01	3.67	4.08	4.37	4.60	4.78	4.94	5.08	5.20	5.31	5.40	5.49	5.58	5.65	5.72	5.79	5.85	5.90	5.96
	.99	4.17	4.83	5.25	5.56	5.80	5.99	6.16	6.31	6.44	6.55	6.66	6.76	6.84	6.93	7.00	7.07	7.14	7.20	7.26
16	.95	3.00	3.65	4.05	4.33	4.56	4.74	4.90	5.03	5.15	5.26	5.35	5.44	5.52	5.59	5.66	5.72	5.79	5.84	5.90
	.99	4.13	4.78	5.19	5.49	5.72	5.92	6.08	6.22	6.35	6.46	6.56	6.66	6.74	6.82	6.90	6.97	7.03	7.09	7.15
17	.95	2.98	3.63	4.02	4.30	4.52	4.71	4.86	4.99	5.11	5.21	5.31	5.39	5.47	5.55	5.61	5.68	5.74	5.79	5.84
	.99	4.10	4.74	5.14	5.43	5.66	5.85	6.01	6.15	6.27	6.38	6.48	6.57	6.66	6.73	6.80	6.87	6.94	7.00	7.05
18	.95	2.97	3.61	4.00	4.28	4.49	4.67	4.82	4.96	5.07	5.17	5.27	5.35	5.43	5.50	5.57	5.63	5.69	5.74	5.79
	.99	4.07	4.70	5.09	5.38	5.60	5.79	5.94	6.08	6.20	6.31	6.41	6.50	6.58	6.65	6.72	6.79	6.85	6.91	6.96
19	.95	2.96	3.59	3.98	4.25	4.47	4.65	4.79	4.92	5.04	5.14	5.23	5.32	5.39	5.46	5.53	5.59	5.65	5.70	5.75
	.99	4.05	4.67	5.05	5.33	5.55	5.73	5.89	6.02	6.14	6.25	6.34	6.43	6.51	6.58	6.65	6.72	6.78	6.84	6.89
20	.95	2.95	3.58	3.96	4.23	4.45	4.62	4.77	4.90	5.01	5.11	5.20	5.28	5.36	5.43	5.49	5.55	5.61	5.66	5.71
	.99	4.02	4.64	5.02	5.29	5.51	5.69	5.84	5.97	6.09	6.19	6.29	6.37	6.45	6.52	6.59	6.65	6.71	6.76	6.82
24	.95	2.92	3.53	3.90	4.17	4.37	4.54	4.68	4.81	4.92	5.01	5.10	5.18	5.25	5.32	5.38	5.44	5.50	5.54	5.59
	.99	3.96	4.54	4.91	5.17	5.37	5.54	5.69	5.81	5.92	6.02	6.11	6.19	6.26	6.33	6.39	6.45	6.51	6.56	6.61
30	.95	2.89	3.49	3.84	4.10	4.30	4.46	4.60	4.72	4.83	4.92	5.00	5.08	5.15	5.21	5.27	5.33	5.38	5.43	5.48
	.99	3.89	4.45	4.80	5.05	5.24	5.40	5.54	5.65	5.76	5.85	5.93	6.01	6.08	6.14	6.20	6.26	6.31	6.36	6.41
40	.95	2.86	3.44	3.79	4.04	4.23	4.39	4.52	4.63	4.74	4.82	4.91	4.98	5.05	5.11	5.16	5.22	5.27	5.31	5.36
	.99	3.82	4.37	4.70	4.93	5.11	5.27	5.39	5.50	5.60	5.69	5.77	5.84	5.90	5.96	6.02	6.07	6.12	6.17	6.21
60	.95	2.83	3.40	3.74	3.98	4.16	4.31	4.44	4.55	4.65	4.73	4.81	4.88	4.94	5.00	5.06	5.11	5.16	5.20	5.24
	.99	3.76	4.28	4.60	4.82	4.99	5.13	5.25	5.36	5.45	5.53	5.60	5.67	5.73	5.79	5.84	5.89	5.93	5.98	6.02
120	.95	2.80	3.36	3.69	3.92	4.10	4.24	4.36	4.48	4.56	4.64	4.72	4.78	4.84	4.90	4.95	5.00	5.05	5.09	5.13
	.99	3.70	4.20	4.50	4.71	4.87	5.01	5.12	5.21	5.30	5.38	5.44	5.51	5.56	5.61	5.66	5.71	5.75	5.79	5.83
∞	.95	2.77	3.31	3.63	3.86	4.03	4.17	4.29	4.39	4.47	4.55	4.62	4.68	4.74	4.80	4.85	4.89	4.93	4.97	5.01
	.99	3.64	4.12	4.40	4.60	4.76	4.88	4.99	5.08	5.16	5.23	5.29	5.35	5.40	5.45	5.49	5.54	5.57	5.61	5.65

Data in this table are extracted from Table 18 from Dixon & Massey, Introduction to Statistical Analysis, McGraw=Hill, 2nd, Ed., 1957 used with permission of McGraw-Hill Book Company.

TABLE A.6. VALUES OF $z = \frac{1}{2}\log_e(1+r)/(1-r)$

r	0.00	0.01	0.02	0.03	0.04	0.05	0.06	0.07	0.08	0.09
.0	0.000	0.010	0.020	0.030	0.040	0.050	0.060	0.070	0.080	0.090
.1	.100	.110	.121	.131	.141	.151	.161	.172	.182	.192
.2	.203	.213	.224	.234	.245	.255	.266	.277	.288	.299
.3	.310	.321	.332	.343	.354	.365	.377	.388	.400	.412
.4	.424	.436	.448	.460	.472	.485	.497	.510	.523	.536
.5	.549	.563	.576	.590	.604	.618	.633	.648	.662	.678
.6	.693	.709	.725	.741	.758	.775	.793	.811	.829	.848
.7	.867	.887	.908	.929	.950	.973	.996	1.020	1.045	1.071
.8	1.099	1.127	1.157	1.188	1.221	1.256	1.293	1.333	1.376	1.422
r	0.000	0.001	0.002	0.003	0.004	0.005	0.006	0.007	0.008	0.009
.90	1.472	1.478	1.483	1.488	1.494	1.499	1.505	1.510	1.516	1.522
.91	1.528	1.533	1.539	1.545	1.551	1.557	1.564	1.570	1.576	1.583
.92	1.589	1.596	1.602	1.609	1.616	1.623	1.630	1.637	1.644	1.651
.93	1.658	1.666	1.673	1.681	1.689	1.697	1.705	1.713	1.721	1.730
.94	1.738	1.747	1.756	1.764	1.774	1.783	1.792	1.802	1.812	1.822
.95	1.832	1.842	1.853	1.863	1.874	1.886	1.897	1.909	1.921	1.933
.96	1.946	1.959	1.972	1.986	2.000	2.014	2.029	2.044	2.060	2.076
.97	2.092	2.109	2.127	2.146	2.165	2.185	2.205	2.227	2.249	2.273
.98	2.298	2.323	2.351	2.380	2.410	2.443	2.477	2.515	2.555	2.599
.99	2.646	2.700	2.759	2.826	2.903	2.994	3.106	3.250	3.453	3.800

TABLE A.7. PERCENTAGE POINTS OF THE DISTRIBUTION OF r_{10}*

n \ 1-α	.80	.90	.95	.98	.99	.995
3	.781	.886	.941	.976	.988	.994
4	.560	.679	.765	.846	.889	.926
5	.451	.557	.642	.729	.780	.821
6	.386	.482	.560	.644	.698	.740
7	.344	.434	.507	.586	.637	.680
8	.314	.399	.468	.543	.590	.634
9	.290	.370	.437	.510	.555	.598
10	.273	.349	.412	.483	.527	.568
11	.259	.332	.392	.460	.502	.542
12	.247	.318	.376	.441	.482	.522
13	.237	.305	.361	.425	.465	.503
14	.228	.294	.349	.411	.450	.488
15	.220	.285	.338	.399	.438	.475
16	.213	.277	.329	.388	.426	.463
17	.207	.269	.320	.379	.416	.452
18	.202	.263	.313	.370	.407	.442
19	.197	.258	.306	.363	.398	.433
20	.193	.252	.300	.356	.391	.425
21	.189	.247	.295	.350	.384	.418
22	.185	.242	.290	.344	.378	.411
23	.182	.238	.285	.338	.372	.404
24	.179	.234	.281	.333	.367	.399
25	.176	.230	.277	.329	.362	.393
26	.173	.227	.273	.324	.357	.388
27	.171	.224	.269	.320	.353	.384
28	.168	.220	.266	.316	.349	.380
29	.166	.218	.263	.312	.345	.376
30	.164	.215	.260	.309	.341	.372

TABLE A.7. (CONTINUED)

$$^{*}\ r_{10} = \frac{Y_{(2)} - Y_{(1)}}{Y_{(n)} - Y_{(1)}} \text{ or } \frac{Y_{(n)} - Y_{(n-1)}}{Y_{(n)} - Y_{(1)}}$$

, where $Y_{(i)}$ is the ith smallest ordered observation in a random sample of size n from a normal distribution.

Values in the table are extracted from W. J. Dixon, "Ratios involving extreme values," The Annals of Mathematical Statistics, Vol. 22, pp. 68-78, (1951).

TABLE A.8. PERCENTAGE POINTS OF T_n OR T_1*

Number of Observations n	1 - α		
	.95	.975	.99
3	1.15	1.15	1.15
4	1.46	1.48	1.49
5	1.67	1.71	1.75
6	1.82	1.89	1.94
7	1.94	2.02	2.10
8	2.03	2.13	2.22
9	2.11	2.21	2.32
10	2.18	2.29	2.41
11	2.23	2.36	2.48
12	2.29	2.41	2.55
13	2.33	2.46	2.61
14	2.37	2.51	2.66
15	2.41	2.55	2.71
16	2.44	2.59	2.75
17	2.47	2.62	2.79
18	2.50	2.65	2.82
19	2.53	2.68	2.85
20	2.56	2.71	2.88
21	2.58	2.73	2.91
22	2.60	2.76	2.94
23	2.62	2.78	2.96
24	2.64	2.80	2.99
25	2.66	2.82	3.01
30	2.75	2.91	
35	2.82	2.98	
40	2.87	3.04	
45	2.92	3.09	
50	2.96	3.13	
60	3.03	3.20	
70	3.09	3.26	
80	3.14	3.31	
90	3.18	3.35	
100	3.21	3.38	

TABLE A.8. (CONTINUED)

* $T_n = \frac{Y_n - \bar{Y}}{s}$; $T_1 = \frac{\bar{Y} - Y_1}{s}$, where $Y_n (Y_1)$ are the largest (smallest) observations in a size n sample from a normal distribution.

Values in the table are from Grubbs, F. E., "Procedures for detecting outlying observations in samples," Technometrics, Vol. 11, p. 4 (1969).

TABLE A.9. COEFFICIENTS $\{a_{n-i+1}\}$ FOR THE W TEST FOR NORMALITY*

i \ n	2	3	4	5	6	7	8	9	10
1	0.7071	0.7071	0.6872	0.6646	0.6431	0.6233	0.6052	0.5888	0.5739
2	-	.0000	.1677	.2413	.2806	.3031	.3164	.3244	.3291
3	-	-	-	.0000	.0875	.1401	.1743	.1976	.2141
4	-	-	-	-	-	.0000	.0561	.0947	.1224
5	-	-	-	-	-	-	-	.0000	.0399

i \ n	11	12	13	14	15	16	17	18	19	20
1	0.5601	0.5475	0.5359	0.5251	0.5150	0.5056	0.4968	0.4886	0.4808	0.4734
2	.3315	.3325	.3325	.3318	.3306	.3290	.3273	.3253	.3232	.3211
3	.2260	.2347	.2412	.2460	.2495	.2521	.2540	.2553	.2561	.2565
4	.1429	.1586	.1707	.1802	.1878	.1939	.1988	.2027	.2059	.2085
5	.0695	.0922	.1099	.1240	.1353	.1447	.1524	.1587	.1641	.1686
6	0.0000	0.0303	0.0539	0.0727	0.0880	0.1005	0.1109	0.1197	0.1271	0.1334
7	-	-	.0000	.0240	.0433	.0593	.0725	.0837	.0932	.1013
8	-	-	-	-	.0000	.0196	.0359	.0496	.0612	.0711
9	-	-	-	-	-	-	.0000	.0163	.0303	.0422
10	-	-	-	-	-	-	-	-	.0000	.0140

i \ n	21	22	23	24	25	26	27	28	29	30
1	0.4643	0.4590	0.4542	0.4493	0.4450	0.4407	0.4366	0.4328	0.4291	0.4254
2	.3185	.3156	.3126	.3098	.3069	.3043	.3018	.2992	.2968	.2944
3	.2578	.2571	.2563	.2554	.2543	.2533	.2522	.2510	.2499	.2487
4	.2119	.2131	.2139	.2145	.2148	.2151	.2152	.2151	.2150	.2148
5	.1736	.1764	.1787	.1807	.1822	.1836	.1848	.1857	.1864	.1870
6	0.1399	0.1443	0.1480	0.1512	0.1539	0.1563	0.1584	0.1601	0.1616	0.1630
7	.1092	.1150	.1201	.1245	.1283	.1316	.1346	.1372	.1395	.1415
8	.0804	.0878	.0941	.0997	.1046	.1089	.1128	.1162	.1192	.1219
9	.0530	.0618	.0696	.0764	.0823	.0876	.0923	.0965	.1002	.1036
10	.0263	.0368	.0459	.0539	.0610	.0672	.0728	.0778	.0822	.0862
11	0.0000	0.0122	0.0228	0.0321	0.0403	0.0476	0.0540	0.0598	0.0650	0.0697
12	-	-	.0000	.0107	.0200	.0284	.0358	.0424	.0483	.0537
13	-	-	-	-	.0000	.0094	.0178	.0253	.0320	.0381
14	-	-	-	-	-	-	.0000	.0084	.0159	.0227
15	-	-	-	-	-	-	-	-	.0000	.0076

TABLE A.9. (CONTINUED)

i \ n	31	32	33	34	35	36	37	38	39	40
1	0.4220	0.4188	0.4156	0.4127	0.4096	0.4068	0.4040	0.4015	0.3989	0.3964
2	.2921	.2898	.2876	.2854	.2834	.2813	.2794	.2774	.2755	.2737
3	.2475	.2463	.2451	.2439	.2427	.2415	.2403	.2391	.2380	.2368
4	.2145	.2141	.2137	.2132	.2127	.2121	.2116	.2110	.2104	.2098
5	.1874	.1878	.1880	.1882	.1883	.1883	.1883	.1881	.1880	.1878
6	0.1641	0.1651	0.1660	0.1667	0.1673	0.1678	0.1683	0.1686	0.1689	0.1691
7	.1433	.1449	.1463	.1475	.1487	.1496	.1505	.1513	.1520	.1526
8	.1243	.1265	.1284	.1301	.1317	.1331	.1344	.1356	.1366	.1376
9	.1066	.1093	.1118	.1140	.1160	.1179	.1196	.1211	.1225	.1237
10	.0899	.0931	.0961	.0988	.1013	.1036	.1056	.1075	.1092	.1108
11	0.0739	0.0777	0.0812	0.0844	0.0873	0.0900	0.0924	0.0947	0.0967	0.0986
12	.0585	.0629	.0669	.0706	.0739	.0770	.0798	.0824	.0848	.0870
13	.0435	.0485	.0530	.0572	.0610	.0645	.0677	.0706	.0733	.0759
14	.0289	.0344	.0395	.0441	.0484	.0523	.0559	.0592	.0622	.0651
15	.0144	.0206	.0262	.0314	.0361	.0404	.0444	.0481	.0515	.0546
16	0.0000	0.0068	0.0131	0.0187	0.0239	0.0287	0.0331	0.0372	0.0409	0.0444
17	-	-	.0000	.0062	.0119	.0172	.0220	.0264	.0305	.0343
18	-	-	-	-	.0000	.0057	.0110	.0158	.0203	.0244
19	-	-	-	-	-	-	.0000	.0053	.0101	.0146
20	-	-	-	-	-	-	-	-	.0000	.0049

TABLE A.9. (CONTINUED)

i \ n	41	42	43	44	45	46	47	48	49	50
1	0.3940	0.3917	0.3894	0.3872	0.3850	0.3830	0.3808	0.3789	0.3770	0.3751
2	.2719	.2701	.2684	.2667	.2651	.2635	.2620	.2604	.2589	.2574
3	.2357	.2345	.2334	.2323	.2313	.2302	.2291	.2281	.2271	.2260
4	.2091	.2085	.2078	.2072	.2065	.2058	.2052	.2045	.2038	.2032
5	.1876	.1874	.1871	.1868	.1865	.1862	.1859	.1855	.1851	.1847
6	0.1693	0.1694	0.1695	0.1695	0.1695	0.1695	0.1695	0.1693	0.1692	0.1691
7	.1531	.1535	.1539	.1542	.1545	.1548	.1550	.1551	.1553	.1554
8	.1384	.1392	.1398	.1405	.1410	.1415	.1420	.1423	.1427	.1430
9	.1249	.1259	.1269	.1278	.1286	.1293	.1300	.1306	.1312	.1317
10	.1123	.1136	.1149	.1160	.1170	.1180	.1189	.1197	.1205	.1212
11	0.1004	0.1020	0.1035	0.1049	0.1062	0.1073	0.1085	0.1095	0.1105	0.1113
12	.0891	.0909	.0927	.0943	.0959	.0972	.0986	.0998	.1010	.1020
13	.0782	.0804	.0824	.0842	.0860	.0876	.0892	.0906	.0919	.0932
14	.0677	.0701	.0724	.0745	.0765	.0783	.0801	.0817	.0832	.0846
15	.0575	.0602	.0628	.0651	.0673	.0694	.0713	.0731	.0748	.0764
16	0.0476	0.0506	0.0534	0.0560	0.0584	0.0607	0.0628	0.0648	0.0667	0.0685
17	.0379	.0411	.0442	.0471	.0497	.0522	.0546	.0568	.0588	.0608
18	.0283	.0318	.0352	.0383	.0412	.0439	.0465	.0489	.0511	.0532
19	.0188	.0227	.0263	.0296	.0328	.0357	.0385	.0411	.0436	.0459
20	.0094	.0136	.0175	.0211	.0245	.0277	.0307	.0335	.0361	.0386
21	0.0000	0.0045	0.0087	0.0126	0.0163	0.0197	0.0229	0.0259	0.0288	0.0314
22	-	-	.0000	.0042	.0081	.0118	.0153	.0185	.0215	.0244
23	-	-	-	-	.0000	.0039	.0076	.0111	.0143	.0174
24	-	-	-	-	-	-	.0000	.0037	.0071	.0104
25	-	-	-	-	-	-	-	-	.0000	.0035

*Reproduced from Shapiro, S. S., and Wilk, M. B., "An analysis of variance test for normality," Biometrika, Vol. 52, pp. 591-612 (1965).

TABLE A.10. PERCENTAGE POINTS OF THE W TEST*

n	α 0.01	0.02	0.05	0.10
3	0.753	0.756	0.767	0.789
4	.687	.707	.748	.792
5	.686	.715	.762	.806
6	0.713	0.743	0.788	0.826
7	.730	.760	.803	.838
8	.749	.778	.818	.851
9	.764	.791	.829	.859
10	.781	.806	.842	.869
11	0.792	0.817	0.850	0.876
12	.805	.828	.859	.883
13	.814	.837	.866	.889
14	.825	.846	.874	.895
15	.835	.855	.881	.901
16	0.844	0.863	0.887	0.906
17	.851	.869	.892	.910
18	.858	.874	.897	.914
19	.863	.879	.901	.917
20	.868	.884	.905	.920
21	0.873	0.888	0.908	0.923
22	.878	.892	.911	.926
23	.881	.895	.914	.928
24	.884	.898	.916	.930
25	.888	.901	.918	.931
30	.900	.912	.927	.939
35	.910	.920	.934	.944
40	.919	.928	.940	.949
45	.926	.934	.945	.953
50	.930	.938	.947	.955

*Reproduced from Shapiro, S. S., and Wilk, M. B., "An analysis of variance test for normality," Biometrika, Vol. 52, pp. 591-612 (1965).

Index

INDEX